Thermal Energy Storage and Energy Conversion Technologies

Thermal Energy Storage and Energy Conversion Technologies

Editor

Ann Lee

MDPI

Basel • Beijing • Wuhan • Barcelona • Belgrade • Novi Sad • Cluj • Manchester

Editor
Ann Lee
School of Engineering
Macquarie University
Sydney
Australia

Editorial Office
MDPI
St. Alban-Anlage 66
4052 Basel, Switzerland

This is a reprint of articles from the Special Issue published online in the open access journal *Energies* (ISSN 1996-1073) (available at: www.mdpi.com/journal/energies/special_issues/thermal_energy_storge).

For citation purposes, cite each article independently as indicated on the article page online and as indicated below:

Lastname, A.A.; Lastname, B.B. Article Title. *Journal Name* **Year**, *Volume Number*, Page Range.

ISBN 978-3-0365-9091-2 (Hbk)
ISBN 978-3-0365-9090-5 (PDF)
doi.org/10.3390/books978-3-0365-9090-5

Contents

About the Editor

Ann Lee

Ann Lee is a distinguished Professor of Mechanical Engineering at Macquarie University, specializing in the field of computational fluid dynamics, with a particular focus on micro- and nano-scale fluid flow and heat transfer. Before his tenure at Macquarie University, Ann Lee embarked on his academic journey as an early-career researcher at the University of New South Wales from 2011 to 2014. Subsequently, he assumed a senior academic role at the Institute of Applied Technology in Abu Dhabi, UAE, from 2014 to 2015. Ann Lee is widely recognized for his pioneering research in three-dimensional investigations of fluid flow and heat transfer within microelectromechanical systems, establishing himself as one of the foremost international researchers in this specialized field. His research interests encompass a diverse range of topics, including fluid-structure interaction, magnetorheological fluids, magnetic drug targeting, refrigeration, and abrasive jet machining. Since 2011, Ann Lee has been a core member of several prestigious international conferences dedicated to fluid mechanics and heat transfer. Notably, he has played integral roles in conferences such as the International Heat Transfer Conference (IHTC), the Australasian Heat and Mass Transfer Conference (AHMTC), and the International Conference on Heat Transfer, Fluid Mechanics, and Thermodynamics (HEFAT). His contributions extend beyond participation, as he has served as a conference chair, demonstrating his leadership within these academic gatherings. Ann Lee is also widely recognized and respected for his role as a meticulous reviewer, providing valuable insights and expertise in the evaluation of research contributions at these conferences.

Preface

Thermal energy storage (TES), also known as heat storage systems, is a technology that accumulates energy when production exceeds demand so that the stored energy can be used later. The stored energy can be used at the user's request for heating and cooling applications or for power generation. TES systems are commonly seen in buildings and industrial processes.

On the other hand, conversion and storage, such as solar and wind energy, help to further increase the share of renewables in the energy mix. TES is becoming crucial for electricity storage in combination with solar power, whereby solar heat can be stored for electricity production when sunlight is absent.

This is a Special Issue dedicated to recent advances in thermal energy storage and energy conversion technologies. All types of research approaches are compiled in this Special Issue: experimental, theoretical, computational, and their mixtures; papers are both of fundamental and applied nature, including industrial case studies.

Finally, I would like to express my gratitude to my family and my Ph.D. students for their enduring patience and support. In particular, Liang Tang has consistently provided understanding, perspective, and love. I would also like to extend my appreciation to my Ph.D. students, Javad and Ryan, who successfully completed their Ph.D. programs between 2022 and 2023, demonstrating unwavering dedication throughout their academic journeys.

Ann Lee

Editor

energies

MDPI

Review

Combined "Renewable Energy–Thermal Energy Storage (RE–TES)" Systems: A Review

Ahmed Elkhatat * and **Shaheen A. Al-Muhtaseb ***

Department of Chemical Engineering, College of Engineering, Qatar University, Doha P.O. Box 2713, Qatar
* Correspondence: ahmed.elkhatat@qu.edu.qa (A.E.); s.almuhtaseb@qu.edu.qa (S.A.A.-M.)

Abstract: Current industrial civilization relies on conventional energy sources and utilizes large and inefficient energy conversion systems. Increasing concerns regarding conventional fuel supplies and their environmental impacts (including greenhouse gas emissions, which contribute to climate change) have promoted the importance of renewable energy (RE) sources for generating electricity and heat. This comprehensive review investigates integrating renewable energy sources (RES) with thermal energy storage (TES) systems, focusing on recent advancements and innovative approaches. Various RES (including solar, wind, geothermal, and ocean energy sources) are integrated with TES technologies such as sensible and latent TES systems. This review highlights the advantages and challenges of integrating RES and TES systems, emphasizing the importance of hybridizing multiple renewable energy sources to compensate for their deficiencies. Valuable outputs from these integrated systems (such as hydrogen production, electric power and freshwater) are discussed. The overall significance of RES–TES hybrid systems in addressing global energy demand and resource challenges is emphasized, demonstrating their potential to substitute fossil-fuel sources. This review provides a thorough understanding of the current state of RES–TES integration and offers insights into future developments in optimizing the utilization of renewable energy sources.

Keywords: renewable energy; thermal energy storage; solar energy; wind energy; biomass energy; geothermal energy; ocean energy

Citation: Elkhatat, A.; Al-Muhtaseb, S.A. Combined "Renewable Energy–Thermal Energy Storage (RE–TES)" Systems: A Review. *Energies* **2023**, *16*, 4471. https://doi.org/10.3390/en16114471

Academic Editor: Ioan Sarbu

Received: 13 May 2023
Revised: 27 May 2023
Accepted: 30 May 2023
Published: 1 June 2023

1. Background

Modern society relies extensively on energy supplies to support its civilization. However, they depend mainly on fossil fuels for heat and power production, as well as liquid fuels for transportation. As a result, they may result in significant greenhouse gas emissions, which contribute to climate change and global warming. To reduce greenhouse gas emissions, many countries are considering using renewable resources to support their needs. This transition from an energy system based primarily on fossil fuels to renewable energy sources is known as the green economy [1].

Over the past few decades, electricity has been generated from low-carbon sources (such as nuclear power, hydroelectricity, and renewable resources). The European Union has committed to reducing greenhouse gas emissions by at least 40% by 2030 [2]. Notably, more than 670 cities have set a renewable energy target worldwide, and most of these cities are located in Europe or North America. Over half of these cities intend to use 100% renewable energy in their transformational efforts. Nevertheless, in the next three decades, 2.5 billion people are expected to become urban dwellers, 90% of whom live in Asia or Africa [3]. Industrial manufacturers are also responsible for planning to reduce global warming. A recent trend among some companies is to commit to achieving sustainable and environmentally conscious operations. This is approached by implementing a differentiated execution strategy and maintaining strong accountability measures to take proactive steps toward a more sustainable future. Ingersoll Rand is a prime example of this trend, as they have pledged to reach net zero greenhouse gas emissions and utilize 100% renewable energy by the year 2050 [4].

Current industrial civilization's energy demand varies daily and relies extensively on conventional sources of energy, which are converted into electric power, oil products, hydrogen, and other energy currency products by heat release. Despite this, energy conversion systems are often large and inefficient because they usually operate below their total capacity due to the commonly intermittent nature of renewable resources. As a result, the global economy can benefit from energy storage (ES) systems to enhance the efficacy of renewable energy facilities and substitute large amounts of fossil fuel by enhancing energy conversion efficiency and supplying peak energy demands [5].

"Thermal energy storage" (TES) is one of the most practical solutions for addressing the shortage of renewable energy resources on electric grids and various power networks. It also offers flexibility and services to facilitate the management of future electricity supply and demand challenges. The storage and release of energy can be accomplished using storage media with high heat capacity, using high latent heat of phase change materials (PCMs), or through the sorption properties of adsorbed species [6–8].

To address future energy demands, it is crucial that energy technologies (1) provide a sustainable and reliable supply of electricity, and (2) integrate effectively renewable energy sources with suitable energy storage technologies. This review distinguishes itself from other reviews that focus on specific technologies such as thermal energy storage integration forms [9], stationary energy storage devices for grid integration [10], and seasonal thermal energy storage technologies [11], applications (such as sensible thermal energy storage for industrial solar applications) [12,13], high-temperature aquifer thermal energy storage [14], ocean thermal energy application technologies for unmanned underwater vehicles [15], or systems such as those exploiting wind energy curtailment [16]. Nonetheless, there is a gap in overviewing the combination of renewable energy sources and thermal energy storage systems. This gap was motivation for this work to give a comprehensive overview of the state-of-the-art research and analysis of the latest studies in this field.

This review aims to address the current gap in the literature on integrating renewable energy sources and thermal energy storage systems. The state-of-the-art research on this topic is reviewed and analyzed, identifying limitations in current studies. To improve stable and grid-friendly operations, practical strategies are presented, including complementary hybrid renewable energy systems, synergy between hybrid thermal/electrical storage systems, and collaboration in multicarrier energy networks. Additionally, recent advances and prospects in renewable and thermal energy storage systems are reviewed, providing insights for policymakers, researchers, and industry professionals. This review offers a practical and holistic approach to address existing limitations and provide guidance for future research and implementation.

2. Types of Renewable Energy Sources

Increasing concerns regarding conventional fuel supplies and environmental impacts have promoted renewable energy sources such as solar, wind, biomass, geothermal, and ocean energy for generating electricity and heat. Many countries have adopted "renewable energy resources" (RESs) to mitigate global warming and dwindling fossil-fuel resources. In a recent study by the International Renewable Energy Agency (IRENA), it has been projected that the global energy landscape will experience a significant shift, with 62% of total energy production coming from renewable sources by the year 2050, gradually replacing fossil fuels, which will decline then to 31%. Around 48% of the total RES contribution will come from solar and wind energies. This prediction is based on the rapid advancements in renewable energy technologies, the decreasing costs of implementing renewable energy systems, and the increasing global awareness and commitment to mitigating climate change effects [3,17].

A statistical review of world energy [18], as shown in Figure 1a,b, illustrates the generation of renewable energy sources (including solar, wind, and other sources) in terawatt-hours (TWh) from the year 1965 to the year 2021. The total renewable energy generation increased significantly from 18 TWh in 1965 to 3657 TWh in 2021. The solar energy generation commenced in 1989 with 0.3 TWh and experienced substantial growth, reaching 1033 TWh in 2021. Similarly, wind energy generation began in 1985 with 0.1 TWh and grew exponentially, becoming the largest renewable energy source, totaling 1862 TWh in 2021. The geothermal, biomass, and other renewable energy sources were the primary contributors until the mid-2000s, and their total generation reached 763 TWh in 2021. The figures also reveal the changing share of each energy source in the total renewable energy generation. In 1965, geothermal, biomass, and other sources constituted 100% of the total renewable energy generation, but their share decreased to 21% in 2021. Solar energy's share started at 0% and grew to 28% by 2021, while wind energy's share increased from 0% to 51% during the same period. These data demonstrate the considerable growth of solar and wind energy generation and the diversification of renewable energy sources over time.

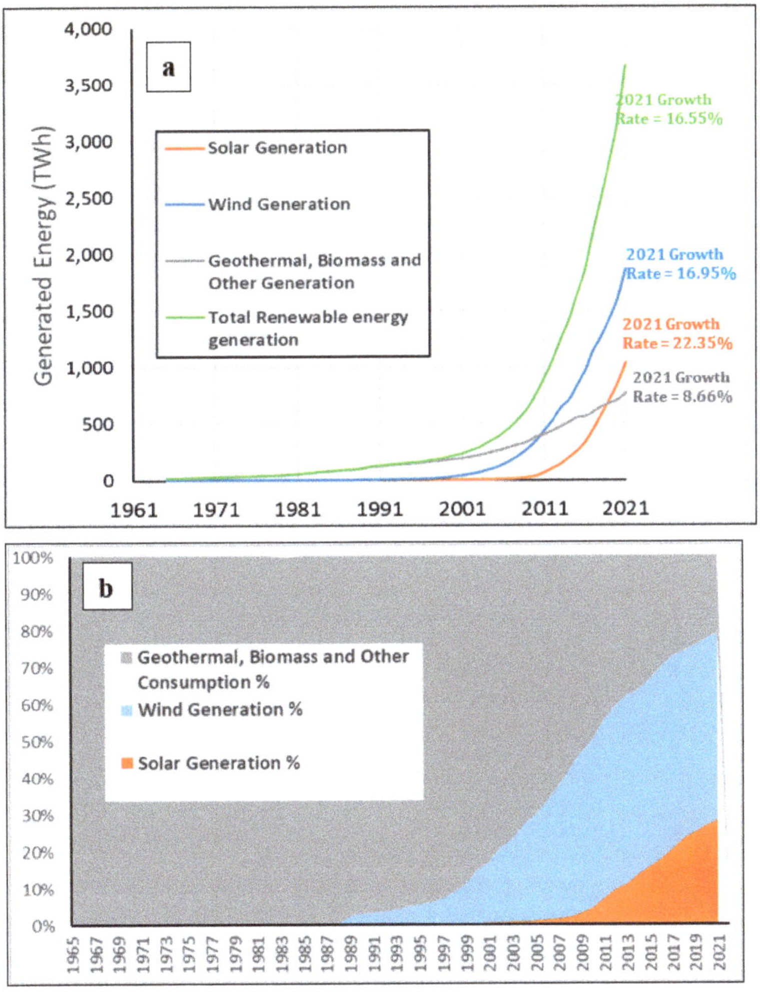

Figure 1. Historical trends and dominance of renewable energy sources (1965–2021), based on data from [18]: (**a**) progression of generated energy by different renewable energy sources; (**b**) relative dominance of various sources of renewable energy (%).

In terms of growth rates, solar energy exhibited the highest growth rate in 2021 at 22%, reflecting ongoing advancements in solar technology and its increasing global adoption. Wind energy followed with a growth rate of 17% in 2021, illustrating its continued significance and expansion due to investments in wind farms and technological improvements in wind turbines. Lastly, other renewables (encompassing geothermal, biomass, and less common sources) grew by 9% in 2021, indicating expansion, albeit slower than solar and wind energy. In conclusion, these data underscore the rapid growth of renewable energy generation, particularly in the solar and wind energy sectors, emphasizing the ongoing global transition toward cleaner and more sustainable energy sources.

An outline of the most dominant energy sources (namely, solar energy, wind energy, biomass energy, geothermal energy, and ocean energy) is given in this section.

2.1. Solar Energy

Sunlight is an abundant source of energy that can be converted into electricity or heat through solar or thermal collectors. Solar energy systems can be utilized in various ways, including portable and industrial-scale applications. Solar energy can be converted to electricity using "photovoltaic" (PV) conversion technologies, which use semiconducting materials. Photovoltaic (PV) or solar cells comprise layers of materials such as silicon. When photons from sunlight strike the cell's surface, they knock electrons out of their atomic orbits, producing electricity. This process is known as the photovoltaic effect. The PV cells are interconnected in a series to form a PV module or solar panel, and multiple panels can be connected to form a PV array. The electrical energy produced by the PV array can be used directly to power devices, stored in batteries, or fed into the electrical grid for later use.

An analysis of the statistics for European Union countries from 2008 to 2019, as depicted in Figure 2, paints a vivid picture of the evolution of solar energy infrastructure over this period. From 2008 to 2013, there was an explosive 700% increase in PV capacity, expanding from 10 GW to 80 GW. This surge was particularly pronounced between 2008 and 2009, when there was a 190% increase in PV capacity, and between 2009 and 2010, with a rise of 76%. The period of 2008 to 2010 was characterized by a substantial investment in solar energy infrastructure, signaling a resolute commitment from European countries to switch to cleaner and more sustainable energy sources. From 2011 to 2012, the growth rate eased somewhat, with a still significant increase of 57%, reaching 80 GW. However, the subsequent years witnessed a notable slowing of this pace. The rise in PV capacity from 2013 to 2015 was only 9% per year, ending with a total of 95 GW in 2015. Interestingly, from 2016 to 2019, the PV capacity remained constant at 95 GW, suggesting a stalling in the expansion of solar energy infrastructure during this period. Over these 4 years, there was a 0% change in PV capacity, which could be attributed to various factors such as policy changes, market saturation, or a shift in investment focus to other renewable energy sources. This period of stagnation contrasts sharply with the dynamic growth seen in the early years of the decade and warrants further investigation [18].

Compared to other renewable energy sources, PV technology offers many advantages. It is fast to install, is easy to adapt, and produces subtle energy [19,20]. However, constructing large-scale PV power plants requires large landscapes, which is rare and costly and might lead to loosing productive agricultural lands [21,22].

Solar energy, widely recognized for its utilization in photovoltaic conversion to generate electricity, also presents opportunities for thermal storage in solar ponds. These ponds can occur naturally, such as saline lakes, or can be artificially constructed in various geographical regions, as shown in Figure 3 [23]. In a typical solar pond, two distinct zones are established: the "upper convective zone" (UCZ) and the "lower convective zone" (LCZ). The LCZ, positioned at the bottom of the pond, acts as the main reservoir for solar energy, storing this energy in the form of heat. It can reach temperatures up to 90 °C and, thus, holds potential for applications in heating and electricity generation. However, as the LCZ heats up, it becomes less dense than the cooler UCZ located above. This disparity in densities can instigate a process known as "natural convection", driven by buoyancy forces where more dense fluids move to the bottom and vice versa. As a result, natural convection can disturb the thermal stability of the pond, mixing warm and cool water layers. This issue can be mitigated by leveraging the characteristic of salt to dissolve more readily in warm water. Given that the LCZ tends to be the warmest part of the pond, it dissolves more salt, thereby establishing a "salinity gradient" within the pond. This gradient results in the highest concentration of salt in the bottom of the LCZ, which gradually decreases toward the UCZ. The higher salinity makes the LCZ denser, which in turn makes it less susceptible to the disruptive effects of natural convection, enhancing the thermal stability of the pond. Solar ponds augmented with salt to create a salinity gradient for improved thermal stability are often referred to as "salinity gradient solar ponds" (SGSP) [24]. Solar ponds are generally classified into two categories: "shallow convective" and "deep non-convective" solar ponds. Among the non-convective variants, SGSPs are frequently favored due to their superior thermal stability [25].

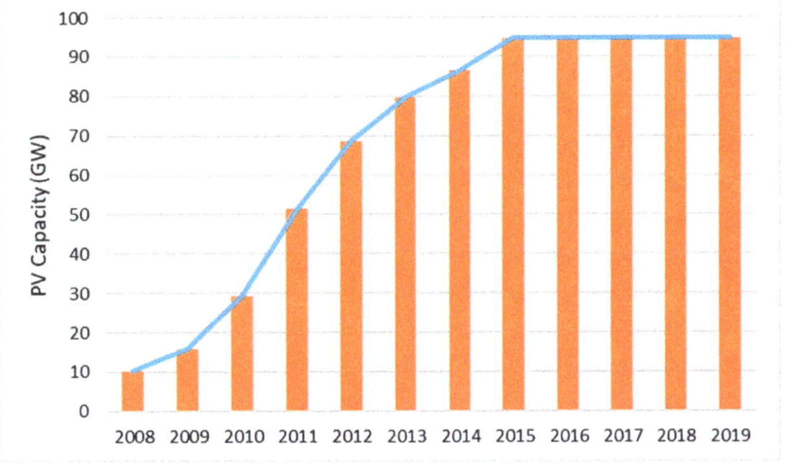

Figure 2. Photovoltaic production capacity in European Union countries (2008–2019), based on data from [1].

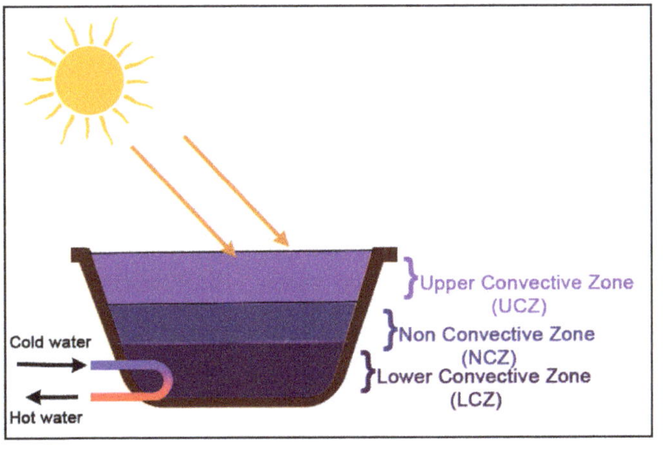

Figure 3. Schematic diagram of a typical solar pond.

Solar flat plate panels and evacuated flat plate solar collectors represent two prominent types of solar collectors harnessing solar energy for diverse applications. Solar flat plate panels are traditional thermal collectors. They absorb solar radiation and transfer this heat to a working fluid, typically water or air. The structural composition of these collectors includes a flat absorber plate, commonly crafted from metal, and a transparent cover that permits sunlight penetration while minimizing heat loss. Once absorbed, the heat is transferred to the working fluid circling through tubes near the absorber plate. In contrast, evacuated flat plate solar collectors serve as an advanced iteration of flat plate collectors. Their design focuses on reducing heat loss and augmenting efficiency. These collectors incorporate a vacuum between the absorber plate and the transparent cover. The created vacuum dramatically curtails heat loss attributable to conduction and convection, contributing to a marked increase in efficiency and heat output [26].

Another approach of residential and industrial interest is to collect solar energy via "solar parabolic trough collectors" (SPTCs), as shown in Figure 4. An SPTC is a concentrated solar power technology device that utilizes a parabolic-shaped reflector, typically constructed from materials such as silvered glass or polished aluminum, to focus sunlight onto a linear receiver tube. The receiver tube, often made of a metal with a selective absorptive coating, contains a heat transfer fluid that absorbs the concentrated solar energy, consequently increasing its temperature. This heated fluid is then utilized to generate steam, which drives a turbine for electric power production. Parabolic trough collectors offer higher energy conversion efficiency and better integration with thermal energy storage systems than other solar energy technologies, such as photovoltaic cells or solar ponds. However, their main limitations include higher upfront costs, the need for precise sun tracking, and the potential for optical losses due to reflector imperfections or misalignment [27–29].

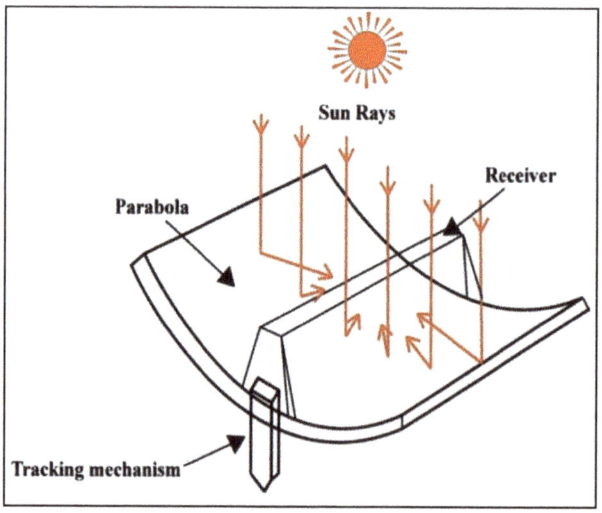

Figure 4. Schematic of a typical parabolic-trough solar concentrator [27].

2.2. Wind Energy

Wind energy is considered to be derived from solar energy, as it is created by the natural convection of air when warm air (that is heated by the sun) rises, creating zones of differential pressures that drive wind formation. In areas that are not sufficiently windy, wind can be accelerated to meet the needs of big cities by certain land formations. The kinetic energy of winds is captured using the rotors of large-scale wind turbines, which create electricity at a few hundred volts that are then amplified (via a transformer) to a few thousand volts to limit electrical loss. The generated electric power is then distributed to nearby towns, residences, and farms [30]. Wind energy generation has received significant attention and investments from several countries. As shown in Figure 5, which presents the global wind power capacity and annual additions in gigawatts (GW) from 2010 to 2020, there was a steady climb in global wind power capacity, moving from 198 GW in 2010 to 743 GW in the year 2020. This reflects an overall growth of 275% over this period, underlining the increasing investment in wind energy infrastructure and the worldwide shift toward cleaner and more sustainable energy sources. In terms of annual percentage growth, the increase was steady between 2010 and 2013. During this period, annual additions oscillated between 36 GW and 45 GW, peaking at 45 GW in 2012. However, 2014 saw a significant surge in annual additions, with 52 GW added, signifying an increase of about 16% from the total capacity of the previous year. The pace of additions quickened from 2015 to 2019, reaching a zenith of 64 GW in the year 2015. The annual additions later stabilized to a range between 51 GW and 61 GW, representing an approximate annual growth of 7% to 9% over the total capacity during these years. However, the year 2020 marked a dramatic rise in annual additions, with 93 GW added, representing a 14% increase from the total capacity in the year 2019. The factors contributing to this substantial growth in the year 2020 include a heightened global emphasis on renewable energy, policy incentives, and advancements in wind turbine technology [31,32]. Notably, the electricity produced by wind energy in China and the United States accounted for more than 74% of the total electricity demand in 2020 [33].

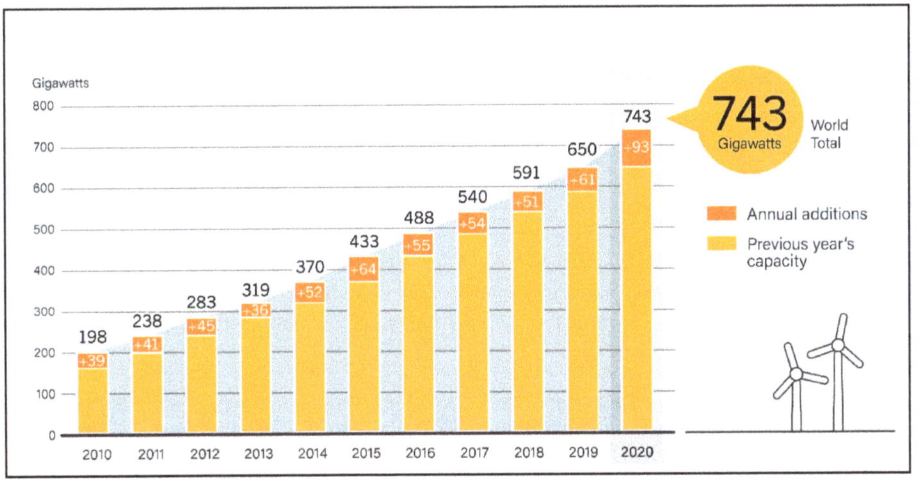

Figure 5. Wind power global capacity and annual additions (2010–2020) [32].

In the field of renewable energy, wind energy has many advantages but also faces several challenges [34]. One of the challenges is the instability of energy generation, as the capacity of wind farms can change seasonally or daily according to the weather conditions [35]. Technically, the wind energy sector also faces some shortages in knowledge, highly qualified supporting staff, and dedicated research ventures. However, embedding electric heating units (e.g., within the wind turbine or outside the turbine near its base) can maximize the efficiency of wind power systems. Furthermore, the overall recovery performance of additional wind energy (i.e., when additional energy is generated from heat produced by the electrical heating units) can be improved by modifying the heating start and end positions, thus benefiting heat treatment quality [36]. There are also several factors to consider in terms of financial, geographic, and environmental challenges. Financial challenges include the need for an initial capital commitment for feasibility studies and ongoing commission and maintenance costs throughout the wind plant lifespan [37]. Geographic and environmental challenges lead to conflicts with fishery groups, changes to the landscape and biodiversity [38], and effects on avian life, including mortality [39] and aerodynamic noise [40].

2.3. Biomass Energy

Biomass energy encompasses a broad spectrum of biofuels that exist in solid, liquid, and gaseous forms, including biomass (solid biofuels), bioethanol and biodiesel (liquid biofuels), and biogas, biohydrogen, and biohythane (gaseous biofuels) [41,42]. However, the storage capacity for biomass is limited compared to that of fossil fuels [43].

First-generation biofuels are produced from food crops such as corn and soy, while second-generation biofuels are derived from nonfood feedstock such as agricultural waste, wood, or grasses [44]. These feedstocks encounter challenges related to food security, productivity, land requirements, costs, and competitive prices [45,46]. Third-generation biofuels have been developed from microalgae and other microorganisms, showing promise due to their remarkable growth rates and high biomass and lipid levels [47–49]. Fourth-generation biofuels improve upon this, using genetically modified microalgae for higher yield, growth rate, and tolerance [50]. The fifth generation of biomass energy production incorporates variable renewable energy sources and energy storage systems, aiming to increase efficiency and reduce emissions during energy conversion processes [51].

Bioethanol is produced through the fermentation of various biomass raw materials, and it can enhance vehicle performance when blended with petrol or gasoline [52–55]. Biodiesel, derived from the transesterification of vegetable or animal fats, microalgae, and cyanobacterial lipids, can replace or supplement diesel to reduce reliance on petroleum-based fuels [44,56]. Biogas, or biomethane, is produced via the anaerobic digestion of biomass, boasting low pollutant emissions and cost-effectiveness [57,58]. Biohydrogen offers high energy density and low pollution levels, making it an attractive energy source [59].

The practical Implementation of biomass power plants faces several challenges, including the need to collect and transport biofuels to power plants, the associated costs, and the competitive edge of other renewable energy sources such as solar or wind power [60].

2.4. Geothermal Energy

Geothermal energy is the thermal energy created from the residual heat from the original formation of the planet (primordial heat) or the decay of natural radioactive materials. Geothermal energy is an RES that has been used to generate electricity for over 80 years and has been proliferating over the past 30 years [30]. In comparison to other renewable sources of energy, geothermal energy is a reliable source that can provide necessities to a community without requiring an external fuel source or grid connection [61]. The global geothermal energy consumption for various applications between 1995 and 2020 (TJ/year) is shown in Figure 6. It is noted that the use of geothermal energy escalated significantly between 2010 and 2020, especially for ground-source heat pumps (GHPs) [62]. GHPs use the Earth's natural heat to warm buildings in the winter and cool them in the summer. They are more efficient than air-source heat pumps and are often used in areas with cold climates.

Sources of geothermal energy have four primary forms: hot rocks, lava, hydrothermal reservoirs, and geopressured brines [63]. However, according to the usage and application, geothermal energy resources can be categorized into three classifications. Firstly, hot rocks (about 7 km deep below the Earth's crust, at temperatures of 200–300 °C) can be used for power generation. Secondly, "shallow geothermal" regions, about 1 km below the Earth's crust, are at ~10–21 °C and can be used for buildings and greenhouse heating. Thirdly, hydrothermal energy reservoirs (>350 °C) are found about 3 km deep below the Earth's crust and can be used for power generation and district heating [62,64]. A "shallow geothermal" system differs from a hydrothermal system in that the thermal energy is trapped in the upper part of the Earth's crust. In contrast, in hydrothermal reservoirs, the groundwater is heated due to geological fractures, faults, or introduced lava [65]. The fluids can be extracted for a variety of applications. Since hot rocks are found deep within the

Earth, they are difficult to exploit as a source of geothermal energy. However, "enhanced geothermal systems" (EGS) can extract heat from artificially fractured hot rocks [66].

The use of geothermal energy for power can take the forms of (1) "flash steam plants", which are primarily used for high-temperature resources, (2) "dry steam plants", which utilize superheated steam extracted from a geothermal reservoir, and (3) "binary plants". Binary plants (i.e., utilizing steam and a secondary fluid with a much lower boiling point) based on medium-temperature resources are becoming increasingly prevalent worldwide [63,67]. The "shallow geothermal" energy is captured through "geothermal heat pumps", which are loop heat exchange systems. Heat pumps using geothermal energy can provide heating, cooling, and hot water to buildings. Agriculture and space heating applications can use geothermal fluids below 100 °C, while industrial processes can use those above 100 °C [68,69].

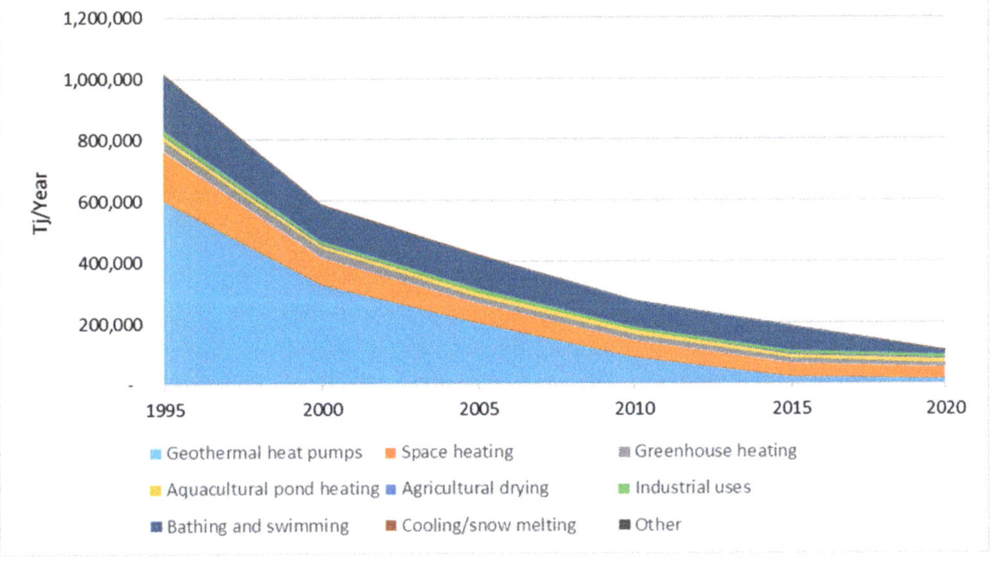

Figure 6. Global geothermal energy consumption between 1995 and 2020 (TJ/year), based on data from [70].

2.5. Ocean Energy

Oceans contain abundant energy that could be harvested over centuries using various means, including mechanical, thermal, and chemical approaches. The ocean's mechanical energy is harvested from waves, tides, and ocean currents. Waves form due to the wind's effect on the ocean's surface layer, and tides form due to the gravitational forces from the sun and moon. In contrast, ocean currents occur due to convective heat transfer between layers of different temperatures. On the other hand, the ocean's thermal energy is harvested from the temperature gradients between the different layers of the ocean, which are exhibited due to solar radiation. The ocean's thermal energy is believed to be an efficient power source for remote islands due to abundant fisheries, tourism, and minerals in the ocean [71,72]. The ocean's chemical energy is harvested from salt concentration gradients via a process called "pressure-retarded osmosis" (PRO) [30,73,74]. In addition, evolving technologies utilizing offshore ocean energies include floating PV and wind technologies [75]. Although the ocean's renewable energy-harvesting technologies have significant potential, they are not extensively implemented due to several factors, including (but not limited to) cost, infrastructure complexity, and lack of knowledge and awareness. In recent years, harvesting technologies of ocean renewable energy have developed rapidly, and many European and Asian countries have explored the potentials of ocean renewable energy where wave power is relatively abundant [76,77]. Notably, the most potent wave power is found in regions with the highest latitude. However, they are susceptible to ice cover formation, which inversely impacts wave power [78].

Wave energy is a promising technology in Southeast Asian countries. China and Japan have vast potential for harnessing wave energy with wave energy fluxes up to around 8 kW/m [79]. Other countries with low energy but a high level of stability (such as Indonesia, Thailand, Myanmar, the Philippines, Malaysia, and Vietnam) have an excellent opportunity to utilize this technology [80]. However, wave energy-harnessing technologies are not always feasible due to their high cost, wave fluctuations, and the possibility of introducing harmful marine species to the ecosystem [81,82]. Three leading technologies have been developed to harvest ocean wave energy. These technologies are mainly the oscillating water column (OWC), overtopping wave (OW), and oscillating bodies (OB) energy converters, as shown in Figure 7 [83,84]. In OWC technology, waves are converted into driven airflow in a submerged column connected to an air turbine at the top, which produces electricity. In OW energy converters, waves are directed into the device's reservoir by a platform supported by ramps. In addition to generating electricity, OW converters can protect the seashore from flooding by intercepting incoming waves and dissipating their energy [85]. OB can be another promising technology for harnessing wave energy by converting it into kinetic energy through the wave-caused movement of the body relative to the platform. Since OB directly utilizes wave energy to generate power without requiring any intermediate steps, it has the highest energy conversion efficiency [86].

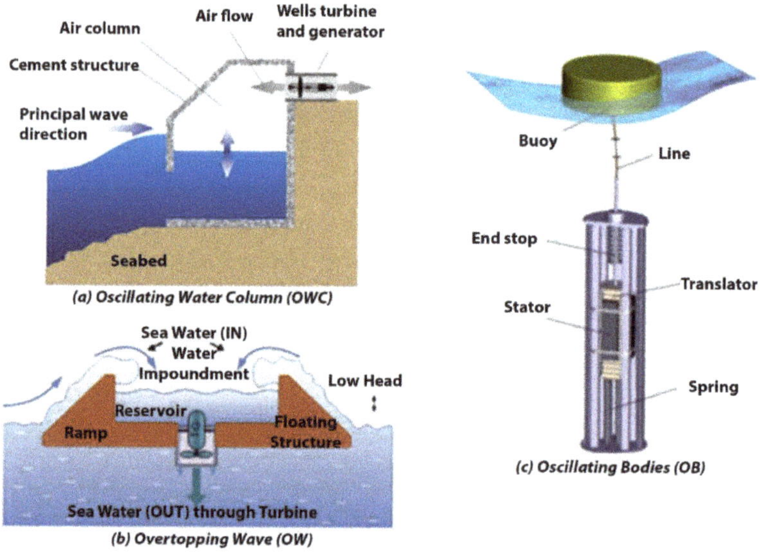

Figure 7. Wave energy-harvesting by (**a**) oscillating water column, (**b**) overtopping wave, and (**c**) oscillating bodies [84].

Tidal energy is another form of ocean energy that is produced from tides. The most commonly used technologies for harnessing tidal energy are "tidal stream generators" and "tidal barrages", shown in Figure 8. Eastern and Southeastern Asia have great potential for developing tidal renewable energy with a possible capability of about 2000 MW [81,87,88]. "Tidal stream generators" are similar to wind turbines because they generate electricity using rotational kinetic energy from moving water. Contrary to "tidal stream generators", "tide barrages" utilize hydraulic heads (potential energy) between high and low tides. Using turbines, this potential energy is converted into mechanical energy and then into electricity.

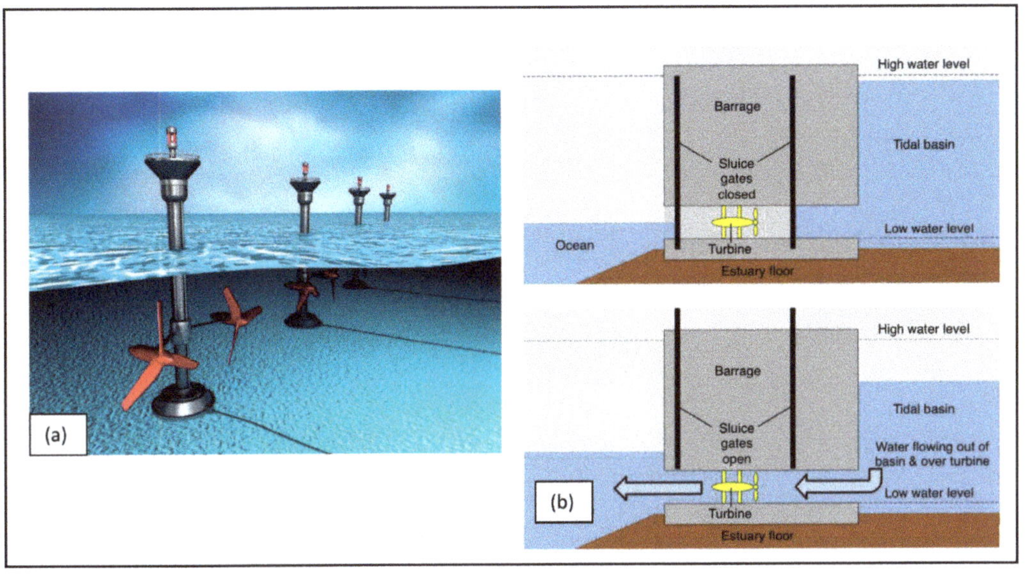

Figure 8. Harnessing tidal energy by (**a**) tidal stream generator and (**b**) tidal barrage. Adapted from [89,90].

Ocean thermal energy conversion (OTEC) generates electricity by utilizing the temperature difference between warm seawater at the ocean's surface and cold deep seawater. Currently, it is used only for small-scale plants, such as the 50 kW unit in Hawaii, 100 MW OTEC plant in India, and 10 MW plant in China. The Philippines has a planned 10 MW plant and is also exploring the feasibility of a 150 MW plant. OTECs are fundamentally flawed due to their low temperature difference, which reduces their net output efficiency. Hence, numerous studies are being conducted to analyze and optimize these systems [91,92]. Several closed cycles have been proposed, including the organic Rankine cycle, the Kalina cycle with ammonia–water mixing, the Uehara cycle, and the two-stage Rankine cycle. Ammonia is the most suitable work gas used to transfer heat to create work in closed cycles and has the highest latent heat and most efficient heat transfer [93,94]. In 2019, Khosravi and his team developed a system that converts ocean thermal energy into hydrogen and electricity for islands by coupling it with a photovoltaic system. Using this configuration was economically more advantageous than other forms of clean energy production. In addition, during off-peak hours, approximately 50% of the electricity generated was stored as ammonia [95].

3. Energy Storage Systems

Recently, energy storage (ES) has become an increasingly important issue and an area of active research to reduce peak energy loads, bridge the gap between energy supply and demand, reduce costs, and mitigate the effects of global warming. During peak energy consumption periods, energy storage systems provide an alternative method of supplying peak energy demands. Many energy storage methods are being investigated, including thermal, mechanical, chemical, electrochemical, magnetic, and electromagnetic storage systems [5]. Figure 9 shows the overall classification of energy storage methods.

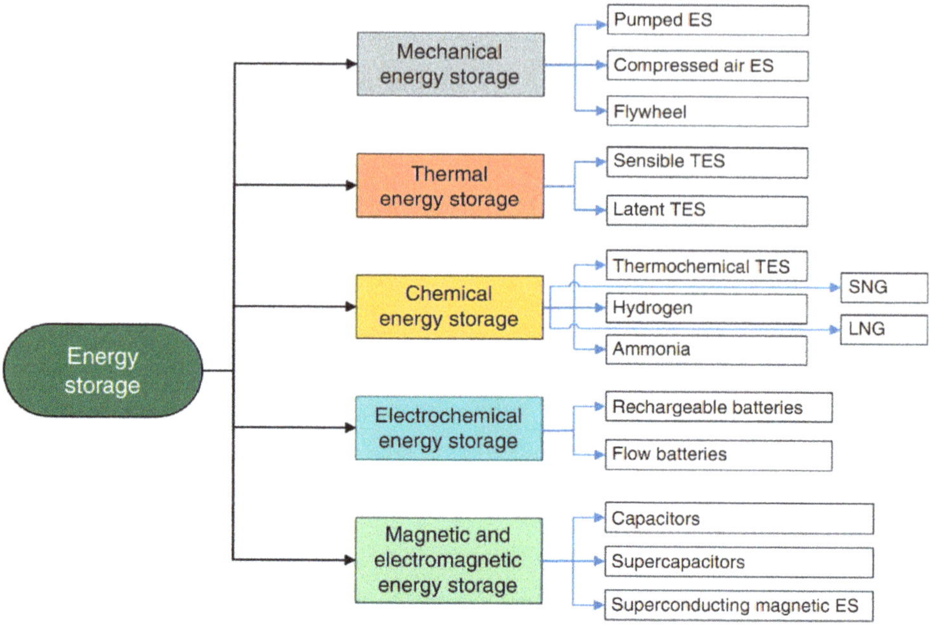

Figure 9. An overview of the different energy storage methods. Adapted from [5].

Energy can be stored in the form of thermal energy (TES). The TES process involves changing the temperature of a substance (sensible TES), changing the phase of the substance (latent TES), combining both, or sorbing energy (sorbing TES). Consequently, winter heating and summer cooling can be attained by using TES [5]. TES and RE–TES systems are detailed in the subsequent sections.

Mechanical energy can be stored in several ways, including pumps, compressed gas, and flywheels. As a simple method of storing electrical energy, pumped energy storage utilizes "pumped water" to produce electricity during the day. The system pumps water upward into a storage reservoir and operates a turbine when the water flows back down. "Pumped water" systems (PWS) are a practical means of storing energy, while a solar or wind power system produces energy in another part of the system. Off-peak power is stored in a dam by "pumping water" for peak load periods. "Compressed-air energy storage" systems (CAES) store energy by compressing air using a gas turbine. In regions where natural caverns exist, CAES can be used to replace PWS. Flywheels are an intelligent way of storing energy in the form of kinetic energy. A flywheel is spun thousands of times per minute by a motor powered by surplus electricity. A frictionless rotation of the flywheel is achieved due to its levitation in an evacuated chamber supported by magnets and highly efficient bearings. The flywheel converts the stored kinetic energy into momentum and can be used to drive an electricity generator [5,96].

In chemical energy storage (ChES) systems, energy can be stored in chemical compounds or mixtures such as solid-state natural gas (SNG) and liquefied natural gas (LNG). These mixtures involve chemical compounds, such as methane and propane, which can store energy by undergoing an endothermic and exothermic process through their synthesis and combustion reactions. Among other chemical compounds, hydrogen and ammonia can be produced using electricity or heat. However, these processes are costly. On the other hand, electrochemical energy storage (EChES) devices (such as batteries, supercapacitors, and fuel cells) are standard in handling electricity loads, operating electric cars and other electric devices, and storing renewable energy. Battery systems are generally preferred over PWS and compressed air systems when storage capacity is not critical.

Magnetic and electromagnetic fields can store energy using capacitors, supercapacitors, and superconducting materials. As a result of the material's superconducting properties, energy can be stored with only a little loss. However, a certain amount of energy is dissipated to cool capacitors and superconducting materials [5].

4. Thermal Energy Storage Systems

Thermal energy storage (TES) can be utilized for sensible thermal processes, such as cooling and heating, latent thermal processes (e.g., melting/freezing and vaporization/condensation), or heat sorption via the interaction between a sorbent and exchange material. A good overview of earlier works in this area is given elsewhere [9,11,13,97]. Thermal energy recovery as heat or coolness can be achieved by reversing the process. In other words, in TES, thermal energy is temporarily stored at high or low temperature for later use. As a result, TES can compensate for the gap between the supply and demand of solar energy, e.g., to provide warming for cold nights and winter by utilizing solar energy from sunny days. Accordingly, TES can manage seasonal temperature fluctuations and meet the demand. Depending on the temperature range and use, there are various thermal energy storage media, such as water (with a high specific heat capacity) and solids (with a compact size and higher specific heat capacity) [98]. TES helps increase generation capacity that helps shift energy purchases to low-cost periods and lowers demand charges by producing and storing energy during off-peak hours. By reducing peak loads for heating and cooling systems, TES can yield up to 70% energy savings [99]. For example, small- and large-scale TES are used to reduce HVAC costs, reducing electric bills. It is also possible to integrate TES into standard energy storage systems depending on the application and specific requirements of the systems.

TES technology is established and economically feasible, where the USA and Canada have used short-term thermal energy storage (TES) and long-term thermal energy storage (which can involve PCMs) since the 1980s [98]. Notwithstanding, it needs to be adopted more widely. Various obstacles hinder TES adoption, including high initial costs, poor infrastructure, and a lack of experience, which make decision-makers only consider using TES for ice storage in air-conditioned buildings. However, recent years have seen significant interest in TES systems, and efforts were made to reduce costs and improve performance by focusing on technologies with local experience and keeping systems as simple as possible. Various studies have addressed technical issues and evaluated the potential for new TES systems, focusing on performance and optimization, design, simulations, applications, and thermal and transport phenomena analyses (including thermodynamics, heat transfer, and mass transfer) [11,51,61,98,100–108].

Heating TES systems is commonly achieved by solar energy or electric-resistance heaters, which produce heat on demand and recuperate it by storing it in TES media. For instance, ceramic bricks are used in new construction, allowing for excellent retention of sensible heat [109], whereas building sites use insulated boxes with small fans in various sizes [12]. In addition, cooling TES systems can shift peak operating times to nighttime by using off-peak electricity to chill a storage medium, such as water or ice, during periods of low electricity demand. This chilled medium can then be used during peak demand periods, such as during hot summer afternoons when air conditioning usage is high, to cool the building or space, which results in reducing the cooling cost. Various TES media can be used for cooling, but ice is the most efficient medium used for this purpose. As a cost-effective alternative, a gas-cooled absorption chiller may replace a vapor-compression refrigeration system during peak cooling [110]. It is essential to consider a number of factors in order to determine the feasibility of TES, including the capacity for daily or seasonal use, the choice of either an individual TES system or an aggregate system, and the compliance with the standards of the American Society of Heating, Refrigerating, and Air Conditioning Engineers (ASHRAE) [5].

An overview of sensible TES, latent TES, and sorption TES is presented in this section.

4.1. Sensible TES

Sensible TES can be illustrated by the sun heating floor tiles during the day, whereafter the tiles release the gained heat during the night. Sensible TES materials can store energy, considering the temperature difference as a driving force for heat gain based on the medium's specific heat capacity (Cp). Water has a high Cp but must be contained in an isolated vessel. Therefore, solid TES media with significant Cp and thermal conductivity values are preferable. For example, iron and iron oxide are excellent thermal storage media due to their high heat capacities and thermal conductivities. On the other hand, although the thermal capacity of rock is half that of water, they are also practical, sensible TES materials due to their abundance and low cost [12]. As a sensible TES medium, concrete bricks can be used at temperatures up to 600 °C [111]. Furthermore, using thermal oils and molten salts in TES systems is highly desirable since they remain stable (no phase change) within a temperature range of 400 °C to 500 °C [111], especially when the application requires higher temperatures and when there is a need for heat transfer fluid to flow through the TES system (where the TES material needs to be circulated through a heat exchanger).

"Aquifer thermal energy storage" (ATES) is a method of storing and recovering thermal energy in the subsurface that uses groundwater to exchange heat with the building. ATES is a seasonal heat storage system that stores and recovers thermal energy by extracting and injecting groundwater from aquifers using groundwater wells connected to heat exchangers. In the winter, groundwater is pumped from the warm well to the building's heat exchanger. Heat is extracted from this source and used as an energy source for the building's heat pumps. During the summer, groundwater is pumped from the cold well to the heat exchanger. Due to the lower temperature of the groundwater, it will be injected into the cold well at a lower rate [12,14,112]. The prescribed approach is considered a low-temperature seasonal ATES system. An overview of a low-temperature seasonal ATES operation is illustrated in Figure 10, which shows the possible directions of heat flow and groundwater flow in different seasons and the modes of operation. For example, in summer, the solar collectors can provide heat to the ATES wells, while the heat exchanger utilizes cold water from the ATES wells for cooling. In winter, the heat pumps can extract heat from the ATES wells and use it for heating, while the solar collectors can provide heat to the soil remediation system. The other geothermal energy systems can also exchange heat with the ATES wells depending on their needs and availability. Several other types of ATES systems, including high-temperature seasonal ATES, typically operate at higher temperatures than low-temperature systems and are used for heating and cooling [113]. Hybrid ATES systems combine ATES with other renewable energy sources, such as solar thermal or geothermal energy, to increase their efficiency and overall performance [114]. Additionally, there are also shallow ATES systems that use a network of underground pipes to store heat or cold in the soil layers close to the surface [115], and deep ATES systems that store heat or cold in deeper aquifers [116].

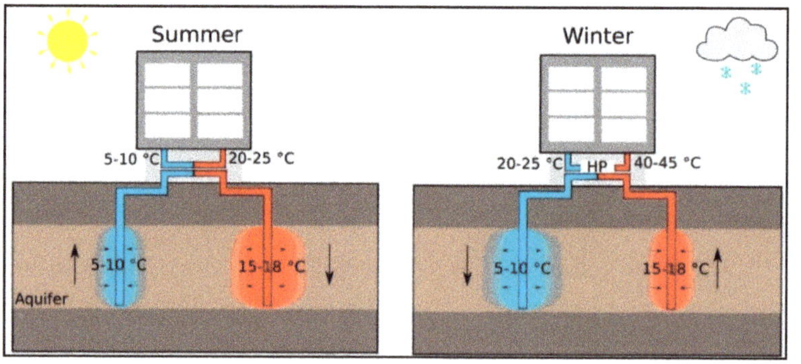

Figure 10. An overview of the essential operation of a low-temperature seasonal ATES system [117].

Notably, the performance and effectiveness of solar ponds depend on the parts integrated into them, including the solar collector, the heat exchanger, and the storage system. Over the years, various technical, developmental, financial, economic, and simulation studies have been conducted to evaluate and improve the efficiency of solar ponds. In addition, many additives, such as nanofluids and "phase change materials" (PCMs), have been used to improve the thermal properties and heat capacity of solar ponds [118]. An earlier work reviewed beneficial developments in solar pond technology between 1980 and 2000 [104].

In addition to ATES and solar ponds as liquid TES systems, recent years have increased the use of stratified TES tanks for heating and air conditioning. The thermal stratification (i.e., layering) phenomenon is the basis for the operation of stratified TES tanks. Water stratification results from the inversely proportional relationship between the temperature and density of water. As a result of stratification, hot water (with a lower density) rises to the top of the tank, while cold water (with a higher density) sinks to its bottom. During the charging and discharging process, the mixing layer, called a thermocline, moves toward the top of the tank, acting as a dynamic natural barrier that keeps warm- and cold-water regions separated. It is generally recommended to reduce the thickness of the thermocline layer to increase the volume of the hot water layer in the tank, which is typically achieved by the optimum design of the stratified TES tank. Heat loss and gain must be prevented in stratified TES tanks, and the thermocline layer and mixing volumes of water of varying temperatures must be minimized, as seen in Figure 11 [119,120]. To recover heat from the hot water in the tank, the hot water is typically pumped from the top of the tank through a heat exchanger, which transfers the heat from the water to a fluid in a closed-loop system. This fluid can then be used to heat a building or provide hot water for domestic use. The cooled water is then returned to the bottom of the tank, which is reheated and rises to the top to complete the cycle. In some cases, a stratified TES tank may also be connected to a heat pump or other heating/cooling system to increase its efficiency. The heat pump can be used to extract heat from the hot water in the tank and transfer it to a higher temperature for use in space heating, while the cold water in the tank can be used for cooling in a similar manner.

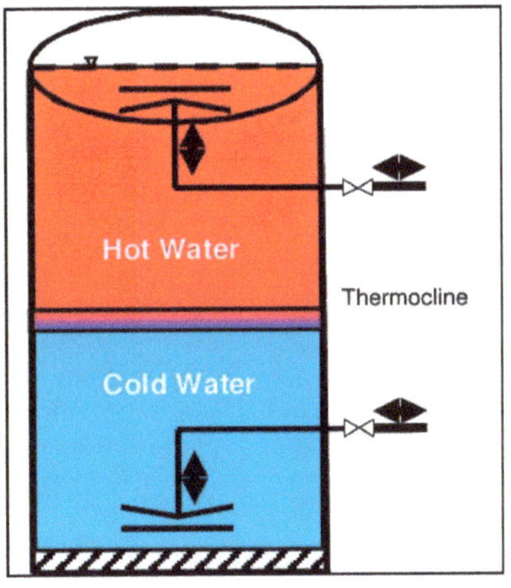

Figure 11. TES with stratified water tanks [119].

"Thermal energy grid storage" (TEGS), influenced by "multijunction photovoltaics" (MPV), is a new emerging technology that utilizes MPVs to provide heat for TEG systems. The thermal energy grid (TEG) system is charged by the excess electricity passed through a refractory heating element with high resistance (such as tungsten or graphite). The heating element then generates heat that is transferred via pumped liquid tin. Tin is preferred because of its high thermal diffusivity and wide range of temperatures in the liquid phase (232–2600 °C). After that, heat is stored in grid storage, consisting of insulated graphite blocks as a low-cost material to store electricity as heat and then convert it back to electricity when required via thermophotovoltaics (TPV) cells [121,122]. The concept of TEGS is illustrated in Figure 12.

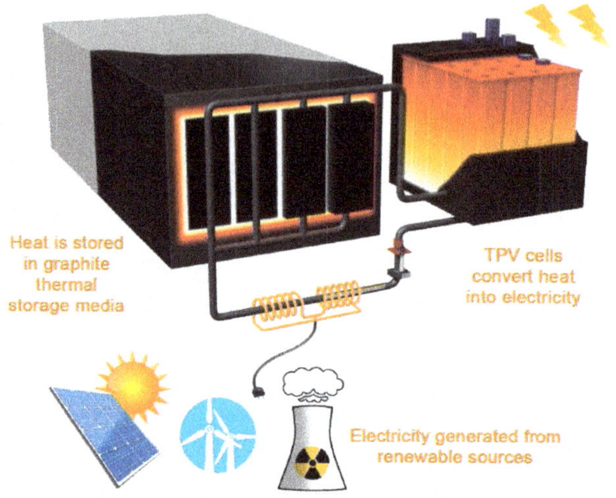

Figure 12. Concept of thermal energy grid storage (TEGS). Adapted from [121].

4.2. Latent TES

Latent thermal energy storage (LTES) is a type of TES system that stores and releases energy through different substances that change phases such as phase change materials (PCMs). These materials undergo phase transitions, such as solid–liquid, solid–solid, or liquid–vapor, which enable them to store or release heat. The advantage of LTES lies in its ability to store and release large amounts of energy at a constant temperature (in the case of pure substances) or a narrow range of temperatures, known as phase transition temperatures. There are several types of LTES systems, which can be broadly categorized into three categories: solid–liquid LTES, solid–solid LTES, and liquid–vapor LTES. In solid–liquid LTES systems, PCMs undergo a solid–liquid phase transition, storing energy as they melt and releasing energy as they solidify. Examples of solid–liquid PCMs include organic materials (e.g., paraffin waxes and fatty acids) and inorganic materials (e.g., salts and salt hydrates). Solid–solid LTES systems involve materials that undergo a solid–solid phase transition, usually involving a change in the crystal structure. This type of LTES is less common than solid–liquid LTES but can be found in applications involving shape memory alloys (SMAs). Liquid–vapor LTES relies on materials that undergo a liquid–vapor phase transitions, such as water or ammonia. While these systems can store large amounts of energy, they typically require high pressures and sophisticated containment systems to maintain the vapor phase [123].

Latent thermal energy storage (LTES) is a promising storage technique that can store and release heat within a small temperature range, and its thermal energy storage capacity is commonly much greater than that obtained with sensible heat for a given medium. This makes LTES systems practical alternatives to sensible heat storage systems for solar energy applications. The essential part of LTES is the phase change material (PCM), which can be organic or inorganic. For example, solutions of salts in water can be considered reliable PCMs if the system is designed to prevent water evaporation and maintain the PCM in its solid or liquid state. These salt compounds, such as sodium sulfate decahydrate and calcium chloride decahydrate, need considerable heat for melting, making them useful for energy storage and drying processes, especially at considerably high temperatures [124]. Another leading PCM is Glauber's salt, which is characterized by a high latent heat that can be used for heat storage, and its required phase-change temperature is well matched with that of solar systems [125] when used for heating and cooling with temperatures around its melting temperature (32.4 °C). Paraffin and zeolites have also been investigated as LTES candidates [126–128]. Most organic and inorganic PCMs melt between 0 and 120 °C and are desirable to possess technical and thermodynamic properties, such as high thermal conductivity, high latent heat of fusion, thermal stability, compatibility with the storage medium and heat transfer fluid, low toxicity, and environmental sustainability. However, some difficulties are found in achieving an optimal match between the phase change temperature range and the operating temperature range concerning the long-term thermal stability of PCMs. Gradient porosity metal foam (GPMF) was suggested as an excellent PCM to overcome the low thermal conductivity associated with PCMs in mid-temperature solar energy storage systems [129]. To enhance robust structure stability, comprehensive energy storage performance, and leakage-proof performance, Zhang's team developed a novel composite PCM using Guar gum (GG), which is a natural polysaccharide used to fabricate carbon aerogel, for encapsulating polyethylene glycol (PEG) [130]. Other polymer-shell microencapsulation methods were studied to prevent PCM leakage.

There is a body of research that has focused extensively on PCMs preparations and potential applications [125,127,128,131–135]. Moreover, an innovative solution was developed by Yan Cao et al. to address leakage and high flammability in PCMs through the creation of a series of leakage-proof phase change composites (PCCs) with excellent solar thermal conversion capabilities and superior flame retardancy. MXene nanosheets were synthesized by the researchers, exhibiting remarkable solar–thermal conversion effects, and MXene/polyimide (PI) aerogel was prepared. MXene/PI aerogels were subsequently impregnated into polyethylene glycol (PEG) using vacuum impregnation, resulting in shape-stable MXene/PI@PEG phase change composites (MPPCCs). MPPCC-4 demonstrated a high PEG loading capacity, high enthalpy, and relative enthalpy efficiency, displaying exceptional flame-retardant properties compared to PEG. These multifunctional PCCs offer a significant potential for use in solar energy utilization systems and the integration of advanced thermal energy management applications with exceptional electromagnetic interference (EMI) shielding [136–138].

PCMs can be stored in "mobile thermal energy storage" (M-TES) systems to use heat in places other than where it is generated. Some M-TES systems use zeolite as a carrier material that can adsorb and desorb water vapor and, thus, release and store the latent heat of adsorption [139]. Other M-TES systems utilize metal-coated microencapsulated PCM that can store sensible heat (by changing temperature) and latent heat (by changing its phase from solid to liquid or vice versa) [140]. M-TES systems can be flexibly adapted to meet the requirements of particular solutions. The approximate lifetime of an M-TES systems is 20 years, with an estimated initial investment cost of 6000 EUR [141]. Furthermore, PCM was suggested as an energy storage medium for ocean-based crewless underwater thermodynamic vehicles. However, some technical issues must be addressed before PCM can be employed widely for storing energy from RES. These problems are caused by PCM's slow heat transfer rate due to their commonly low thermal conductivities, low energy storage efficiency, small energy storage density, and lack of synergy between the motion and heat transfer profiles [15].

A transcritical cycle (TRCC) is another efficient option for converting low- and medium-grade heat sources into electricity. The TRCC working fluid (mainly CO_2, but also including R134a, R123, and ammonia) passes through subcritical and supercritical states, where an evaporator converts the fluid into a superheated vapor. Then, a compression process increases the pressure and energy of the superheated vapor. When the pressure exceeds the critical point of the used fluid, it transforms into an undefined gas (supercritical fluid) with combined liquid and gas properties. The energy can be recovered from the supercritical fluid through the expansion process, where the fluid is expanded through a turbine, which drives a generator to produce electricity. Utilizing CO_2 as the TRCC working fluid is cheaper, safer, and more environmentally friendly than using organic fluids and is recommended for "waste heat recovery" (WHR) due to its relatively low operation temperature [142].

4.3. Sorption TES

Sorption TES is a leading-edge technology that depends on thermal adsorption–desorption cycles between a sorbent material and a refrigerant. Two major energy storage processes are involved in sorption TES: "endothermic decomposition" (i.e., desorption step) and "exothermic synthesis" (i.e., adsorption step). Adsorption TES has the advantage of controlling the humidity and storing heat and cold through these endothermic and exothermic steps, making it a promising technology for air conditioning in buildings. However, some challenging phenomena in the sorption bed, including deliquescence, agglomeration, swelling, and low thermal conductivity, hinder the broad application of sorption TES systems [8,143].

A summary of TES forms, media, usage, development status, and advantages and disadvantage is demonstrated in Figure 13. The development and application of the three TES technologies—sensible, latent, and sorption—vary significantly. Sensible TES, being more advanced, is utilized in large-scale district heating/cooling plants, while latent and sorption TES are still in laboratory study stages. Sensible TES requires efficient thermal stratification to optimize these systems, whereas latent- and sorption TES demand heat transfer enhancement techniques. Factors such as cost-effectiveness, environmental sustainability, reliability, and stability must be considered for practical applications [9].

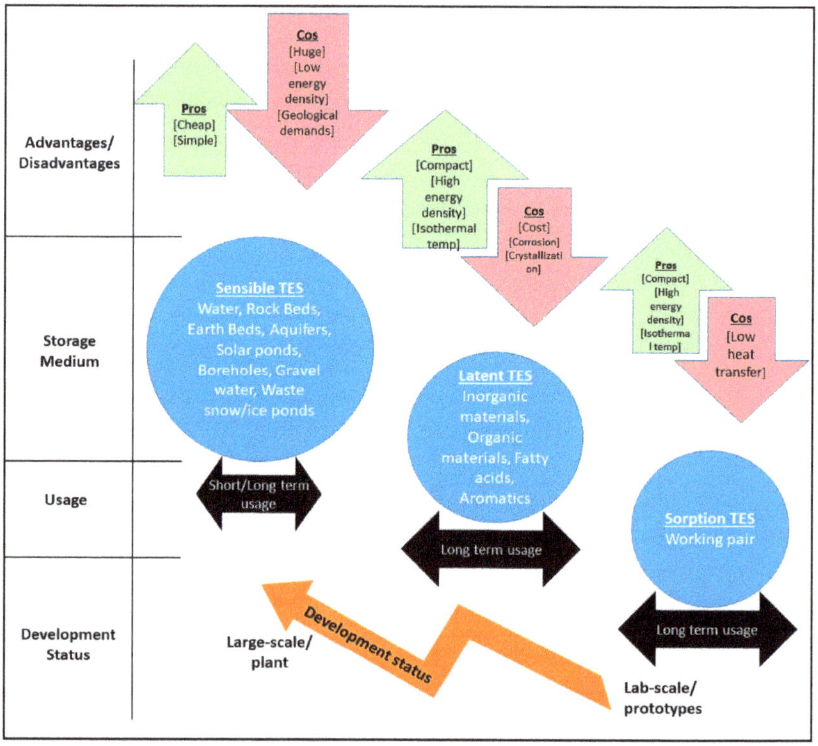

Figure 13. Summary of TES forms, media, usage, development status, advantages, and disadvantage, based on data from [9,97].

5. Combined Renewable Energy–Thermal Energy Storage Systems

As the world's energy demand increases, there is an increasing cognition that nonrenewable resources are limited and can drastically impact the environment. On the other hand, renewable energy sources are commonly unstable and/or intermittent in nature. To mitigate the impacts of the intermittence of renewable energy sources, TES may be a viable option for addressing future electricity demand challenges and assuring a sustainable and steady power supply. Despite the wealth of literature available in the renewable energy and energy storage systems field, previous reviews typically investigated only specific topics related to renewable energy sources or energy storage systems, including thermal ones. This work reviews the state-of-the-art literature that combines renewable energy and thermal storage systems (RE–TES). Table 1 summarizes publications on combined renewable energy and thermal energy storage systems, while Table 2 compares the different RE–TES options on the basis of their efficiency, cost-effectiveness, scalability, and environmental impact. A summary of various simulation tools and research methods used in RE and TES systems is also given in Table 3.

Table 1. Summary of the literature on combined renewable energy and thermal energy storage systems.

Combined Renewable/TES System	Renewable System	Thermal Energy System	Work Type	Opportunities and Constraints	Reference
Solar/sensible TES	Parabolic-trough solar concentrator	Compressed CO_2,	Original research/modeling	***Opportunities***: rising fossil-fuel prices, growing renewable energy demand, technological advancements ***Constraints***: high investment cost, space requirement, weather vulnerability	[106]
Solar/sensible TES	"Concentrated solar power" (CSP)	"geothermal-based plate tectonic boundaries"	Original research/modeling	***Opportunities***: increasing demand for hydrogen, dual purpose of CSP systems (electricity and heat), reliable and sustainable heat from geothermal systems ***Constraints***: high cost of CSP systems, limited geothermal resources, environmental impacts of geothermal systems	[61]
Solar/sensible TES	Ambient solar radiation	Concrete mixtures/molten salt	Original research/modeling	***Opportunities***: efficient solar energy harnessing, concrete's heat capacity and thermal conductivity, scalability and adaptability ***Constraints***: requirement of a heat transfer system, limited scope of study (focuses on comparing the heat capacity and thermal conductivity of concrete as a filler material)	[111]
Solar/sensible TES	Ambient solar radiation	Aquifer	Original research/modeling	***Opportunities***: energy cost savings, carbon footprint reduction, improved comfort and wellbeing ***Constraints***: lack of generalizability (conducted in a single hospital in Turkey), cost consideration, did not consider the environmental impact	[144]

Table 1. *Cont.*

Combined Renewable/TES System	Renewable System	Thermal Energy System	Work Type	Opportunities and Constraints	Reference
Solar/latent TES	"Visible solar storage fabric" (VSSF)	"Azo-PCM@PS" nanocapsule and "$Cs_{0.32}WO_3$" nanoparticle	Original research/experiment	***Opportunities***: sustainable energy for wearable devices, versatile applications, environmental data collection ***Constraints***: laboratory setting, early development stage, limited application suitability	[145]
Solar/latent TES	"Concentric solar power" (CSP)	Thermochemical water-splitting cycles	Original research/experiment	***Opportunities***: hydrogen production from renewable sources, diverse hydrogen applications, amelioration of climate change ***Constraints***: theoretical model, early development stage, climate and location limitations	[146]
Solar/latent TES	Shell-and-tube heat exchanger with a transparent silica glass shell	Copper foam embedded in PCM	Original research/experiment	***Opportunities***: improved heat transfer capacity, better temperature uniformity, reduced melting time ***Constraints***: scalability, availability, further research needed	[129]
Solar/sensible and latent TES	Hot air collectors	Sensible TES: packed bed TES (PBTES) using pebble stones Latent-TES: PCM using paraffin wax	Original research/experiment	***Opportunities***: increased energy efficiency, reduced drying time, preserved product quality, value of computational numerical modeling ***Constraints***: research needs, implementation challenges	[101]
Wind/sensible TES	Wind power with wind turbines	"Thermal energy grid storage influenced by multijunction photovoltaics" (TEGS-MPV)	Original research/modeling	***Opportunities***: environmental benefits, optimization and sensitivity analysis, renewable energy utilization ***Constraints***: validation, real-world implementation, geographic specificity	[147]
Wind and solar/variable TES and other energy storage systems	Wind power with wind turbines	"Compressed air energy storage" (CAES), "pumped hydroelectric storage", and "sodium–sulfur batteries"	Original research/experiment	***Opportunities***: potential for wind and solar energy, various energy storage methods ***Constraints***: technical limitations, fluctuating wind turbine output	[16]
Wind/sensible TES	Wind power with wind turbines	TES embedded inside wind turbine nacelles, with	Original research/ thermodynamic analysis	***Opportunities***: increased energy efficiency, sustainable energy solutions, effective utilization of renewable energy ***Constraints***: technical challenges and uncertainties in the development, need for detailed assessments	[148]

Table 1. *Cont.*

Combined Renewable/TES System	Renewable System	Thermal Energy System	Work Type	Opportunities and Constraints	Reference
Biomass/latent TES	"Guar gum"	carbon aerogel for encapsulating "polyethylene glycol" (PEG)	Original research/experiment	***Opportunities***: improved solar–thermal energy conversion and storage, use of biomass materials, cost-effective energy solutions ***Constraints***: achieving homogeneity, optimization of carbon aerogels	[130]
Biomass/sensible TES	Bioproducts	Different TES	Review	***Opportunities***: sustainable production of electricity, fuels, and chemicals, flexible renewable-based utility plants, economic and technological feasibility ***Constraints***: limitation of biomass resources, complex conversion processes, environmental impacts, carbon release	[51]
Biomass/sensible TES	Wood pellet	TES reboiler tank	Original research/modeling	***Opportunities***: improving district heating systems, reducing costs and emissions ***Constraints***: technical barriers, economic barriers, geographical and temporal barriers	[149]
Biomass and other sources/sensible TES	Biomass-powered combined heat and power systems	Different TES	Review	***Opportunities***: enhanced energy efficiency, reduced greenhouse gas emissions, new revenue streams, energy security, sustainable energy transition ***Constraints***: sustainable biomass supply, efficient conversion technology	[43]
Geothermal/latent TES	"Geothermal district heating system" (GDHS)	Undefined TES	Original research/case study analysis	***Opportunities***: sustainable and renewable energy source, enhanced performance and reliability, informed decision making with multicriteria decision analysis ***Constraints***: dimensioning the district heating system, efficiency and cost of thermal energy storage technologies	[150]
Geothermal/sensible TES	"High-temperature aquifer thermal energy storage" (HT-ATES)	Undefined TES	Original research/risk analysis	***Opportunities***: risk identification and mitigation, promoting sustainable energy solutions ***Constraints***: limited data on HT-ATES systems, uncertainties in subsurface conditions, potential environmental impacts	[14]
Geothermal/sensible TES	"Mobile thermal storage system" (M-TES)	Undefined PMC	Original research/ techno-economic assessment	***Opportunities***: renewable and stable energy source, high-capacity factors, mobile thermal energy storage ***Constraints***: climate factors	[141]

Table 1. *Cont.*

Combined Renewable/TES System	Renewable System	Thermal Energy System	Work Type	Opportunities and Constraints	Reference
Ocean/sensible TES	"Ocean thermal energy conversion" system (OTEC)	Undefined TES	Original research/modeling	*Opportunities:* improved control strategies, renewable energy for islands, enhanced efficiency and effectiveness ***Constraints*:** limited controller comparison	[94]
Ocean/sensible TES	Ocean energy	Undefined TES	Review	*Opportunities:* advanced ocean energy converters, diversified ocean energy systems and hybrid energy storage, artificial intelligence integration, complementary hybrid renewable system integrations ***Constraints*:** power frequency fluctuation	[151]
Ocean/latent TES	Ocean thermal energy	Different PMC	Review	*Opportunities:* utilization of ocean thermal energy, PCM thermal-harvesting systems ***Constraints*:** slow heat transfer rates, low conversion efficiency, low energy storage density, conceptual design phase	[15]
Poly: solar-geothermal/sensible TES	"Enhanced geothermal energy" (EGS) and "concentrated solar power" (CSP)	Undefined TES	Original research/case study analysis	*Opportunities:* continuous energy production, integration of thermal energy storage, sustainable and green alternative, improved techno-economic performances ***Constraints*:** harnessing geothermal energy, solar intermittency, large-scale land use	[152]
Poly: ocean and solar/latent TES	"Concentrated solar plant" (CSP), a "bifacial photovoltaic" (BiPV), a "cascaded heat pump", and a "multieffect desalination" process	"Polymer electrolyte membrane" (PEM) electrolyzer, fuel cell systems, and thermal energy storage systems	Original research/case study analysis	*Opportunities:* renewable energy utilization, integrated hydrogen production and thermal energy storage, combining food production systems, quality of life improvement ***Constraints*:** harsh arctic conditions, system optimization	[107]
Poly: solar, wind, and ocean/TES	"Ocean thermal energy convertor" (OTEC), a wind turbine, and a "solar flat plate panel"	Undefined TES	Original research/case study analysis	*Opportunities:* sustainable and efficient energy system, reduced dependence on fossil fuels, contribution to energy security ***Constraints*:** geographical specificity, substantial initial investment and infrastructure development, intermittency of renewable energy sources	[153]

Table 1. *Cont.*

Combined Renewable/TES System	Renewable System	Thermal Energy System	Work Type	Opportunities and Constraints	Reference
Geothermal and solar/latent TES	A solar-geothermal hybrid system	CO_2 cycle with an organic Rankine cycle and heat exchangers	Original research/ thermodynamic analysis	*Opportunities:* efficient utilization of renewable resources, reliability and flexibility in energy supply, improved thermodynamic performance, optimization of system configurations ***Constraints***: high initial capital costs, geographical and climatic dependency, availability of suitable materials and technologies for thermal energy storage, technical challenges with the transcritical CO_2 Cycle	[114]
Geothermal and solar/sensible TES	ATES combined with solar collectors	Undefined TES	Original research/case study analysis	*Opportunities:* enhanced energy efficiency, reduced environmental impact, cost-effective integration of renewable energy sources ***Constraints***: technical challenges, regulatory barriers, dependence on local conditions	[112]
Geothermal and solar/sensible TES	Solar energy system powered by geothermal	Organic Rankine cycle	Original research/ thermodynamic analysis	*Opportunities:* enhanced energy efficiency, reduced dependence on fossil fuels, improved ORC performance ***Constraints***: geographical limitations, high initial costs, intermittency of solar energy	[105]

Table 2. A comprehensive comparison of the different RE–TES options based on their efficiency, cost-effectiveness, scalability, and environmental impact.

RE–TES Option	Efficiency	Cost-Effectiveness	Scalability	Environmental Impact	References
Solar systems with battery storage	Moderate (15–20%)	Moderate–high (depending on battery type)	Moderate–low (limited by battery capacity)	Low (no emissions during operation, but batteries require rare Earth metals and other materials)	[61,101,105,106,111,112, 114,129,144–146,153]
Wind turbines with pumped hydro storage	High (30–50%)	High (long lifespan of hydro storage infrastructure)	High (can be scaled up to meet large energy demands)	Moderate (construction of hydro storage infrastructure can have environmental impacts, but the operation is low-impact)	[16,147,148,153]
Geothermal power with thermal energy storage	High (70–90%)	Moderate–high (depending on location and drilling costs)	Low–moderate (limited by geothermal resources in certain areas)	Low–moderate (minimal emissions during operation, but drilling can have environmental impacts)	[14,105,112,114,141,150]
Biomass power with thermal energy storage	Moderate (20–30%)	Low–moderate (depending on feedstock availability and cost)	Low–moderate (limited by feedstock supply)	Moderate–high (emissions from biomass combustion can contribute to air pollution and climate change, but sustainable sourcing can mitigate some impacts)	[43,51,130,137,149]
Ocean energy with thermal energy storage	Low–moderate (10–20%)	Moderate (infrastructure can be costly, but operational costs are low)	High (especially in coastal regions with consistent wave/tidal action)	Low–moderate (potential impact on marine ecosystems during construction, but operation is low-impact)	[15,94,107,151]

Table 3. Summary of various simulation tools and research methods used in RE and TES systems.

Simulation/Method	Description	References
Artificial neural networks (ANNs)	This machine learning tool is used to model and predict the behavior of systems based on training data. It was used in some studies to develop predictive models for the performance of solar collectors and TES systems.	[154]
EnergyPlus	This building energy simulation program is used to model and simulate the energy performance of buildings. It was used in some studies to analyze the performance of building integrated renewable energy systems and TES systems.	[155,156]
MATLAB/Simulink	This numerical computing tool is used for mathematical modeling, simulation, and analysis of dynamic systems. It was used in some studies to model and simulate the behavior of solar thermal systems and TES systems.	[153,157,158]
System dynamics modeling	This simulation tool is used to model complex systems' behavior over time. It was used in some studies to simulate the behavior of renewable energy systems and thermal energy storage systems.	[159,160]
TRNSYS (Transient System Simulation Tool)	This dynamic simulation tool is used to model and simulate the performance of renewable and thermal energy storage systems. It was used in some studies to analyze the performance of different solar collectors and TES systems.	[161,162]

Table 1 summarizes various studies in the literature that focused on combined renewable energy sources and thermal energy storage (RES–TES) systems. The studies span a range of renewable energy sources, including solar, wind, biomass, and geothermal power, as well as ocean-based systems. The TES systems considered in these studies include sensible and latent storage technologies, with some studies examining integrating multiple types of storage in a single system. The research methodologies employed in these studies are diverse, with some employing original research through modeling, experimentation, thermodynamic analysis, or case study analysis. Other studies also take the form of comprehensive reviews, providing an overview of the current knowledge in the field.

The wide range of studies demonstrates the growing interest in developing innovative combined RES–TES systems to address energy supply and storage challenges sustainably. Several studies that investigate solar energy combined with sensible TES employ technologies such as parabolic trough solar concentrators, concentrated solar power (CSP) systems, and ambient solar radiation collectors [61,106,111,138]. Other studies focus on solar energy combined with latent TES, utilizing materials such as phase change materials (PCMs) and copper foam [128,139,140]. Wind energy is also considered in combination with various TES systems, including sensible storage using thermal energy grid storage and variable TES systems such as compressed air energy storage (CAES), pumped hydroelectric storage, and sodium–sulfur batteries [14,141,142]. Moreover, biomass energy is explored with sensible and latent TES systems, often involving bioproducts and different storage materials [35,51,129,143]. Geothermal energy is frequently combined with sensible TES, often in the form of high-temperature aquifer thermal energy storage (HT-ATES) and mobile thermal storage systems (M-TES) [15,135,144]. Ocean energy, another renewable source, is also examined in combination with both sensible and latent TES systems [16,94,145]. Some studies investigate hybrid renewable energy systems incorporating multiple renewable sources, such as solar–geothermal, ocean–solar, and solar–wind–ocean systems [105,107,112,114,146,147]. These hybrid systems typically involve sensible or latent TES to optimize the combined use of different renewable resources.

5.1. Combined Solar/TES System

The combined use of solar energy and thermal energy storage systems has been examined in several studies. Many researchers investigated solar energy storage in the form of sensible energy. Soni and his team [106] devised a pioneering mathematical model to simulate the process of thermal energy storage (TES) drawn from solar energy, purposed explicitly for winter residential heating. This renewable system employed a parabolic trough solar concentrator with a novel design for the TES medium, utilizing compressed CO_2. This design aimed to minimize heat loss by integrating several layers of highly insulating materials, thereby maximizing solar energy storage within the gas medium. The study resulted in an innovative solar/TES system for home heating that offers a compelling solution for year-round heating at a cost that is 42.5% less than traditional methods. Promising opportunities abound for the adoption and enhancement of such technology. With the steady rise in fossil-fuel prices, renewable energy alternatives such as solar-powered TES systems could become increasingly appealing economically, despite their substantial initial costs. Furthermore, as societies become more aware and responsive to climate changes, the demand for renewable energy sources is projected to grow. This creates a more conducive environment for the implementation of TES technologies. The relentless pace of technological advancement, particularly in material science, also offers the prospect of addressing current limitations. As more efficient, resilient, and cost-effective TES solutions are developed, the potential of TES as a residential heating option is likely to grow. However, there are significant constraints. A substantial initial financial outlay is required for the installation of TES systems, including both the cost of the system and the installation. In addition, deploying solar-powered TES systems often requires large land areas to house the solar collectors. This requirement could present a significant hurdle, particularly in urban or densely populated areas with limited space. Weather vulnerability also poses a substantial challenge, as TES systems are susceptible to extreme weather events. This vulnerability is particularly pertinent in regions prone to such conditions, where damage could compromise system performance or cause complete failure. Despite these challenges, the potential benefits and opportunities presented by TES systems could serve as compelling reasons for further research and development.

Temiz and Dincer [61] crafted an innovative model, merging solar and geothermal systems to create an array of valuable outputs, including hydrogen, electric power, and freshwater. This was accomplished through comprehensive thermodynamic analysis. Their approach incorporated concentrated solar power (CSP) as a renewable energy source and harnessed geothermal-based plate tectonic boundaries for thermal energy storage (TES) to generate electricity, hydrogen generation, space heating, and freshwater production within a hypothetical community near the Geysers area in California. This sophisticated system was thermodynamically scrutinized using Aspen Plus, NREL's SAM, and HOMER Pro software packages. TES and CSP systems were modeled using tanks of 15 m height and absorber tubes of 8 cm diameter. The CueCl hydrogen production plant, when combined with desalination processes, generated approximately 160,390 tons of fresh water. Additionally, it contributed 48 GWh of electricity, 453 GWh of heat, and 297 tons of hydrogen. CSP and geothermal systems, despite their complexities, offer unique opportunities in the renewable energy realm. One significant opportunity lies in the burgeoning demand for hydrogen. The rising interest in clean energy sources has highlighted hydrogen's potential as a green fuel in various sectors. Furthermore, the dual-purpose nature of CSP systems is an attractive prospect. These systems generate electricity and produce and store heat for future use, thereby enhancing energy use efficiency and reducing dependency on the electrical grid. Geothermal systems also present valuable opportunities. They provide a stable, reliable heat source, making them an ideal choice for district heating and industrial processes. Their heat can also be harnessed for electricity generation, presenting a sustainable and steady power source. However, several constraints demand a critical attention. The high cost associated with CSP systems is a significant limitation. These systems necessitate considerable initial investments, primarily due to the requirement of large-scale solar collector installations, energy storage systems, and associated infrastructure. The expense can be particularly daunting for smaller-scale projects. Geothermal energy, while promising, is limited by the availability of suitable geological conditions—specifically, heat emanating from the Earth's crust. This inherent constraint restricts the scalability of geothermal energy. Moreover, geothermal systems, despite being renewable, pose potential environmental impacts. These include the release of trapped greenhouse gases, surface instability due to geothermal fluid extraction, and a risk of water contamination if the extraction process is not meticulously managed. Balancing these opportunities and constraints is critical for successfully integrating and advancing CSP and geothermal systems in the renewable energy sector.

Boretti [146] proposed an innovative approach of amalgamating solar–thermal energy storage (TES) systems with thermochemical water-splitting cycles to produce cost-effective hydrogen. The envisaged system anticipated the generation of 0.1 GW of continuous electricity, thereby facilitating the production of approximately 2750 kg/h of hydrogen daily by the year 2030. The model utilized concentric solar power (CSP), deploying photovoltaic (PV) panels for solar energy-harvesting, and advanced ultra-supercritical CO_2 as the medium for storing the produced energy. Opportunities for this initiative are abundant. The system presents a promising path toward hydrogen production from renewable energy sources, potentially reducing the dependence on fossil fuels. Moreover, the generated hydrogen could find myriad applications, including its use in fuel cells, ammonia production, and rocket fuel, thereby encouraging cleaner and more efficient energy sources. The system also offers the prospect of mitigating climate change by enabling cleaner energy production and reducing carbon emissions. Beyond environmental benefits, the broad application of the produced hydrogen could extend to various sectors, including transportation, power generation, and industrial processes, thus broadening the reach of renewable energy utilization. Furthermore, developing and implementing this system could stimulate job creation and economic growth within the clean energy sector. Despite these opportunities, certain constraints demand thoughtful consideration. One such limitation is that the study is based on a theoretical model; hence, its effectiveness and efficiency in real-world scenarios still need to be verified. Moreover, as the system is in its infancy, vital factors such as cost and performance still need to be defined, potentially discouraging immediate adoption. Additionally, the system's effectiveness may depend on geographical location and climatic conditions, limiting its applicability. The high initial investment could also present a significant barrier to implementing such systems for potential users or applications. Moreover, the system may necessitate regular maintenance, potentially inflating the total cost of ownership over time. Lastly, the system's susceptibility to external factors, such as dust, corrosion, or extreme weather conditions, could compromise its durability and efficiency. Balancing these opportunities and constraints will be crucial for the successful implementation and broader acceptance of this promising technology.

In solar/TES systems, concrete can be used as a TES medium to store the heat generated by solar collectors during the day and release it at night or during periods of low sunlight. However, in order for concrete to be an effective storage medium, it must be able to withstand high temperatures without significant degradation or loss of structural integrity. In a study investigating concrete as a thermal energy storage medium, John and his coworker [111] undertook a novel endeavor, developing 26 unique concrete mixtures using sandstone, limestone, and syenite, combined with washed river sand in varied proportions. These mixtures underwent casting, curing, and a month-long immersion in water at room temperature. Subsequently, they were cured in saturated air at 90 °C for a couple of days. The endurance of these concrete mixtures was then tested in molten salt at 585 °C for 500 h, followed by exposure to 30 cycles in the air. A notable outcome of thermal cycling in molten salt was a substantial reduction in the compressive strength of both concrete and mortar when subjected to temperatures up to 600 °C. This research offers intriguing opportunities for applying concrete in thermocline solar energy storage systems. The methodology introduced by John and his team reveals an innovative pathway toward more efficient solar energy harnessing. Utilizing concrete as a storage medium, solar thermal energy can be effectively absorbed and held for subsequent use. The inherent properties of concrete, such as its commendable heat capacity and thermal conductivity, position it as an ideal candidate for solar thermal energy storage systems. This implies that concrete can absorb, retain, and transmit heat effectively. Furthermore, the concept of a modular cement-based solid–liquid heat storage system suggests a potential for scalability and adaptability. This system offers the flexibility to be easily modified or expanded based on specific requirements. While the opportunities are substantial, a few constraints necessitate careful consideration. A concrete-based storage medium requires a heat transfer system, generally a fluid, for optimal functioning. This requirement introduces an additional component into the system, increasing its complexity and cost. Moreover, the limited scope of the study presents another constraint. The focus was primarily on comparing the heat capacity and thermal conductivity of concrete used as a filler material, leaving out other potentially influential factors such as durability, long-term performance, and environmental impact, which indicates the necessity for a more comprehensive research approach in this domain. Balancing these opportunities and constraints will be integral to effectively use concrete as a thermal energy storage medium in solar energy systems.

Paksoy and coworkers [144] employed the CONFLOW simulation program to conceive a system that optimizes energy conservation by harnessing solar power and an aquifer for seasonal thermal energy storage. Explicitly envisioned for a hospital in Adana, Turkey, this system served for heating and cooling purposes. It exhibited the capacity to store 7000 MW/year at an average temperature exceeding 98 °C. During winter, the proposed system ingeniously utilized hospital ventilation air and proximal surface water to cool the aquifer, storing thermal energy. Concurrently, the system prewarmed the ventilation air with the assistance of two heat exchangers. This operation was reversed during the summer. Despite the promising aspects, some constraints exist and could be improved. One concern is the lack of generalizability, as the study was conducted in a specific hospital setting in Turkey. Thus, the applicability of its findings to hospitals located in different regions or countries could be questioned due to variations in climate, architectural design, and pre-existing energy infrastructure. A noteworthy omission in the study was the cost of implementing such a system. Considering that hospitals function within prescribed budget constraints, the financial implications of adopting this system could play a crucial role in its broader usage. The study also overlooked the environmental impact of deploying this system, a significant consideration given the escalating concern about sustainability across all sectors, including healthcare. On the flip side, there are significant opportunities to be explored. Foremost is the potential for significant energy cost savings. The savings could be substantial, considering the high-energy demands typically associated with hospital operations. The system also provides an opportunity to reduce reliance on nonrenewable energy sources, potentially aiding hospitals in reducing their carbon emissions and aligning with global climate change mitigation efforts. Furthermore, establishing a well-regulated and efficient energy system could create a more comfortable environment for patients and staff. This could improve patient satisfaction and enhance staff productivity, thereby underscoring the system's holistic benefits. Balancing these opportunities against the constraints will be essential in determining the system's broader applicability and effectiveness.

Many researchers investigated the storage of solar energy in the form of latent energy. In an experimental study, Wang and colleagues [129] embarked on a study exploring the potential of latent thermal energy storage by leveraging n-tetradecane as a phase change material (PCM) embedded in gradient porosity copper foam. This was aimed at the efficient storage of mid-temperature solar energy. Solar energy was harnessed via a shell-and-tube heat exchanger, where a transparent silica glass shell was utilized to capture solar radiation. Simultaneously, the TES system was endowed with PCM encapsulated within the copper foam to amplify its hardness, thermal conductivity, and stability. The team devised a gradient porosity metal foam (GPMF) to circumvent the issue of low conductivity typically associated with PCM in mid-temperature solar energy storage systems. The experimental results indicated a substantial improvement in heat transfer with the employment of GPMF. The study, however, presents several constraints. A critical constraint is the scalability of the production process for gradient porosity copper foam. Given the specificity of this material, it could be challenging to mass-produce, potentially impeding the broad adoption of this technology. Additionally, the availability of gradient porosity copper foam could be restricted, making it challenging to source and potentially hindering the technology's widespread deployment. Furthermore, the findings necessitate more exhaustive research to validate the conclusions drawn and explore potential long-term impacts across various applications. The initial study may not fully address the practical realities and challenges that emerge in real-world scenarios and applications. Conversely, the study also highlights several promising opportunities. Employing gradient porosity copper foam could drastically enhance heat transfer capacity in mid-temperature solar energy storage systems, improving their efficiency and effectiveness. Another significant benefit is the potential for better temperature uniformity achieved through this material, which could lead to the more consistent and efficient performance of solar energy storage systems. Lastly, the study indicates that using gradient porosity copper foam could reduce the melting time by 37.6%. This could lead to solar energy storage systems becoming quicker and more efficient in capturing and storing solar energy.

For a solar energy-assisted drying process, Atalay [101] represents a crucial step toward understanding the energy and cost implications of the packed bed and phase change material (PCM) thermal energy storage systems. By comprehensively evaluating these media for storing sensible and latent energies, Atalay underscores the economic advantage of packed bed due to its lower initial investment cost but points to the superior thermal storage capacity of paraffin wax PCM. Atalay's investigation opens up promising opportunities. Integrating solar dryers and thermal energy storage techniques can enhance energy efficiency, leading to more sustainable and potentially cost-effective drying processes. Moreover, adopting these energy storage systems could significantly reduce drying time, boosting productivity. An intriguing insight from the study suggests that these systems aid in preserving product quality during the drying process. This potential benefit could result in higher-quality dried goods, enhancing their marketability. Furthermore, the study underscores the utility of computational numerical modeling for optimizing drying systems, thus spotlighting an avenue for further improvement in system efficiency and performance. Despite these promising opportunities, the study also presents considerable constraints that warrant further attention. The research underlines the need for more in-depth exploration to identify optimal materials and configurations for these energy storage systems, highlighting that the full potential of these systems is yet to be realized. The challenge of implementing these systems under varying climates and conditions is also identified. This suggests that customization is necessary to suit specific environmental factors, which could amplify complexity and cost. Fei and his team [145] innovatively developed a wearable solar energy management system that leverages visible solar thermal energy storage for complete solar spectrum utilization. Their invention termed the "visible solar storage fabric" (VSSF), employs a unique combination of "Azo-PCM@PS" nanocapsules and "$Cs_{0.32}WO_3$" nanoparticles. The outcome is a "thermochemical–thermophysical coupled energy" storage system with visible to near-infrared (Vis–NIR) light-harvesting capabilities. Remarkably, the designed wearable solar energy system has demonstrated the ability to release significant solar heat up to about 85 °C, which could provide notable protection against cold injuries to the human body. Fei's team's pioneering work presents several exciting opportunities. First, it has the potential to provide a sustainable and reliable energy source for a wide array of wearable devices, including but not limited to smartwatches, fitness trackers, and medical devices. The technology's versatility powers a broad spectrum of wearable tech applications. Additionally, the system can be harnessed to collect environmental data such as temperature, humidity, and air quality, thus providing valuable insights for various applications. Beyond these notable points, the solar energy management system could extend its applications to power other portable devices such as smartphones and laptops, further enhancing its usefulness. This technology could be a crucial renewable energy source in off-grid situations where traditional power infrastructure is lacking. Furthermore, if these systems gain widespread acceptance, the data collected could enhance weather forecasting accuracy and refine climate modeling. Despite these promising opportunities, the study by Fei et al. also uncovers potential constraints that require careful attention. For one, the testing environment was a laboratory, and it remains to be seen how the system will perform under real-world conditions, which include weather fluctuations, outdoor wear and tear, and varied user behaviors. Secondly, being in the early stages of development, the cost implications and total system performance still need to be defined, potentially posing a challenge to immediate implementation. A few more concerns also arise. The system's application may be restricted by certain factors, such as location and climate, which can influence the availability of sufficient solar energy. Additionally, wearability and durability questions surface; the system might be too heavy or bulky, negatively affecting user comfort, or fragile, leading to concerns about its durability with frequent use. Lastly, regular maintenance might be necessary for optimal system functioning, potentially contributing to higher total ownership costs.

Overall, the combined use of solar energy and thermal energy storage systems presents several opportunities, including the potential for cost-effective hydrogen production, significant energy cost savings, and reduced reliance on nonrenewable energy sources. It also offers the prospect of mitigating climate change and creating a more comfortable environment for patients and staff. However, several constraints demand critical attention, including the high cost associated with CSP systems, the lack of generalizability of some studies, and the need for regular maintenance. Additionally, the effectiveness of the system may depend on geographical location and climatic conditions, and the cost implications and total system performance still need to be defined.

5.2. Combined Wind/TES System

Several studies have explored the integration of wind energy and thermal energy storage (TES) systems to enhance the efficiency and performance of renewable energy applications. Many researchers investigated the storage of wind energy in the form of sensible energy, but no research was found on storing wind energy in the form of latent energy. Al-Mashakbeh and colleagues [147] devised a compelling combination of a simulated wind farm with "thermal energy grid storage multijunction photovoltaics" (TEGS-MPV) in Jordan. Their approach harnessed the capabilities of HOMER® software to simulate a 60 MW wind power production scenario involving wind turbines, inverters, and TEGS-MPV, resulting in an attractive energy cost of 0.04252 USD/kWh. The study opens up many opportunities, beginning with its potential to generate significant environmental benefits. It could substantially curb greenhouse gas emissions, with an estimated annual reduction of 293,764 tons of carbon emissions. Another highlight of the study is the importance placed on optimization and sensitivity analysis, achieved through the application of HOMER® software. This process enables precise calculation of levelized cost of energy (LCOE) values, assisting in identifying optimal scenarios for the TEGS-MPV system deployment. Moreover, the research promotes an innovative strategy for integrating wind energy with thermal energy grid storage. This creative approach fosters broader acceptance and use of renewable energy sources while enhancing energy efficiency. Despite these promising opportunities, the study also brings to light several constraints that merit attention. One notable constraint is the need for practical validation of the proposed system. Given the study's theoretical nature, it is crucial to test the feasibility of the TEGS-MPV system in real-world settings. Another constraint lies in translating this theoretical approach into practical application. There may be technical, financial, or regulatory challenges during the implementation phase of the TEGS-MPV technology that need to be surmounted. Additionally, the study is geographically specific, focusing on Mafraq, Jordan. Therefore, the transferability and effectiveness of the TEGS-MPV system in different regions or countries might be influenced by various factors, including local climatic conditions and regulations.

Caralis and his team [16] delve into the evaluation of potential energy storage systems in Crete, designed to harness the fluctuations of wind energy. They put a spotlight on a "compressed air energy storage" (CAES) system, integrated with a thermal energy storage (TES), in comparison to other energy storage systems. Several storage solutions, including CAES, "pumped hydroelectric storage", and "sodium–sulfur batteries", were scrutinized for their efficacy in storing electrical energy derived from wind power during periods of curtailment exploitation. In terms of economics, CAES was slightly more advantageous, costing 0.21 EUR/kWh. However, despite being economically less viable, sodium–sulfur batteries emerged as strong contenders for positively balancing generated loads. Starting with opportunities, it is clear that Crete boasts significant untapped potential for exploiting wind and solar energy. With the right infrastructure and energy storage systems, this latent renewable energy could be harnessed more efficiently. Energy storage systems are key in storing excess energy wind turbines produce. Utilizing these technologies could reduce, if not eliminate, the need for energy curtailment. This would augment the wind energy system's overall efficiency and economic viability. Furthermore, this study presents a comprehensive examination of different energy storage technologies, assessing their potential to mitigate the issue of wind energy curtailment. This analysis is a valuable foundation for future research and developments to maximize the benefits of renewable energy utilization in Crete. Turning to constraints, the study points out potential technical limitations with existing thermal units. These issues may impede the effective harvesting of wind energy, necessitating appropriate measures to overcome these limitations. Another challenge lies in the inconsistency of wind turbine output, which can fluctuate due to varying wind speeds. This could lead to challenging, often curtailed energy surpluses, reducing the wind energy system's overall efficiency and economic appeal.

Karasu and Dincer [148] evaluated the effectiveness of a hybrid system combining electromagnetic induction and thermal energy storage (TES), designed to convert wind energy into heat directly. A distinctive feature of their system is the placement of the TES cycle within wind turbine nacelles, which allows the system to operate continuously. Using induction heating to heat the TES within wind turbine nacelles empowers the system to operate without interruptions, facilitating the direct conversion of wind energy to heat. Overall energy efficiencies of 7.0% and 8.6% were achieved when using electromagnetic induction and TES, respectively. The innovative wind energy/TES system is a potential substitute for traditional fossil-fuel-based facilities and nuclear power plants. Opportunities present themselves in this study with the possibility of enhanced energy efficiency. The direct transformation of wind energy into heat sidesteps unnecessary intermediary conversion steps, which can reduce overall energy losses. This approach is in harmony with a broader global shift toward renewable energy and could be instrumental in curtailing greenhouse gas emissions. Moreover, the innovative system proposes a more effective way of harnessing renewable energy resources. As the global demand for cleaner, renewable energy surges, technologies such as this wind energy/TES system could significantly affect our future energy landscape. Despite these promising prospects, there are a few constraints that need consideration. The research underscores the need for a comprehensive understanding of thermal energy storage and the cost-effectiveness of wind energy-TES technology. A thorough economic and technical feasibility assessment is crucial for its large-scale implementation. Furthermore, given the relative novelty of this approach, technical challenges and uncertainties might arise during the development, deployment, and operation of these systems. Hence, further research and development are imperative to refine this technology and ensure its reliability and efficiency.

Overall, the opportunities of combining wind energy and thermal energy storage systems include the potential for more efficient use of renewable energy resources, reduction in greenhouse gas emissions, and the possibility of replacing traditional fossil-fuel-based facilities and nuclear power plants. However, constraints include technical limitations with existing thermal units, the inconsistency of wind turbine output, and the need for a comprehensive understanding of thermal energy storage and the cost-effectiveness of wind energy/TES technology. Further research and development are necessary to refine this technology and ensure its reliability and efficiency.

5.3. Combined Biomass/TES System

The combined use of biomass energy and thermal energy storage systems has been examined in several studies. Some researchers investigated the storage of biomass energy in the form of sensible thermal energy. Fushimi [51] offers an insightful review of biomass power generation systems, evaluating their economic worth and inherent limitations. He delves into a thorough examination of combustion technologies, the issues of lower energy conversion rates from solar power to electricity, and high fuel costs. To mitigate the mismatch between the sporadic electricity supply from various renewable energy sources and the demand for electricity, the review proposes integrating next-generation biomass energy systems with other fluctuating renewable energy sources such as solar and wind power and energy storage systems. This approach could potentially address the non-steady operations induced by an intermittent power supply. The opportunities provided by the study illuminate the potential for biomass to serve as a sustainable alternative for producing electricity, fuels, and chemicals. Substituting fossil fuels with biomass opens up promising prospects for a greener future. Moreover, the study emphasizes the potential of establishing flexible renewable-based utility plants. Such plants could generate a stable stream of renewable energy, significantly enhancing the power grid's reliability. Moreover, it prompts a discussion on the economic and technological feasibility of power-generating systems that utilize biomass resources. This comprehensive analysis could inform policymaking decisions, advocating for the broader adoption of biomass power generation. However, several constraints need to be considered. The limitation of biomass resources emerges as a critical challenge. The availability of these resources can vary significantly across regions, demanding sustainable management strategies to prevent adverse environmental effects such as deforestation and biodiversity loss. The complexity and potential expense of the biomass-to-energy conversion processes can also pose challenges to the widespread deployment of biomass power generation. Furthermore, environmental impacts associated with improper resource management can lead to ecological degradation, highlighting the need for careful consideration of the environmental implications. Additionally, the carbon released during biomass combustion, although often deemed part of a closed carbon cycle, can contribute to short-term spikes in atmospheric carbon levels, posing another potential constraint.

Rezaei and colleagues [43] undertake a thorough review of the integration of biomass-powered combined heat and power (BCHP) systems with thermal energy storage (TES) in district heating. They scrutinize many optimization models, including economic and environmental ones incorporating parametric or sensitivity analyses. Various parameters related to thermodynamics, hydraulics, chemistry, costing, and decision-making variables are considered. Since the heat value of syngas produced from biomass is lower than fossil fuels, the researchers investigate the alternatives of combining syngas with natural gas or utilizing it in a coal-fired subsystem. To deal with uncertainties in the biomass supply chain, they propose a post-optimization evaluation to estimate the probability of the optimal scenarios. Rezaei and colleagues spotlight a significant opportunity to augment energy efficiency by integrating BCHP systems with district heating and thermal energy storage. Their approach opens up the potential for substantial reductions in greenhouse gas emissions, offering a significant step forward in combating climate change. Moreover, the installation and upkeep of BCHP systems have the potential to stimulate local job creation, contributing to economic development. There is also an enticing prospect of opening up new revenue streams by exploiting biomass waste, which adds an economic incentive to adopting BCHP systems. Furthermore, BCHP integration could boost energy security by diversifying the energy portfolio and diminishing dependence on fossil fuels. Integrating BCHP systems could expedite the transition toward a more sustainable energy infrastructure by increasing the utilization of renewable resources. However, several constraints emerge in this field. One of the significant challenges lies in securing a consistent and sustainable supply of biomass, which is subject to regional and seasonal fluctuations. There is also a pressing need for the evolution and broad acceptance of efficient and environmentally friendly technologies for converting biomass to energy. Integrating BCHP systems into the existing energy infrastructure presents a significant challenge due to potential technical and regulatory barriers. Issues related to biomass combustion emissions, regulatory obstacles, and public acceptance are crucial and need to be addressed adequately.

In their study, Wang and colleagues [149] implemented a simulation of a biomass-fueled boiler based on field data harvested from a wood pellet boiler equipped with radiant floor heating. Key variables such as thermal energy storage (TES) tank discharge efficiency, maximum product temperature, and boiler on/off times are considered. Using these parameters, the simulation provides insights into the optimal TES reboiler volume, capacity, and heat demand profiles. Findings suggest that a reboiler operating on a medium heat demand profile requires the smallest TES tank volume. Conversely, an intermittently used reboiler demands a considerably larger TES tank. Interestingly, a reboiler with a high heat demand profile leads to the highest TES tank discharge efficiency but also calls for a substantial storage volume. For economic efficiency, the team recommends sizing the thermal energy storage system such that the boiler's nominal capacity is roughly 45% of the average building heat demand. The research by Wang et al. elucidates the potential of thermal energy storage tanks in the context of biomass boiler heating systems, offering an avenue for optimizing the operation and improving the overall energy efficiency of these systems. Integrating TES systems into district heating (DH) infrastructure represents a significant opportunity. These systems can provide several benefits, including peak load leveling, enhanced operational flexibility and reduced production costs; which can lead to more sustainable and economically viable heating solutions. In their comparison of different heat demand scenarios, the researchers highlight the potential of TES in minimizing costs and emissions. This is especially noteworthy in the context of sustainable industrial operations and aligns with the global push for reducing the environmental impact of energy production and consumption. Despite these promising opportunities, the study also uncovers some constraints. Technically, challenges related to the design, implementation, and integration of TES systems into existing infrastructures might pose barriers to their adoption. Economic factors also present significant constraints. The capital and operational costs of implementing and maintaining TES systems might be high. They could deter widespread adoption, especially in areas where the initial investment is not immediately apparent or outweighed by the economic benefits. The study also points to geographical and temporal constraints that can influence the effectiveness and feasibility of TES systems. Factors such as the availability and type of waste heat, heat demand patterns, and seasonal fluctuations can all impact the overall system efficiency.

Zhang and colleagues [130] embarked on a trailblazing investigation into biomass's latent energy storage potential. The researchers innovatively developed a multifunctional form-stable composite PCM that leverages Guar gum, a natural polysaccharide, to create a carbon aerogel encapsulating polyethylene glycol. The biomass-derived composite PCMs demonstrated exceptional characteristics such as robust structural stability, comprehensive energy storage performance, and leakproof quality. When these synthesized PCMs were applied to solar-thermal energy conversion and storage, they performed admirably, marking a substantial advancement in the renewable energy landscape. A key opportunity highlighted by this study is the potential to enhance solar-thermal energy conversion and storage significantly. Developing such innovative carbon aerogels can escalate solar energy systems' efficiency and storage capacity, rendering them more attractive for broader, large-scale energy production. This could revolutionize the renewable energy sector and expedite the global transition toward a more sustainable energy economy. Additionally, the successful utilization of biomass materials in creating aerogels showcases a promising approach to developing eco-friendly energy solutions. This novel use of biomass supports the sustainability agenda and provides a pathway to efficiently utilize biomass waste, thereby contributing to the circular economy concept. This approach also paves the way for more cost-effective energy solutions, assuming that the production cost of these carbon aerogels is economically viable. However, despite these promising opportunities, the study also reveals significant constraints. One of the primary challenges is the difficulty in achieving homogeneity in the biomass materials used to create carbon aerogels. Maintaining consistency in these materials is a critical factor that directly influences the performance and efficiency of the aerogels in their solar-thermal energy conversion and storage roles. Another significant constraint is the need to optimize the properties of carbon aerogels to boost solar-thermal energy conversion efficiency. Determining the proper parameters for this optimization is a complex task and can affect the overall efficacy of the aerogels. It highlights the need for additional research and development to fully unlock the potential of these innovative materials.

Overall, the combined use of biomass energy and thermal energy storage systems presents several opportunities, including reducing greenhouse gas emissions, stimulating local job creation, opening up new revenue streams, diversifying the energy portfolio, and increasing the utilization of renewable resources. Thermal energy storage systems can optimize the operation and improve the overall energy efficiency of biomass boiler heating systems, as well as provide peak load leveling, enhanced operational flexibility, and reduced production costs. Additionally, the use of biomass materials in creating carbon aerogels can escalate solar energy systems' efficiency and storage capacity, rendering them more attractive for broader, large-scale energy production. However, there are several constraints, including securing a consistent and sustainable supply of biomass, developing efficient and environmentally friendly technologies for converting biomass to energy, technical and regulatory barriers to integrating BCHP systems into the existing energy infrastructure, and high capital and operational costs of implementing and maintaining thermal energy storage systems. Additionally, geographical and temporal constraints can influence the effectiveness and feasibility of thermal energy storage systems, and achieving homogeneity in the biomass materials used to create carbon aerogels is a critical factor that directly influences their performance and efficiency.

5.4. Combined Geothermal/TES System

The combined use of geothermal energy and thermal energy storage systems has been examined in several studies. Some researchers investigated the storage of geothermal energy in the form of latent energy. The research conducted by Matuszewska and her team [141] provides an insightful analysis of using a mobile thermal energy storage (M-TES) system for delivering geothermal heat to individual recipients, specifically in Polish conditions. They present an innovative solution using a phase change material (PCM) energy storage container of 55 kWh capacity, demonstrating its potential to overcome critical logistical challenges in the application of geothermal energy. One of the primary opportunities this study highlights is the potential of geothermal energy as a renewable and stable energy source with high utilization efficiency. This sustainable energy resource holds significant potential in terms of environmental benefits, making it an appealing solution for energy generation. Another significant opportunity lies in the high-capacity factors achievable through geothermal resources. The study illustrates that up to 60% of capacity factors are viable, offering a promising avenue for economically and environmentally efficient use of geothermal energy. Moreover, the researchers propose a mobile thermal energy storage (M-TES) system as a novel solution to tackle the challenges associated with the long-distance transportation of geothermal heat. This M-TES system enhances the versatility and feasibility of geothermal energy usage, allowing for efficient storage and transportation of geothermal heat to individual recipients over longer distances, thereby improving the economic viability of such operations. However, alongside these opportunities, the study also acknowledges a notable constraint. Climate factors are recognized as a significant challenge that can limit the effective use of geothermal resources in certain regions. This necessitates developing and implementing additional technical measures to optimize system performance and maintain efficient energy delivery in varying climate conditions.

In the study conducted by Fleuchaus and colleagues [14], they address the risks associated with high-temperature aquifer thermal energy storage (HT-ATES) systems and present a risk assessment framework to mitigate those risks, thereby bridging the seasonal gap between the demand and supply of thermal energy. A key focus of their work lies in the exploration of potential opportunities while also acknowledging inherent constraints. The significant opportunities highlighted by this study involve risk identification and mitigation strategies for HT-ATES systems. The proposed risk assessment framework paves the way for a more robust and secure approach toward designing and operating these systems. Moreover, the study also stresses the importance of promoting sustainable energy solutions. Their research encourages the integration of renewable energy sources and enhances thermal energy storage system efficiency, contributing a valuable dimension to sustainable energy. On the other hand, the study recognizes the constraints associated with these systems. A notable challenge is the limited availability of data on HT-ATES systems, which poses an obstacle to a comprehensive understanding and efficient design. Uncertainties inherent in subsurface conditions also pose a considerable challenge to the development and operation of these systems, impacting their reliability and performance. Furthermore, the potential environmental impacts associated with the heat storage and extraction processes of HT-ATES systems are also a cause for concern, necessitating careful risk assessment and mitigation strategies.

In their compelling study, Arslan and Arslan [150] investigated the potential for integrating residential-scale latent heat thermal energy storage (RS-LTES) using phase change material (PCM) energy storage containers in a geothermal district heating system. Located in Simav, Turkey, the system exhibited numerous opportunities and certain constraints that are imperative to consider for potential implementation. On the positive side, Arslan and Arslan's research illuminates several opportunities. Their investigation underscores the promise of geothermal energy as a sustainable and renewable energy source for district heating systems. This energy choice is environmentally friendly and a viable path to reduce greenhouse gas emissions substantially. They further highlight the enhanced performance and reliability of integrating thermal energy storage into the district heating system. This feature contributes to the system's long-term sustainability and operational efficiency. Lastly, the study introduces an important tool: a multicriteria decision analysis. This framework supports informed decision making, thoroughly evaluating different design options and technologies. This approach is instrumental in fostering the development of efficient, environmentally friendly district heating systems. Despite these promising opportunities, Arslan and Arslan also shed light on the constraints accompanying the use of geothermal energy. These include the difficulty of accurately calculating the size and capacity of the generator plant, distribution system, and substations using geothermal energy as the source. Precise dimensioning of the district heating system is essential for optimal performance and economic feasibility. They underscore the challenge of selecting the appropriate thermal energy storage technologies for district heating systems. The efficiency and cost of these technologies can pose limitations, which need careful consideration during the design and implementation phases.

Overall, the combined use of geothermal energy and thermal energy storage systems offers promising opportunities such as sustainable and renewable energy sources for district heating systems, reduced greenhouse gas emissions, enhanced performance and reliability of the system, and a multicriteria decision analysis framework for informed decision making. However, there are also constraints, such as the difficulty of accurately calculating the size and capacity of the generator plant, distribution system, and substations, and the challenge of selecting appropriate thermal energy storage technologies for district heating systems due to their efficiency and cost limitations.

5.5. Combined Ocean/TES System

The combined use of ocean energy and thermal energy storage systems has been examined in several studies. Some researchers investigated the storage of ocean energy in the form of sensible energy. In their notable study, Li and colleagues [94] developed a thermodynamic model for the ocean thermal energy conversion system (OTEC), offering a load-following control strategy. Their analysis quantitatively evaluated the impact of manipulated variables on the system's power output and the superheating of the evaporator outlet. Despite certain constraints, this study also illuminated several opportunities that could profoundly influence the effectiveness and efficiency of the OTEC systems. The study opened the door to several positive prospects. Firstly, it presented the opportunity for developing advanced control strategies for OTEC systems, enhancing the power output and mitigating temperature fluctuations in the evaporator outlet superheating degree. Such improvements could significantly improve the overall efficiency of OTEC systems. Moreover, their findings offer valuable insights into renewable energy sourcing for islands. Reducing reliance on fossil fuels could substantially decrease greenhouse gas emissions, contributing to environmental conservation efforts. Lastly, the study provided valuable insights, which could enhance the overall efficiency and effectiveness of OTEC systems, further establishing its potential as a robust and sustainable energy solution. However, the study by Li et al. also brings certain constraints to light. One major limitation lies in comparing only two controllers—the model predictive control (MPC) and the proportional integral (PI) controller. This focus on their performance leaves room for exploring other potential control strategies for OTEC systems. It indicates a need for broadening the scope to evaluate a more comprehensive range of control strategies, potentially leading to more optimized OTEC systems.

In a comprehensive heuristic review, Zhou [151] scrutinized the integration of ocean energy into intelligent energy systems, focusing on diversified ocean energy systems for coastal residential communities. This work identified potential limitations and illuminated remarkable opportunities that could shape the future of ocean energy applications. In terms of opportunities, Zhou's review uncovered several promising prospects. The review explores the potential utilization of advanced ocean energy converters to increase energy extraction efficiency from ocean resources. Furthermore, the study pointed toward the benefits of diversified ocean energy systems and hybrid energy storages, which could bolster energy production and storage capacity and consequently improve the reliability and effectiveness of the energy system. The study also highlighted the integration of artificial intelligence as a significant opportunity. This integration could enhance the sustainability and efficiency of these systems through improved prediction, optimization, and control mechanisms. Lastly, Zhou proposed strategies for complementary hybrid renewable system integrations. This effective strategy could help overcome constraints and seize opportunities, leading toward a carbon-neutral transition. Nevertheless, the study has its limitations. The primary challenge identified is the fluctuating power frequency, a side-effect of vertical cascade ocean energy systems. Ensuring stable and grid-friendly operations necessitates identifying practical solutions to this challenge. This constraint underscores the need for further exploration and innovation in this realm to ensure that the efficiency and reliability of these systems are not compromised.

Similarly, in an insightful review, Wang and colleagues [15] critically assessed the potential of harnessing ocean energy as latent energy for crewless underwater vehicles (UUVs) operation. The researchers explored the mechanics of ocean thermal energy formation, the development of ocean-based thermodynamic UUVs, and the current challenges faced in their application. This comprehensive review lends itself to opportunities and constraints in ocean thermal energy storage and application. Wang and colleagues identified significant opportunities for advancing ocean thermal energy technology. The potential utilization of ocean thermal energy, a vast yet largely untapped renewable energy source, was emphasized for powering UUVs. Furthermore, the review identified PCM-based thermal energy-harvesting systems as the most promising technology in the context of these vehicles. This recognition suggests room for considerable advancements and innovation in PCM-based systems, thus paving the way for more efficient and sustainable energy solutions in UUVs. However, the application of these technologies has its limitations. The review noted that the slow heat transfer rates of the phase change material (PCM) thermal-harvesting systems impede their efficiency. Moreover, ocean thermal energy conversion technologies, including PCM-based systems, are grappling with low conversion efficiency. Energy storage technologies used in ocean thermal UUVs, such as PCM thermal-harvesting systems, were highlighted as having a low energy storage density, which presents an obstacle to their operational effectiveness. Lastly, Wang and colleagues pointed out that many ocean thermal UUV concepts, such as shape memory alloys, thermoelectric generators, and thermodynamic cycles, are still in the conceptual design phase, indicating an urgent need for further research and development. Despite these constraints, the study by Wang and his team provides a valuable resource for understanding the current state and future potential of ocean thermal energy-powered UUVs. As the study explores opportunities for innovation, it also outlines the areas where substantial improvements are needed, thereby contributing significantly to the development of more efficient, sustainable, and advanced ocean thermal energy applications.

Overall, the combined use of ocean energy and thermal energy storage systems presents significant opportunities for advancing ocean thermal energy technology, particularly in powering crewless underwater vehicles (UUVs). PCM-based thermal energy-harvesting systems are identified as the most promising technology in this context. However, the slow heat transfer rates of PCM thermal-harvesting systems and the low conversion efficiency of ocean thermal energy conversion technologies present constraints to their operational effectiveness. Additionally, many ocean thermal UUV concepts are still in the conceptual design phase, indicating an urgent need for further research and development.

5.6. Renewable Polygeneration/TES System

Renewable resources can be hybridized to complement the individual deficiencies found in these resources. For instance, solar and geothermal energies can be used as a di-generation system in some areas to maximize the generated power. Similarly, solar, ocean, and wind energies can be used as a di- or tri-generation system to enhance electricity and heat [163,164].

In their noteworthy research, Temiz and Dincer [107], introduce an ocean and solar-based multigeneration system tailored to the challenging conditions of Arctic communities. This comprehensive system incorporates a variety of components, such as a concentrated solar plant (CSP), bifacial photovoltaic (BiPV) cells, cascaded heat pumps, a multi-effect desalination process, and a polymer electrolyte membrane (PEM) electrolyzer. Additionally, the system utilizes fuel cell systems and thermal energy storage for energy retention. Through an in-depth analysis using multiple methodologies and software, the researchers explore the potential of this system in providing essential services, including food and energy production, for Arctic communities. In terms of opportunities, Temiz and Dincer's work highlights the potential of renewable energy technologies, such as ocean thermal energy conversion, concentrated solar plants, and bifacial photovoltaic systems, in addressing the energy needs of isolated Arctic communities. Their proposed system also pioneers the integration of hydrogen production and thermal energy storage, providing promising avenues for enhancing the efficiency and sustainability of energy solutions in these harsh climates. Moreover, the study advocates a comprehensive approach to food production by amalgamating a fish farm, a greenhouse, and a food drying facility, all powered by their innovative energy system. Through these innovative strategies, this research opens the door to significant improvements in the quality of life for Arctic communities by addressing their perennial food and energy shortages. Despite the promising opportunities, certain constraints are identified in the study. The harsh and often unpredictable Arctic conditions pose substantial challenges to implementing and operating such a sophisticated and integrated system. These conditions require robust and resilient systems that withstand freezing, ice, and limited sunlight. Another hurdle lies in system optimization; achieving maximum efficiency and effectiveness necessitates extensive analysis and simulations, which can be time-consuming and complex.

Additionally, Assareh and colleagues [153] proffer an intriguing model of an integrated energy system combining an ocean thermal energy convertor (OTEC), a wind turbine, and a solar flat plate panel to cater to the electricity needs of Iranian households. Using Rankine cycles for power generation and wind turbines for backup, the system offers a synergistic approach to harnessing multiple renewable energy sources. This innovative setup underwent thorough thermo-economic analysis, was gauged against an Iranian household's annual electricity consumption, and was further honed using the NSGA-II optimization algorithm. Their system could feasibly supply electricity to 38 Iranian households annually, operating with an energy conversion efficiency of 12.53% and producing a net power output of 448 kW. Furthermore, the proposed system promises an affordable electricity supply of 57.6 USD per hour. Commencing with the opportunities, Assareh and his team's study highlights the potential of crafting a sustainable and efficient energy system through the concerted utilization of diverse renewable energy sources such as solar, wind, and ocean energy. This integrated approach offers a promising pathway to reducing dependence on fossil fuels, thus curbing greenhouse gas emissions—an increasingly crucial goal in the face of global warming. Moreover, their proposed system enhances energy security by leveraging locally available renewable energy resources, thereby safeguarding against geopolitical uncertainties and supply chain disruptions. Conversely, several constraints come into play. Foremost among these is the geographical specificity of the study, which focuses on Bandar Abas, Iran. This localization could restrict the direct applicability of the findings to regions with different climatic conditions and resource availability. Additionally, deploying such an integrated system necessitates substantial initial investment and infrastructure development, which could pose challenges in regions with limited resources. Lastly, the inherent intermittency of renewable energy sources, particularly solar and wind, raises the need for further exploration into effective energy storage solutions to ensure the system's reliability and stability.

Similarly, solar and geothermal energies can be used as a di-generation system in some areas to maximize the generated power. A recent study by Li, Tao, Zhang, and Fu [114] illuminates a suite of opportunities inherent in their innovative solar–geothermal hybrid system. Foremost among these is the efficient utilization of renewable resources. The proposed system leverages both strengths by seamlessly integrating solar and geothermal energy sources, mitigating dependence on fossil fuels and significantly curtailing greenhouse gas emissions. In addition, implementing thermal energy storage bolsters the system's reliability, delivering unwavering flexibility in power generation, hydrogen production, and freshwater supply, regardless of the intermittent nature of the solar input. Furthermore, pioneering a transcritical CO_2 cycle heightens thermodynamic performance while concurrently diminishing environmental impact, providing a compelling alternative to conventional power cycles. The two-objective optimization process delineated by the researchers also opens up the prospect of optimizing system configurations. This method uncovers the most efficient configurations, paving the way for future renewable energy projects by gleaning invaluable insights. However, a critical appraisal of Li's research also brings to light some constraints that could impede widespread adoption. A fundamental limitation lies in the high initial capital costs associated with the complex nature of the hybrid system. This considerable financial barrier could limit the accessibility of the technology and slow its adoption. Geographical and climatic dependency presents another hurdle. The effectiveness of the solar and geothermal elements is heavily contingent upon the geographical location and local climate, which can vary markedly and consequently influence the system's overall performance. The efficacy of the thermal energy storage component also hinges on the availability of suitable materials and technologies. As these might still be in the nascent stages of development, this could stymie the practical deployment of the system. Lastly, technical challenges in optimizing the transcritical CO_2 cycle require advanced control strategies and high-level technical acumen. Given the limited pool of expertise, this might present an obstacle to widespread implementation.

Picone and team [112] meticulously explored the plethora of opportunities in integrating Aquifer Thermal Energy Storage (ATES) with solar collectors, soil remediation, and other geothermal energy systems. Doing so paves the way for enhanced energy efficiency, a crucial asset in today's energy-demanding era. Their simulations at sites in Belgium and the Netherlands demonstrated that such an innovative combination increases efficiency and energy savings. This, in turn, enables ATES to be used even under water scarcity conditions, thereby expanding its application spectrum. Furthermore, the promising results showing reduced dechlorination of chlorinated ethenes attest to the potential for reducing environmental impact, a critical consideration in today's climate-sensitive scenario. Picone's research not only introduces a novel combination of energy systems but also presents a blueprint for the cost-effective integration of renewable energy sources. This provides a creative solution to ongoing challenges in sustainable energy production and management, further underlining the value of the study. However, even with the positive strides made in this direction, it is important to recognize emerging constraints. Integrating diverse energy systems efficiently is riddled with technical challenges that must be overcome to function optimally. Additionally, legal and regulatory barriers might present obstacles, potentially limiting the widespread adoption of these innovative combinations. Of equal significance is the fact that the effectiveness and feasibility of these combinations are intrinsically tied to local conditions, particularly the geothermal and hydrogeological factors. Therefore, despite the myriad benefits these integrated systems offer, addressing these hurdles to unlock their full potential remains paramount.

The research conducted by Senturk Acar and Arslan [105], along with the insights gleaned from Boretti's [152] investigation, presents a compelling case for the integration of solar and geothermal energy to power an organic Rankine cycle (ORC). By incorporating these renewable sources, there is significant potential for enhanced energy efficiency and a subsequent reduction in dependence on fossil fuels. Consequently, this transition could drastically cut greenhouse gas emissions, contributing to a more sustainable energy landscape. Moreover, using both solar and geothermal energy has been shown to optimize the performance of the ORC, paving the way for improved energy conversion rates. In particular, Boretti's exploration of the considerable geothermal resources within Saudi Arabia's Earth's crust demonstrates the potential for "enhanced geothermal systems" (EGS) to increase thermal cycle efficiencies by 40% more than traditional methods. The good coupling of "concentrated solar power" (CSP) and EGS with "thermal energy storage" (TES) could boost thermal cycle efficiency by over 50%, particularly in regions abundant in CSP and EGS resources, such as Saudi Arabia. Despite the promising prospects, however, certain constraints could impede the wide-scale deployment of these hybrid systems. For one, the availability and suitability of geothermal resources are primarily confined to specific geographical regions, which could limit the overall adoption. High initial costs associated with the necessary geothermal and solar energy infrastructure can also pose a significant financial barrier, potentially deterring investments in these systems. Lastly, the inherently intermittent nature of solar energy may impact the hybrid system's overall efficiency and reliability, adding to its implementation's complexities. Hence, while these studies underscore the potential of integrating solar and geothermal energy sources, they also illustrate the need to address these challenges further to maximize the potential of these promising renewable energy systems.

Overall, the combined use of renewable polygeneration and thermal energy storage systems offers opportunities such as reducing dependence on fossil fuels, enhancing energy security, and mitigating greenhouse gas emissions. However, there are constraints such as geographical specificity, high initial investment and infrastructure development, intermittency of renewable energy sources, technical challenges, legal and regulatory barriers, and dependence on local conditions. Overcoming these constraints is necessary to unlock the full potential of these integrated systems.

In conclusion, the comprehensive review of various studies into integrating renewable energy and thermal energy storage (RES–TES) systems presents an optimistic outlook for the future of energy production. Harnessing the synergies between disparate renewable energy sources and TES systems can significantly contribute to the global transition toward more sustainable and low-carbon energy systems. Such a transition ensures energy efficiency and mitigates our reliance on fossil fuels, easing the ongoing pressures of climate change. Nevertheless, several constraints must be considered, such as geographical and climatic dependencies, high initial capital costs, and technical challenges in optimizing these systems. Continued research, development, and investment are paramount to address these challenges and unlock the full potential of these pioneering energy solutions. Moreover, the document underscores the value of various simulation tools and research methodologies used in studying RES–TES systems, serving as a rich resource for researchers and practitioners committed to ensuring a sustainable future for future generations. A summary of the crucial parameters and terms mentioned in RE and TES systems is given in Table 4.

Table 4. Key terms for designing combined renewable energy and thermal energy storage systems.

Parameters/Term	Description
Biomass	Organic matter, such as wood, crops, or animal waste, can be used as fuel for heating or electricity generation.
Concentrated solar power (CSP)	A technology that uses mirrors or lenses to concentrate sunlight onto a small area, which heats a fluid to produce steam and generate electricity.
Di-generation	A system that generates two forms of energy, usually electricity and heat, from a single energy source.
Enhanced geothermal systems (EGS)	A type of geothermal energy production involves creating artificial fractures in hot rock formations to extract heat from the Earth.
Latent thermal energy storage (LTES)	A type of TES that stores energy by changing the phase of a material, such as melting or solidifying.
Model predictive control (MPC)	A control strategy that uses a mathematical model to predict future behavior of a system and optimize control actions accordingly.
Ocean energy	Energy derived from the ocean, including ocean waves, tides, ocean currents, thermal ocean energy, and chemical ocean energy.
Ocean thermal energy conversion (OTEC)	A technology that harnesses the temperature difference between warm surface water and cold deep water to generate electricity.
Organic Rankine cycle (ORC)	A thermodynamic cycle that uses an organic fluid as the working fluid in a closed loop to generate electricity.
Oscillating water columns	A type of ocean wave energy converter that uses waves to compress and decompress air in a chamber, which then drives a turbine to generate electricity.
Overtopping wave energy converters	A type of ocean wave energy converter that uses waves to fill a reservoir with water, which is then released through a turbine to generate electricity.
Phase change materials (PCMs)	Substances that store and release thermal energy during phase transitions commonly used in thermal energy storage systems for heating, cooling, and air conditioning applications.
Polymer electrolyte membrane (PEM) electrolyzer	A device that uses an electric current to split water into hydrogen and oxygen gases.
Proportional integral (PI) control	A control strategy that adjusts a system's output based on the difference between the desired set point and the actual value, using proportional and integral terms.
Renewable energy sources (RES)	Energy sources such as solar, wind, hydro, geothermal, and biomass replenished naturally and sustainably.
Sensible thermal energy storage (STES)	A type of TES that stores energy by changing the temperature of a material without changing its phase.
Solar parabolic trough collectors	A type of solar thermal technology that uses curved mirrors to concentrate sunlight onto a pipe containing a heat transfer fluid, which is then used to generate steam and produce electricity.
Thermal energy storage (TES)	A method of storing thermal energy by heating or cooling material to retrieve the energy later for heating or cooling purposes.
Tri-generation	A system that generates three forms of energy, usually electricity, heat, and cooling, from a single energy source.

6. Conclusions

Renewable energy sources (RES) are vital in maintaining the wellbeing of today's civilization. Nonetheless, RES often suffer from their intermittent nature, which limits their utilization. In summarizing the array of studies in this review, it is evident that mitigating the shortcomings of RES with thermal energy storage systems (TES) presents promising opportunities and significant constraints. While these systems bring to light an avenue toward sustainable and renewable energy, their practical effectiveness and efficiency require empirical substantiation. Cost, performance, geographical influences, and climatic conditions necessitate careful consideration. Furthermore, the complexities and cost of these systems, coupled with potential environmental ramifications, signify formidable challenges. Nevertheless, through meticulous examination and exhaustive research, these systems possess the capability to revolutionize the clean energy sector, paving the way toward a more sustainable future. Moreover, this review underscores the benefits of hybridizing diverse renewable energy resources, such as di-generation solar and geothermal systems. These hybrid systems amplify the efficiency and reliability of renewable energy production, thereby fostering sustainable development by reducing fossil-fuel dependence and curtailing greenhouse gas emissions.

Considering the potential of RES–TES hybrid systems to supplant fossil fuels, exploring interdisciplinary collaborations in these systems is crucial. Encouraging partnerships amongst academia, industry, and policymakers can facilitate developing and deploying state-of-the-art RES–TES technologies. Innovation can be expedited by uniting expertise from various sectors, where potential challenges can be identified and addressed, and an environment conducive to the investment and adoption of these sustainable energy solutions can be fostered. Establishing standardized performance metrics and benchmarks is imperative to effectively compare diverse systems, stimulate innovation, and guide future research and development endeavors. By adopting a holistic and cooperative approach, the field of RES–TES hybrid systems can advance, contributing significantly to a more sustainable, reliable, and low-carbon energy future.

Author Contributions: Conceptualization, A.E. and S.A.A.-M.; methodology, A.E. and S.A.A.-M.; formal analysis, A.E.; investigation, A.E.; resources, S.A.A.-M.; data curation, A.E.; writing—original draft preparation, A.E.; writing—review and editing, S.A.A.-M.; visualization, A.E.; supervision, S.A.A.-M.; project administration, S.A.A.-M. All authors have read and agreed to the published version of the manuscript.

Funding: This research received no external funding.

Data Availability Statement: No new data were created or analyzed in this study. Data sharing is not applicable to this article.

Conflicts of Interest: The authors declare no conflict of interest.

References

1. Wolniak, R.; Skotnicka-Zasadzien, B. Development of Photovoltaic Energy in EU Countries as an Alternative to Fossil Fuels. *Energies* **2022**, *15*, 662. [CrossRef]
2. Amanatidis, G. *Report: European Policies on Climate and Energy towards 2020, 2030 and 2050*; Policy Department for Economic, Scientific and Quality of Life Policies; Directorate-General for Internal Policies: Brussels, Belgium, 2019.
3. IRENA. Renewable Energy Solutions for Climate-Safe Cities. Available online: https://www.irena.org/News/articles/2020/Oct/Renewable-Energy-Solutions-for-Climate-Safe-Cities (accessed on 11 November 2022).
4. Kini, V. Ingersoll Rand 2030 and 2050 Environmental Goals Set to Mitigate Climate Change and Make Life Better for Generations to Come. Available online: https://investors.irco.com/news/news-details/2021/Ingersoll-Rand-2030-and-2050-Environmental-Goals-Set-to-Mitigate-Climate-Change-and-Make-Life-Better-for-Generations-to-Come/default.aspx (accessed on 11 November 2022).
5. Dincer, I.; Rosen, M.A. Energy Storage Systems. In *Thermal Energy Storage: Systems and Applications*, 3rd ed.; John Wiley & Sons: Hoboken, NJ, USA, 2021; pp. 59–123.
6. Li, G. Sensible heat thermal storage energy and exergy performance evaluations. *Renew. Sustain. Energy Rev.* **2016**, *53*, 897–923. [CrossRef]
7. Li, G. Energy and exergy performance assessments for latent heat thermal energy storage systems. *Renew. Sustain. Energy Rev.* **2015**, *51*, 926–954. [CrossRef]

8. Liu, Y.L.; Wang, R.Z.; Xia, Z.Z. Experimental study on a continuous adsorption water chiller with novel design. *Int. J. Refrig.-Rev. Int. Du Froid* **2005**, *28*, 218–230. [CrossRef]
9. Li, G.; Zheng, X.F. Thermal energy storage system integration forms for a sustainable future. *Renew. Sustain. Energy Rev.* **2016**, *62*, 736–757. [CrossRef]
10. Kebede, A.A.; Kalogiannis, T.; Van Mierlo, J.; Berecibar, M. A comprehensive review of stationary energy storage devices for large scale renewable energy sources grid integration. *Renew. Sustain. Energy Rev.* **2022**, *159*, 112213. [CrossRef]
11. Xu, J.; Wang, R.Z.; Li, Y. A review of available technologies for seasonal thermal energy storage. *Sol. Energy* **2014**, *103*, 610–638. [CrossRef]
12. Koçak, B.; Fernandez, A.I.; Paksoy, H. Review on sensible thermal energy storage for industrial solar applications and sustainability aspects. *Sol. Energy* **2020**, *209*, 135–169. [CrossRef]
13. Alva, G.; Liu, L.K.; Huang, X.; Fang, G.Y. Thermal energy storage materials and systems for solar energy applications. *Renew. Sustain. Energy Rev.* **2017**, *68*, 693–706. [CrossRef]
14. Fleuchaus, P.; Schuppler, S.; Bloemendal, M.; Guglielmetti, L.; Opel, O.; Blum, P. Risk analysis of High-Temperature Aquifer Thermal Energy Storage (HT-ATES). *Renew. Sustain. Energy Rev.* **2020**, *133*, 110153. [CrossRef]
15. Wang, G.H.; Yang, Y.A.; Wang, S.X. Ocean thermal energy application technologies for unmanned underwater vehicles: A comprehensive review. *Appl. Energy* **2020**, *278*, 115752. [CrossRef]
16. Caralis, G.; Christakopoulos, T.; Karellas, S.; Gao, Z.Q. Analysis of energy storage systems to exploit wind energy curtailment in Crete. *Renew. Sustain. Energy Rev.* **2019**, *103*, 122–139. [CrossRef]
17. Ali, U. Bloomberg New Energy Outlook 2019: The Future of the Energy Sector: Bloomberg New Energy Finance (BNEF) Held Its Annual New Energy Outlook (NEO). Available online: https://www.power-technology.com/news/bloomberg-new-energy-outlook-2019-2/ (accessed on 23 November 2022).
18. BP, P.L.C. Statistical Review of World Energy: Primary Energy. Available online: https://www.bp.com/en/global/corporate/energy-economics/statistical-review-of-world-energy/primary-energy.html (accessed on 23 November 2022).
19. Lichołai, L.; Kruzel, R.; Helbrych, P.; Dębska, B.; Miąsik, P.; Szyszka, J.; Krasoń, J.; Szalacha, A. Analysis of the profitability of a photovoltaic installation in the context of sustainable development of construction. *E3S Web Conf.* **2018**, *49*, 61.
20. Mitscher, M.; Ruther, R. Economic performance and policies for grid-connected residential solar photovoltaic systems in Brazil. *Energy Policy* **2012**, *49*, 688–694. [CrossRef]
21. Dinesh, H.; Pearce, J.M. The potential of agrivoltaic systems. *Renew. Sustain. Energy Rev.* **2016**, *54*, 299–308. [CrossRef]
22. Patel, M.T.; Khan, M.R.; Sun, X.S.; Alam, M.A. A worldwide cost-based design and optimization of tilted bifacial solar farms. *Appl. Energy* **2019**, *247*, 467–479. [CrossRef]
23. Saxena, A.; Goel, V. A Technical Note on Fabrication and Thermal Performance Studies of a Solar Pond Model. *J. Renew. Energy* **2013**, *2013*, 475282. [CrossRef]
24. Perumal, P.; Dharmalingam, M. Solar ponds-a mini review. *Environ. Sci. Pollut. Res. Int.* **2022**, *29*, 45063–45069. [CrossRef]
25. Ganguly, S.; Jain, R.; Date, A.; Akbarzadeh, A. On the addition of heat to solar pond from external sources. *Sol. Energy* **2017**, *144*, 111–116. [CrossRef]
26. Moss, R.W.; Henshall, P.; Arya, F.; Shire, G.; Hyde, T.; Eames, P.C. Performance and operational effectiveness of evacuated flat plate solar collectors compared with conventional thermal, PVT and PV panels. *Appl. Energy* **2018**, *216*, 588–601. [CrossRef]
27. Joardder, M.U.H.; Halder, P.K.; Rahim, M.A.; Masud, M.H. Solar Pyrolysis. In *Clean Energy for Sustainable Development*; Rasul, M.G., Azad, A.K., Sharma, S.C., Eds.; Academic Press: Cambridge, MA, USA, 2017; pp. 213–235.
28. Fernández-García, A.; Zarza, E.; Valenzuela, L.; Pérez, M. Parabolic-trough solar collectors and their applications. *Renew. Sustain. Energy Rev.* **2010**, *14*, 1695–1721. [CrossRef]
29. Gautam, A.; Saini, R.P. A review on technical, applications and economic aspect of packed bed solar thermal energy storage system. *J. Energy Storage* **2020**, *27*, 101046. [CrossRef]
30. Dincer, I.; Rosen, M.A. Renewable Energy Systems with Thermal Energy Storage. In *Thermal Energy Storage: Systems and Applications*, 3rd ed.; John Wiley & Sons: Hoboken, NJ, USA, 2021; pp. 521–555.
31. Jones, D. Global Electricity Review 2021. 2021. Available online: https://ember-climate.org/insights/research/global-electricity-review-2021/ (accessed on 1 February 2023).
32. REN21. Renewables 2021: Global Status Report. 2021, p. 146. Available online: https://www.ren21.net/wp-content/uploads/2019/05/GSR2021_Full_Report.pdf (accessed on 1 February 2023).
33. Komusanac, I.; Brindley, G.; Fraile, D.; Ramirez, L. Report: Wind Energy in Europe—2020 Statistics and the Outlook for 2021–2025. 2021, pp. 1–40. Available online: https://windeurope.org/intelligence-platform/product/wind-energy-in-europe-2020-statistics-and-the-outlook-for-2021-2025/ (accessed on 1 February 2023).
34. Msigwa, G.; Ighalo, J.O.; Yap, P.S. Considerations on environmental, economic, and energy impacts of wind energy generation: Projections towards sustainability initiatives. *Sci. Total Environ.* **2022**, *849*, 157755. [CrossRef]
35. Ulazia, A.; Nafarrate, A.; Ibarra-Berastegi, G.; Saenz, J.; Carreno-Madinabeitia, S. The Consequences of Air Density Variations over Northeastern Scotland for Offshore Wind Energy Potential. *Energies* **2019**, *12*, 2635. [CrossRef]
36. Islam, S.; Dincer, I.; Yilbas, B.S. Analysis and Assessment of a Biomass Energy-Based Multigeneration System with Thermoelectric Generators. *Energy Fuels* **2017**, *31*, 10901–10915. [CrossRef]

37. Tazi, N.; Bouzidi, Y. Evolution of wind energy pricing policies in France: Opportunities and new challenges. *Energy Rep.* **2020**, *6*, 687–692. [CrossRef]
38. Diffendorfer, J.E.; Dorning, M.A.; Keen, J.R.; Kramer, L.A.; Taylor, R.V. Geographic context affects the landscape change and fragmentation caused by wind energy facilities. *PeerJ* **2019**, *7*, e7129. [CrossRef]
39. Perold, V.; Ralston-Paton, S.; Ryan, P. On a collision course? The large diversity of birds killed by wind turbines in South Africa. *Ostrich* **2020**, *91*, 228–239. [CrossRef]
40. KaradaĞ, İ.; KuruÇAy, E. Noise Emission from Building Integrated Wind Turbines: A Case Study of a Tall Building. *Sak. Univ. J. Sci.* **2021**, *25*, 563–570. [CrossRef]
41. Kandasamy, S.; Devarayan, K.; Bhuvanendran, N.; Zhang, B.; He, Z.X.; Narayanan, M.; Mathimani, T.; Ravichandran, S.; Pugazhendhi, A. Accelerating the production of bio-oil from hydrothermal liquefaction of microalgae via recycled biochar-supported catalysts. *J. Environ. Chem. Eng.* **2021**, *9*, 105321. [CrossRef]
42. Shanmugam, S.; Sekar, M.; Sivaramakrishnan, R.; Raj, T.; Ong, E.S.; Rabbani, A.H.; Rene, E.R.; Mathimani, T.; Brindhadevi, K.; Pugazhendhi, A. Pretreatment of second and third generation feedstock for enhanced biohythane production: Challenges, recent trends and perspectives. *Int. J. Hydrogen Energy* **2021**, *46*, 11252–11268. [CrossRef]
43. Rezaei, M.; Sameti, M.; Nasiri, F. Biomass-fuelled combined heat and power: Integration in district heating and thermal-energy storage. *Clean Energy* **2021**, *5*, 44–56. [CrossRef]
44. Whangchai, K.; Mathimani, T.; Sekar, M.; Shanmugam, S.; Brindhadevi, K.; Van Hung, T.; Chinnathambi, A.; Alharbi, S.A.; Pugazhendhi, A. Synergistic supplementation of organic carbon substrates for upgrading neutral lipids and fatty acids contents in microalga. *J. Environ. Chem. Eng.* **2021**, *9*, 105482. [CrossRef]
45. Voloshin, R.A.; Rodionova, M.V.; Zharmulzhamedou, S.K.; Veziroglu, T.N.; Allakhverdiev, S.I. Review: Biofuel production from plant and algal biomass. *Int. J. Hydrogen Energy* **2016**, *41*, 17257–17273. [CrossRef]
46. Saravanan, A.P.; Mathimani, T.; Deviram, G.; Rajendran, K.; Pugazhendhi, A. Biofuel policy in India: A review of policy barriers in sustainable marketing of biofuel. *J. Clean. Prod.* **2018**, *193*, 734–747. [CrossRef]
47. Alagumalai, A.; Mathimani, T.; Pugazhendhi, A.; Atabani, A.E.; Brindhadevi, K.; Canh, N.D. Experimental insight into co-combustion characteristics of oxygenated biofuels in modified DICI engine. *Fuel* **2020**, *278*, 118303. [CrossRef]
48. Deviram, G.; Mathimani, T.; Anto, S.; Ahamed, T.S.; Ananth, D.A.; Pugazhendhi, A. Applications of microalgal and cyanobacterial biomass on a way to safe, cleaner and a sustainable environment. *J. Clean. Prod.* **2020**, *253*, 119770. [CrossRef]
49. Pal, P.; Chew, K.W.; Yen, H.W.; Lim, J.W.; Lam, M.K.; Show, P.L. Cultivation of Oily Microalgae for the Production of Third-Generation Biofuels. *Sustainability* **2019**, *11*, 5424. [CrossRef]
50. Shokravi, H.; Heidarrezaei, M.; Shokravi, Z.; Ong, H.C.; Lau, W.J.; Din, M.F.M.; Ismail, A.F. Fourth generation biofuel from genetically modified algal biomass for bioeconomic development. *J. Biotechnol.* **2022**, *360*, 23–36. [CrossRef] [PubMed]
51. Fushimi, C. Valorization of Biomass Power Generation System: Noble Use of Combustion and Integration with Energy Storage. *Energy Fuels* **2021**, *35*, 3715–3730. [CrossRef]
52. Das, P.; Chandramohan, V.P.; Mathimani, T.; Pugazhendhi, A. Recent advances in thermochemical methods for the conversion of algal biomass to energy. *Sci. Total Environ.* **2021**, *766*, 144608. [CrossRef] [PubMed]
53. Mathimani, T.; Baldinelli, A.; Rajendran, K.; Prabakar, D.; Matheswaran, M.; van Leeuwen, R.P.; Pugazhendhi, A. Review on cultivation and thermochemical conversion of microalgae to fuels and chemicals: Process evaluation and knowledge gaps. *J. Clean. Prod.* **2019**, *208*, 1053–1064. [CrossRef]
54. Mathimani, T.; Mallick, N. A review on the hydrothermal processing of microalgal biomass to bio-oil—Knowledge gaps and recent advances. *J. Clean. Prod.* **2019**, *217*, 69–84. [CrossRef]
55. Verma, D.; Fortunati, E.; Jain, S.; Zhang, X. *Biomass, Biopolymer-Based Materials, and Bioenergy: Construction, Biomedical, and Other Industrial Applications*; Woodhead Publishing: Duxford, UK, 2019.
56. Lin, E.Y.; Lu, C.R.E. Development perspectives of promising lignocellulose feedstocks for production of advanced generation biofuels: A review. *Renew. Sustain. Energy Rev.* **2021**, *136*, 110445. [CrossRef]
57. Glivin, G.; Sekhar, S.J. Experimental and Analytical Studies on the Utilization of Biowastes Available in an Educational Institution in India. *Sustainability* **2016**, *8*, 1128. [CrossRef]
58. Glivin, G.; Sekhar, S.J. Waste Potential, Barriers and Economic Benefits of Implementing Different Models of Biogas Plants in a Few Indian Educational Institutions. *Bioenergy Res.* **2020**, *13*, 668–682. [CrossRef]
59. Kumar, G.; Cho, S.K.; Sivagurunathan, P.; Anburajan, P.; Mahapatra, D.M.; Park, J.H.; Pugazhendhi, A. Insights into evolutionary trends in molecular biology tools in microbial screening for biohydrogen production through dark fermentation. *Int. J. Hydrogen Energy* **2018**, *43*, 19885–19901. [CrossRef]
60. Chambon, C.L.; Karia, T.; Sandwell, P.; Hallett, J.P. Techno-economic assessment of biomass gasification-based mini-grids for productive energy applications: The case of rural India. *Renew. Energy* **2020**, *154*, 432–444. [CrossRef]
61. Temiz, M.; Dincer, I. Concentrated solar driven thermochemical hydrogen production plant with thermal energy storage and geothermal systems. *Energy* **2021**, *219*, 119554. [CrossRef]
62. Hu, Y.; Cheng, H.; Tao, S. Opportunity and challenges in large-scale geothermal energy exploitation in China. *Crit. Rev. Environ. Sci. Technol.* **2021**, *52*, 3813–3834. [CrossRef]
63. Moya, D.; Aldas, C.; Kaparaju, P. Geothermal energy: Power plant technology and direct heat applications. *Renew. Sustain. Energy Rev.* **2018**, *94*, 889–901. [CrossRef]

64. Zhang, L.; Chen, S.; Zhang, C. Geothermal power generation in China: Status and prospects. *Energy Sci. Eng.* **2019**, *7*, 1428–1450. [CrossRef]
65. Dhar, A.; Naeth, M.A.; Jennings, P.D.; El-Din, M.G. Geothermal energy resources: Potential environmental impact and land reclamation. *Environ. Rev.* **2020**, *28*, 415–427. [CrossRef]
66. Pan, S.Y.; Gao, M.Y.; Shah, K.J.; Zheng, J.M.; Pei, S.L.; Chiang, P.C. Establishment of enhanced geothermal energy utilization plans: Barriers and strategies. *Renew. Energy* **2019**, *132*, 19–32. [CrossRef]
67. Sircar, A.; Bist, N.; Yadav, K.J.M.S. A comprehensive review on exploration and exploitation of offshore geothermal energy. *Mar. Syst. Ocean Technol.* **2022**, *17*, 135–146. [CrossRef]
68. Lund, J.W.; Boyd, T.L. Direct utilization of geothermal energy 2015 worldwide review. *Geothermics* **2016**, *60*, 66–93. [CrossRef]
69. Dalsgaard, J.; Lund, I.; Thorarinsdottir, R.; Drengstig, A.; Arvonen, K.; Pedersen, P.B. Farming different species in RAS in Nordic countries: Current status and future perspectives. *Aquac. Eng.* **2013**, *53*, 2–13. [CrossRef]
70. Lund, J.W.; Toth, A.N. Direct utilization of geothermal energy 2020 worldwide review. *Geothermics* **2021**, *90*, 101915. [CrossRef]
71. Morales-Nin, B.; Moranta, J.; Garcia, C.; Tugores, M.P.; Grau, A.M.; Riera, F.; Cerda, M. The recreational fishery off Majorca Island (western Mediterranean): Some implications for coastal resource management. *Ices J. Mar. Sci.* **2005**, *62*, 727–739. [CrossRef]
72. Lim, C.C.; Cooper, C. Beyond Sustainability: Optimising Island Tourism Development. *Int. J. Tour. Res.* **2009**, *11*, 89–103. [CrossRef]
73. Song, C.; Zhu, X.; Wang, M.; Yang, P.; Chen, L.; Hong, L.; Cui, W. Recent advances in ocean energy harvesting based on triboelectric nanogenerators. *Sustain. Energy Technol. Assess.* **2022**, *53*, 102767. [CrossRef]
74. Alsvik, I.L.; Hagg, M.B. Pressure Retarded Osmosis and Forward Osmosis Membranes: Materials and Methods. *Polymers* **2013**, *5*, 303–327. [CrossRef]
75. Li, M.; Luo, H.J.; Zhou, S.J.; Kumar, G.M.S.; Guo, X.M.; Law, T.C.; Cao, S.L. State-of-the-art review of the flexibility and feasibility of emerging offshore and coastal ocean energy technologies in East and Southeast Asia. *Renew. Sustain. Energy Rev.* **2022**, *162*, 112404. [CrossRef]
76. Quirapas, M.A.J.R.; Lin, H.; Abundo, M.L.S.; Brahim, S.; Santos, D. Ocean renewable energy in Southeast Asia: A review. *Renew. Sustain. Energy Rev.* **2015**, *41*, 799–817. [CrossRef]
77. Magagna, D.; Uihlein, A. Ocean energy development in Europe: Current status and future perspectives. *Int. J. Mar. Energy* **2015**, *11*, 84–104. [CrossRef]
78. Reguero, B.G.; Losada, I.J.; Mendez, F.J. A global wave power resource and its seasonal, interannual and long-term variability. *Appl. Energy* **2015**, *148*, 366–380. [CrossRef]
79. Zhang, Y.-l.; Lin, Z.; Liu, Q.-l. Marine renewable energy in China: Current status and perspectives. *Water Sci. Eng.* **2014**, *7*, 288–305.
80. Samrat, N.H.; Ahmad, N.B.; Choudhury, I.A.; Taha, Z. Prospect of wave energy in Malaysia. In Proceedings of the 2014 IEEE 8th International Power Engineering and Optimization Conference (PEOCO2014), Langkawi, Malaysia, 24–25 March 2014.
81. Kim, G.; Lee, M.E.; Lee, K.S.; Park, J.S.; Jeong, W.M.; Kang, S.K.; Soh, J.G.; Kim, H. An overview of ocean renewable energy resources in Korea. *Renew. Sustain. Energy Rev.* **2012**, *16*, 2278–2288. [CrossRef]
82. Inger, R.; Attrill, M.J.; Bearhop, S.; Broderick, A.C.; James Grecian, W.; Hodgson, D.J.; Mills, C.; Sheehan, E.; Votier, S.C.; Witt, M.J.; et al. Marine renewable energy: Potential benefits to biodiversity? An urgent call for research. *J. Appl. Ecol.* **2009**, *46*, 1145–1153. [CrossRef]
83. Babarit, A. Working Principles and Technologies of Wave Energy Conversion. In *Wave Energy Conversion*; Babarit, A., Ed.; Elsevier: Amsterdam, The Netherlands, 2017; pp. 99–151.
84. Hendroko Setyobudi, R.; Syarif Arief, I.; Aria Pria Utama, I.K.; Hantoro, R.; Prananda, J.; Muhammad Megawan, A.; Danang Birowosuto, M.; Kala Mahaswa, R.; Jani, Y.; Vincēviča-Gaile, Z. Computational Fluid Dynamics (CFD) Simulation for Designing Mooring Bitts Position at the Barge for Wave Energy Conversion (WEC). *E3S Web Conf.* **2020**, *190*, 17.
85. Barbosa, D.V.E.; Santos, A.L.G.; dos Santos, E.D.; Souza, J.A. Overtopping device numerical study: Openfoam solution verification and evaluation of curved ramps performances. *Int. J. Heat Mass Transf.* **2019**, *131*, 411–423. [CrossRef]
86. Kim, S.S.; Lee, J.C.; Kang, D.; Lee, S.S. Motion characteristics of a floating wave energy converter with wave activating body type. *Int. J. Nav. Archit. Ocean Eng.* **2019**, *11*, 244–255. [CrossRef]
87. Bricker, J.D.; Esteban, M.; Takagi, H.; Roeber, V. Economic feasibility of tidal stream and wave power in post-Fukushima Japan. *Renew. Energy* **2017**, *114*, 32–45. [CrossRef]
88. Liu, Y.J.; Li, Y.; He, F.L.; Wang, H.F. Comparison study of tidal stream and wave energy technology development between China and some Western Countries. *Renew. Sustain. Energy Rev.* **2017**, *76*, 701–716. [CrossRef]
89. Woo, M. Tidal Energy: The New Sustainable Resource. Available online: https://www.woodharbinger.com/tidal-energy-sustainable-resource/ (accessed on 1 February 2023).
90. Roberts, A.; Thomas, B.; Sewell, P.; Khan, Z.; Balmain, S.; Gillman, J. Current tidal power technologies and their suitability for applications in coastal and marine areas. *J. Ocean Eng. Mar. Energy* **2016**, *2*, 227–245. [CrossRef]
91. Rajagopalan, K.; Nihous, G.C. Estimates of global Ocean Thermal Energy Conversion (OTEC) resources using an ocean general circulation model. *Renew. Energy* **2013**, *50*, 532–540. [CrossRef]
92. Faizal, M.; Ahmed, M.R. Experimental studies on a closed cycle demonstration OTEC plant working on small temperature difference. *Renew. Energy* **2013**, *51*, 234–240. [CrossRef]

93. Liu, W.M.; Xu, X.J.; Chen, F.Y.; Liu, Y.J.; Li, S.Z.; Liu, L.; Chen, Y. A review of research on the closed thermodynamic cycles of ocean thermal energy conversion. *Renew. Sustain. Energy Rev.* **2020**, *119*, 109581. [CrossRef]
94. Li, D.M.; Fan, C.C.; Zhang, C.B.; Chen, Y.P. Control strategy of load following for ocean thermal energy conversion. *Renew. Energy* **2022**, *193*, 595–607. [CrossRef]
95. Khosravi, A.; Syri, S.; Assad, M.E.H.; Malekan, M. Thermodynamic and economic analysis of a hybrid ocean thermal energy conversion/photovoltaic system with hydrogen-based energy storage system. *Energy* **2019**, *172*, 304–319. [CrossRef]
96. Arabkoohsar, A.; Sadi, M. Flywheel energy storage. In *Mechanical Energy Storage Technologies*; Arabkoohsar, A., Ed.; Academic Press: Cambridge, MA, USA, 2021; pp. 101–124.
97. Dincer, I.; Rosen, M.A. Thermal Energy Storage Methods. In *Thermal Energy Storage: Systems and Applications*, 3rd ed.; John Wiley & Sons: Hoboken, NJ, USA, 2021; pp. 125–259.
98. Dincer, I.; Rosen, M.A. *Thermal Energy Storage: Systems and Applications*, 3rd ed.; John Wiley & Sons: Hoboken, NJ, USA, 2021.
99. Zhou, D.; Zhao, C.Y.; Tian, Y. Review on thermal energy storage with phase change materials (PCMs) in building applications. *Appl. Energy* **2012**, *92*, 593–605. [CrossRef]
100. AlShafi, M.; Bicer, Y. Thermodynamic performance comparison of various energy storage systems from source-to-electricity for renewable energy resources. *Energy* **2021**, *219*, 119626. [CrossRef]
101. Atalay, H. Assessment of energy and cost analysis of packed bed and phase change material thermal energy storage systems for the solar energy-assisted drying process. *Sol. Energy* **2020**, *198*, 124–138. [CrossRef]
102. Başer, T.; Dong, Y.; Moradi, A.M.; Lu, N.; Smits, K.; Ge, S.; Tartakovsky, D.; McCartney, J.S. Role of Nonequilibrium Water Vapor Diffusion in Thermal Energy Storage Systems in the Vadose Zone. *J. Geotech. Geoenvironmental Eng.* **2018**, *144*, 1. [CrossRef]
103. Diyoke, C.; Wu, C.F. Thermodynamic analysis of hybrid adiabatic compressed air energy storage system and biomass gasification storage (A-CAES plus BMGS) power system. *Fuel* **2020**, *271*, 117572. [CrossRef]
104. Saxena, A.; Cuce, E.; Singh, D.B.; Cuce, P.M.; Gupta, P.; Suryavanshi, A.; Farrokhi, M.; El-Sebaii, A.A. A thermodynamic review on solar ponds. *Sol. Energy* **2022**, *242*, 335–363. [CrossRef]
105. Senturk Acar, M.; Arslan, O. Energy and exergy analysis of solar energy-integrated, geothermal energy-powered Organic Rankine Cycle. *J. Therm. Anal. Calorim.* **2019**, *137*, 659–666. [CrossRef]
106. Soni, N.; Sharma, D.; Rahman, M.M.; Hanmaiahgari, P.R.; Reddy, V.M. Mathematical Modeling of Solar Energy based Thermal Energy Storage for House Heating in Winter. *J. Energy Storage* **2021**, *34*, 102203. [CrossRef]
107. Temiz, M.; Dincer, I. A unique ocean and solar based multigenerational system with hydrogen production and thermal energy storage for Arctic communities. *Energy* **2022**, *239*, 122126. [CrossRef]
108. Zhang, X.F.; Yan, R.S.; Zeng, R.; Zhu, R.L.; Kong, X.Y.; He, Y.C.; Li, H.Q. Integrated performance optimization of a biomass-based hybrid hydrogen/thermal energy storage system for building and hydrogen vehicles. *Renew. Energy* **2022**, *187*, 801–818. [CrossRef]
109. Dincer, I.; Dost, S. A perspective on thermal energy storage systems for solar energy applications. *Int. J. Energy Res.* **1996**, *20*, 547–557. [CrossRef]
110. Bindra, H.; Betz, A.R.; Stallbaumer, E.M.; Ross, M.; Riley, G.A.; Morrow, J.A.; Huber, R.A.; Haque, M.R.; Chakraborty, P.P.; Adams, A.N.; et al. Heat and Mass Transfer in the Food, Energy, and Water Nexus—A Review. *J. Heat Transf.* **2020**, *142*, 090801.
111. John, E.; Hale, M.; Selvam, P. Concrete as a thermal energy storage medium for thermocline solar energy storage systems. *Sol. Energy* **2013**, *96*, 194–204. [CrossRef]
112. Picone, S.; Bloemendal, M.; Pellegrini, M.; Hoekstra, N.; Andreu Gallego, A.; Rodriguez Comins, J.; Murrel, A. Novel combinations of aquifer thermal energy storage with solar collectors, soil remediation and other types of geothermal energy systems. In Proceedings of the European Geothermal Congress, Hague, The Netherlands, 11–14 June 2019.
113. Bu, X.; Jiang, K.; Guo, Z. Storing high temperature solar thermal energy in shallow depth artificial reservoir for space heating. *Sci. Rep.* **2022**, *12*, 19549. [CrossRef] [PubMed]
114. Li, H.; Tao, Y.; Zhang, Y.; Fu, H. Two-objective optimization of a hybrid solar-geothermal system with thermal energy storage for power, hydrogen and freshwater production based on transcritical CO cycle. *Renew. Energy* **2022**, *183*, 51–66. [CrossRef]
115. Ahmed, A.A.; Assadi, M.; Kalantar, A.; Sliwa, T.; Sapińska-Śliwa, A. A critical review on the use of shallow geothermal energy systems for heating and cooling purposes. *Energies* **2022**, *15*, 4281. [CrossRef]
116. Romanov, D.; Leiss, B. Geothermal energy at different depths for district heating and cooling of existing and future building stock. *Renew. Sustain. Energy Rev.* **2022**, *167*, 112727. [CrossRef]
117. Aquifer Thermal Energy Storage for Decarbonisation of Heating and Cooling: Overcoming Technical, Economic and Societal Barriers to UK Deployment (ATESHAC). Available online: https://www.imperial.ac.uk/earth-science/research/research-groups/ateshac/ (accessed on 24 February 2023).
118. Khodabandeh, E.; Safaei, M.R.; Akbari, S.; Akbari, O.A.; Alrashed, A.A. Application of nanofluid to improve the thermal performance of horizontal spiral coil utilized in solar ponds: Geometric study. *Renew. Energy* **2018**, *122*, 1–16. [CrossRef]
119. Sayegh, M.A.; Zwierzchowski, R.; Malicki, M.; Lipka, M.; Danielewicz, J.; Jouhara, H.; Kaźmierczak, B.; Kutyłowska, M.; Piekarska, K. Innovative model of trigeneration system generating desalinated water, hot and cold by using low grade heat recovery from nuclear reactor set in cascade of sorption devices. *E3S Web Conf.* **2019**, *116*, 00108.
120. Ghaddar, N.K. Stratified Storage Tank Influence on Performance of Solar Water-Heating System Tested in Beirut. *Renew. Energy* **1994**, *4*, 911–925. [CrossRef]

121. Atomistic Simulation & Energy Research Group. Thermal Energy Grid Storage (TEGS) Concept. Available online: https://ase.mit.edu/projects/thermal-energy-grid-storage-tegs/ (accessed on 1 December 2022).
122. Amy, C.; Seyf, H.R.; Steiner, M.A.; Friedman, D.J.; Henry, A. Thermal energy grid storage using multi-junction photovoltaics. *Energy Environ. Sci.* **2019**, *12*, 334–343. [CrossRef]
123. Cárdenas, B.; León, N. High temperature latent heat thermal energy storage: Phase change materials, design considerations and performance enhancement techniques. *Renew. Sustain. Energy Rev.* **2013**, *27*, 724–737. [CrossRef]
124. Ndukwu, M.C.; Bennamoun, L.; Abam, F.I.; Eke, A.B.; Ukoha, D. Energy and exergy analysis of a solar dryer integrated with sodium sulfate decahydrate and sodium chloride as thermal storage medium. *Renew. Energy* **2017**, *113*, 1182–1192. [CrossRef]
125. García-Romero, A.; Diarce, G.; Ibarretxe, J.; Urresti, A.; Sala, J.M. Influence of the experimental conditions on the subcooling of Glauber's salt when used as PCM. *Sol. Energy Mater. Sol. Cells* **2012**, *102*, 189–195. [CrossRef]
126. Chaouch, W.B.; Khellaf, A.; Mediani, A.; Slimani, M.E.; Loumani, A.; Hamid, A. Experimental investigation of an active direct and indirect solar dryer with sensible heat storage for camel meat drying in Saharan environment. *Sol. Energy* **2018**, *174*, 328–341. [CrossRef]
127. Akgün, M.; Aydın, O.; Kaygusuz, K.J.E.C. Experimental study on melting/solidification characteristics of a paraffin as PCM. *Energy Convers. Manag.* **2007**, *48*, 669–678. [CrossRef]
128. Degues, K.d.M.; Cypriano, M.G.; Coelho, K.B.; Luza, A.L.; Montedo, O.R.K.; de Castro, L.C.; Angioletto, E. Assessment of PCM-impregnated zeolite as a matrix for latent heat storage. *Mater. Sci. Forum* **2018**, *912*, 87–92. [CrossRef]
129. Wang, Z.F.; Wu, J.N.; Lei, D.Q.; Liu, H.; Li, J.P.; Wu, Z.Y. Experimental study on latent thermal energy storage system with gradient porosity copper foam for mid-temperature solar energy application. *Appl. Energy* **2020**, *261*, 114472. [CrossRef]
130. Zhang, Q.F.; Xia, T.F.; Zhang, Q.H.; Zhu, Y.C.; Zhang, H.Z.; Xu, F.; Sun, L.X.; Wang, X.D.; Xia, Y.P.; Lin, X.C.; et al. Biomass Homogeneity Reinforced Carbon Aerogels Derived Functional Phase-Change Materials for Solar-Thermal Energy Conversion and Storage. *Energy Environ. Mater.* **2023**, *6*, 1. [CrossRef]
131. Al-Shannaq, R.; Farid, M.; Al-Muhtaseb, S.; Kurdi, J. Emulsion stability and cross-linking of PMMA microcapsules containing phase change materials. *Sol. Energy Mater. Sol. Cells* **2015**, *132*, 311–318. [CrossRef]
132. Al-Shannaq, R.; Kurdi, J.; Al-Muhtaseb, S.; Dickinson, M.; Farid, M. Supercooling elimination of phase change materials (PCMs) microcapsules. *Energy* **2015**, *87*, 654–662. [CrossRef]
133. Al-Shannaq, R.; Kurdi, J.; Al-Muhtaseb, S.; Farid, M. Innovative method of metal coating of microcapsules containing phase change materials. *Sol. Energy* **2016**, *129*, 54–64. [CrossRef]
134. Ansari, J.A.; Al-Shannaq, R.; Kurdi, J.; Al-Muhtaseb, S.A.; Ikutegbe, C.A.; Farid, M.M. A Rapid Method for Low Temperature Microencapsulation of Phase Change Materials (PCMs) Using a Coiled Tube Ultraviolet Reactor. *Energies* **2021**, *14*, 7867. [CrossRef]
135. Roberts, N.S.; Al-Shannaq, R.; Kurdi, J.; Al-Muhtaseb, S.A.; Farid, M.M. Efficacy of using slurry of metal-coated microencapsulated PCM for cooling in a micro-channel heat exchanger. *Appl. Therm. Eng.* **2017**, *122*, 11–18. [CrossRef]
136. Cao, Y.; Weng, M.; Mahmoud, M.H.H.; Elnaggar, A.Y.; Zhang, L.; El Azab, I.H.; Chen, Y.; Huang, M.; Huang, J.; Sheng, X. Flame-retardant and leakage-proof phase change composites based on MXene/polyimide aerogels toward solar thermal energy harvesting. *Adv. Compos. Hybrid Mater.* **2022**, *5*, 1253–1267. [CrossRef]
137. Shen, R.; Weng, M.; Zhang, L.; Huang, J.; Sheng, X. Biomass-based carbon aerogel/Fe3O4@PEG phase change composites with satisfactory electromagnetic interference shielding and multi-source driven thermal management in thermal energy storage. *Compos. Part A Appl. Sci. Manuf.* **2022**, *163*, 107248. [CrossRef]
138. Cao, Y.; Zeng, Z.; Huang, D.; Chen, Y.; Zhang, L.; Sheng, X. Multifunctional phase change composites based on biomass/MXene-derived hybrid scaffolds for excellent electromagnetic interference shielding and superior solar/electro-thermal energy storage. *Nano Res.* **2022**, *15*, 8524–8535. [CrossRef]
139. Fujii, S.; Nakagaki, T.; Kanematsu, Y.; Kikuchi, Y. Prospective life cycle assessment for designing mobile thermal energy storage system utilizing zeolite. *J. Clean. Prod.* **2022**, *365*, 132592. [CrossRef]
140. Liu, C.; Cao, H.; Jin, S.; Bao, Y.; Cheng, Q.; Rao, Z. Synthesis and characterization of microencapsulated phase change material with phenol-formaldehyde resin shell for thermal energy storage. *Sol. Energy Mater. Sol. Cells* **2022**, *243*, 111789. [CrossRef]
141. Matuszewska, D.; Kuta, M.; Olczak, P. Techno-Economic Assessment of Mobilized Thermal Energy Storage System Using Geothermal Source in Polish Conditions. *Energies* **2020**, *13*, 3404. [CrossRef]
142. Meng, F.X.; Wang, E.H.; Zhang, B.; Zhang, F.J.; Zhao, C.L. Thermo-economic analysis of transcritical CO_2 power cycle and comparison with Kalina cycle and ORC for a low-temperature heat source. *Energy Convers. Manag.* **2019**, *195*, 1295–1308. [CrossRef]
143. ElBahloul, A.A.; Zeidan, E.-S.B.; El-Sharkawy, I.I.; Hamed, A.M.; Radwan, A. Recent advances in multistage sorption thermal energy storage systems. *J. Energy Storage* **2022**, *45*, 103683. [CrossRef]
144. Paksoy, H.O.; Andersson, O.; Abaci, S.; Evliya, H.; Turgut, B. Heating and cooling of a hospital using solar energy coupled with seasonal thermal energy storage in an aquifer. *Renew. Energy* **2000**, *19*, 117–122. [CrossRef]
145. Fei, L.; Yin, Y.J.; Yang, M.F.; Zhang, S.F.; Wang, C.X. Wearable solar energy management based on visible solar thermal energy storage for full solar spectrum utilization. *Energy Storage Mater.* **2021**, *42*, 636–644. [CrossRef]
146. Boretti, A. Concentrated Solar Energy with Thermal Energy Storage for Hydrogen Production by Three-Step Thermochemical Water-Splitting Cycles. *Energy Fuels* **2021**, *35*, 10832–10840. [CrossRef]

147. Al-Mashakbeh, A.S.; Al-Mashaqbeh, S.F.; Awwad, A.M.E.; Almaita, E.K. Design and Optimization of Wind Energy System installed in Rehab Gas Power Station Combined with Thermal Energy Grid Storage Multi-Junction Photovoltaics Mean in Mafraq, Jordan. *Int. J. Renew. Energy Res.* **2022**, *12*, 39–47.
148. Karasu, H.; Dincer, I. Analysis and Efficiency Assessment of Direct Conversion of Wind Energy into Heat Using Electromagnetic Induction and Thermal Energy Storage. *J. Energy Resour. Technol.-Trans. Asme* **2018**, *140*, 71201. [CrossRef]
149. Wang, K.; Satyro, M.A.; Taylor, R.; Hopke, P.K. Thermal energy storage tank sizing for biomass boiler heating systems using process dynamic simulation. *Energy Build.* **2018**, *175*, 199–207. [CrossRef]
150. Arslan, O.; Arslan, A.E. Performance evaluation and multi-criteria decision analysis of thermal energy storage integrated geothermal district heating system. *Process Saf. Environ. Prot.* **2022**, *167*, 21–33. [CrossRef]
151. Zhou, Y. Ocean energy applications for coastal communities with artificial intelligence a state-of-the-art review. *Energy AI* **2022**, *10*, 100189. [CrossRef]
152. Boretti, A. The perspective of enhanced geothermal energy integration with concentrated solar power and thermal energy storage. *Energy Storage* **2022**, *4*, 1. [CrossRef]
153. Assareh, E.; Assareh, M.; Alirahmi, S.M.; Jalilinasrabady, S.; Dejdar, A.; Izadi, M. An extensive thermo-economic evaluation and optimization of an integrated system empowered by solar-wind-ocean energy converter for electricity generation—Case study: Bandar Abas, Iran. *Therm. Sci. Eng. Prog.* **2021**, *25*, 100965. [CrossRef]
154. Li, L.; Gao, Y.; Ning, D.Z.; Yuan, Z.M. Development of a constraint non-causal wave energy control algorithm based on artificial intelligence. *Renew. Sustain. Energy Rev.* **2021**, *138*, 110519. [CrossRef]
155. Harish, V.S.K.V.; Kumar, A. A review on modeling and simulation of building energy systems. *Renew. Sustain. Energy Rev.* **2016**, *56*, 1272–1292. [CrossRef]
156. Dahanayake, K.W.D.K.C.; Chow, C.L. Studying the potential of energy saving through vertical greenery systems: Using EnergyPlus simulation program. *Energy Build.* **2017**, *138*, 47–59. [CrossRef]
157. Djamel, I.; Camara, M.B.; Dakyo, B. Offshore wind farms energy injection in the electrical grid-Lithium battery to mitigate power fluctuations. *Int. J. Renew. Energy Res.* **2015**, *5*, 1049–1061.
158. Yu, H.; Helland, H.; Yu, X.; Gundersen, T.; Sin, G. Optimal design and operation of an Organic Rankine Cycle (ORC) system driven by solar energy with sensible thermal energy storage. *Energy Convers. Manag.* **2021**, *244*, 114494. [CrossRef]
159. Bamooeifard, A. Future studies in Iran development plans for wind power, a system dynamics modeling approach. *Renew. Energy* **2020**, *162*, 1054–1064. [CrossRef]
160. Zhou, Y.; Cao, S. Quantification of energy flexibility of residential net-zero-energy buildings involved with dynamic operations of hybrid energy storages and diversified energy conversion strategies. *Sustain. Energy Grids Netw.* **2020**, *21*, 100304. [CrossRef]
161. Stritih, U.; Butala, V. Optimization of a thermal storage unit combined with a biomass boiler for heating buildings. *Renew. Energy* **2004**, *29*, 2011–2022. [CrossRef]
162. Terziotti, L.T.; Sweet, M.L.; McLeskey, J.T. Modeling seasonal solar thermal energy storage in a large urban residential building using TRNSYS 16. *Energy Build.* **2012**, *45*, 28–31. [CrossRef]
163. Elghamry, R.; Hassan, H. Impact a combination of geothermal and solar energy systems on building ventilation, heating and output power: Experimental study. *Renew. Energy* **2020**, *152*, 1403–1413. [CrossRef]
164. Ciani Bassetti, M.; Consoli, D.; Manente, G.; Lazzaretto, A. Design and off-design models of a hybrid geothermal-solar power plant enhanced by a thermal storage. *Renew. Energy* **2018**, *128*, 460–472. [CrossRef]

energies

MDPI

Review

Nano-Enhanced Phase Change Materials for Thermal Energy Storage: A Bibliometric Analysis

Javad Mohammadpour [1], Ann Lee [1,*], Victoria Timchenko [2] and Robert Taylor [2]

[1] School of Engineering, Macquarie University, Sydney, NSW 2109, Australia; javad.mohammadpour@mq.edu.au
[2] School of Mechanical and Manufacturing Engineering, The University of New South Wales, Sydney, NSW 2052, Australia; v.timchenko@unsw.edu.au (V.T.); robert.taylor@unsw.edu.au (R.T.)
* Correspondence: ann.lee@mq.edu.au

Abstract: The high latent heat thermal energy storage (LHTES) potential of phase change materials (PCMs) has long promised a step-change in the energy density for thermal storage applications. However, the uptake of PCM systems has been limited due to their relatively slow charging response, limited life, and economic considerations. Fortunately, a concerted global research effort is now underway to remove these remaining technical challenges. The bibliometric analysis of this review reveals that a major focus is now on the development of nano-enhanced phase change materials (NePCM), which have the potential to mitigate many of these technical challenges for PCM-based thermal energy storage systems. As such, our bibliometric analysis has zeroed in on research in the field of thermal energy storage using NePCMs since 1977. It was found that journal articles were the most frequently used document type, representing 79% of the records and that the pace of new work in this specific area has increased exponentially over these two decades, with China accounting for the highest number of citations and the most publications (168), followed by India and Iran. China has also played a central role in the collaboration network among the most productive countries, while Saudi Arabia and Vietnam show the highest international collaboration level.

Keywords: nano-enhanced phase change material; NePCM; latent heat thermal energy storage; LHTES; bibliometric analysis; solar energy; electronic devices; cooling

Citation: Mohammadpour, J.; Lee, A.; Timchenko, V.; Taylor, R. Nano-Enhanced Phase Change Materials for Thermal Energy Storage: A Bibliometric Analysis. *Energies* **2022**, *15*, 3426. https://doi.org/10.3390/en15093426

Academic Editor: Antonio Lecuona

Received: 8 March 2022
Accepted: 5 May 2022
Published: 7 May 2022

1. Introduction

Fossil fuels as primary resources are non-renewable and have severe impacts on the environment, such as the emission of greenhouse gases and global warming. On the other hand, energy demands rapidly grow, while generating conventional energy sources is much lower than their depletion. Hence, renewable energy sources have drawn extensive attention as potential alternatives [1]. Renewable energy with lower costs can impact energy supply and demand.

1.1. Thermal Energy Storage

Since the output of many renewable energy sources is unpredictable, they require energy storage systems. Solar energy is an excellent example of renewable resources coming with low negative impacts. Given that this energy is intermittent and dependent on weather conditions, efficient energy storage is essential to compensate for the inherent limitations. Thermal energy storage (TES) systems are widely studied as an effective technique to store solar energy all day and later utilise it overnight or on overcast days [2]. Recent advancements in TES systems are classified into two main parts: (i) physical storage and (ii) chemical storage. Physical storage is based on heat transfer mechanisms, while reversible chemical reactions play a significant role in chemical storage. As shown in Figure 1, thermal energy can be stored in physical storage by sensible and latent techniques [3].

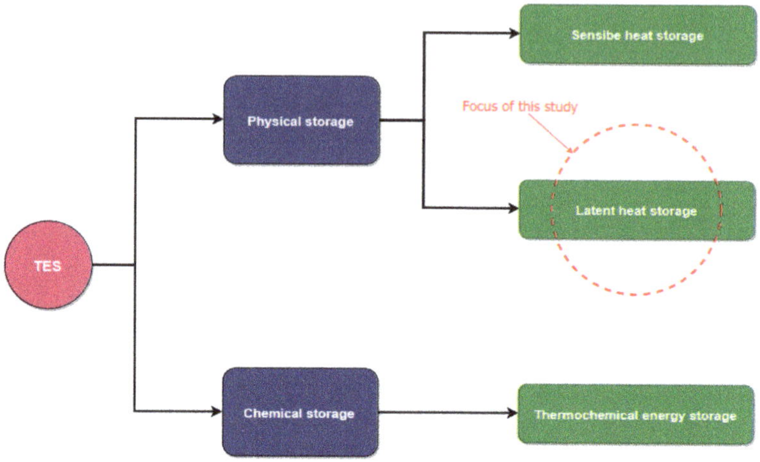

Figure 1. Classification of TESs. Adapted with permission from Ref. [3]. Copyright 2019 International Journal of Heat and Mass Transfer.

1.2. PCMs in LHTES

PCMs in LHTES units represent a compact medium to absorb and release energy [4]. The phase transition of PCMs is the key factor in charge and discharge TES systems, which occurs from the solid or liquid phase to the others [3]. Various materials are adopted for PCMs, such as organic, inorganic, or eutectics with different melting temperatures (−100 to 1000 °C). The performance of PCMs is highly dependent on the melting point. In this regard, PCMs are divided into three categories: low, middle, and high melting temperatures [2]. Huang et al. [5] assumed that the low temperature for melting temperatures is up to 120 °C, the middle for temperatures is between 120 °C and 300 °C, and the high melting temperature for temperatures is over 300 °C. Since PCMs are highly beneficial for storing thermal energy, they have a wide range of applications in solar systems, save energy in buildings, lithium-ion battery cooling, and efficient electronic cooling systems [6–12].

However, the serious challenge of PCMs is their low intrinsic thermal conductivity, increasing phase change process time, especially in solidification, since natural convection is low and conduction heat transfer is dominant [2]. Hence, various passive solutions have been proposed to tackle this issue and improve PCM functionality [11,13–15]. A simple approach is to work on the arrangement of heat transfer fluid (HTF) tubes surrounded by PCMs. Senthil [13] studied the position of HTF tubes on the melting of PCM. The inclined configuration was superior to concentric and eccentric arrangements. The following solution is HTF tubes with embedded fins to increase the contact area with PCMs and enhance the heat transfer rate [16,17]. Another method is to combine multiple PCMs with different melting temperatures to resolve the long process of phase changing [11,14,15]. Mozafari et al. [11] compared the performance of a single PCM to multiple PCMs in a heat sink. They obtained the best thermal performance when n-Eicosane and RT44 were combined. Further, the operational time was accelerated by 3.3–12%. Another method is the nano encapsulated phase change material (NEPCM), which can be adopted in base fluids [2,18,19]. Eventually, one of the most efficient techniques to overcome the remaining challenge is nano-enhanced phase change materials (NePCMs), achieved by adding nanoparticles with high thermal conductivity into PCMs [2,20].

Although a large number of NePCM articles have been published in recent years, no bibliometric study has been reported. The present study provides an overview of the application of NePCMs in LHTES systems. In addition, it aims to review publications on nanostructure additives into PCMs systematically and identify the scientific research direction, productive authors, sources, and countries in this field using bibliometric analysis.

The study is synthesised into four sections in this paper. Section 2 presents the overview of NePCMs in LHTES. The literature findings using bibliometric analysis are introduced in Section 3. A full bibliographic analysis is presented in Section 4, aiming to explore the literature based on different bibliometrics. Finally, conclusions are drawn in Section 5. By identifying the research direction, active researchers, leading sources, and productive countries, the findings in this paper are beneficial for the research communities seeking literary sources, insight into research trends and focus on NEPCM as well as international research collaboration opportunities.

2. Overview of NePCMs

Nanotechnology has paved the way to produce particles with higher thermal conductivity in nanoscales, including metal, metal oxide, single-walled carbon nanotube (SWCNT), multi-walled carbon nanotube (MWCNT), graphene, and graphite [21]. The addition of these highly thermal conductive nanoparticles in PCMs improves heat transfer performance. However, the thermal conductivity of NePCMs is affected by factors such as the size, shape, and type of particles, temperature, surfactant, and particle concentration, as shown in Figure 2 [21].

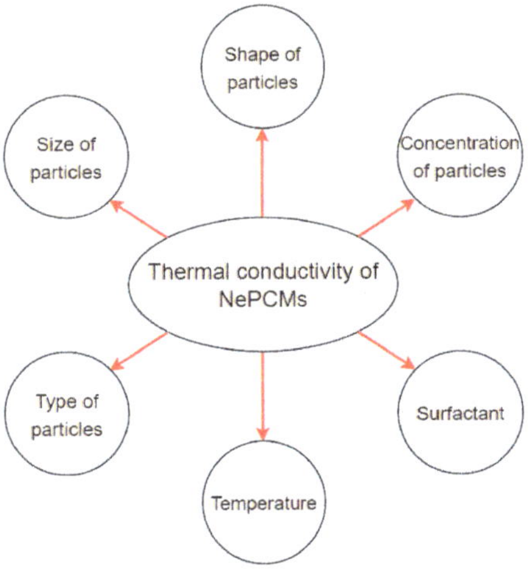

Figure 2. Influential factors affecting NePCMs' thermal conductivity. Adapted with permission from Ref. [21]. Copyright 2019 Journal of Energy Storage.

The use of nanostructures with high thermal conductivity in PCMs enhances the functionality of LHTES significantly [22]. A few contributions were reported using nanoparticles to enhance the thermal conductivity of PCMs before 2005 [23,24]. Later, a significant number of publications embraced nanotechnology to improve the performance of LHTES systems. In 2005, Elgafy and Lafdi [25] examined the effect of nanofibers in paraffin wax and enhanced thermal conductivity remarkably in a highly-cited study. As a result, increasing the carbon additives reduced the solidification time and enhanced the output power. Then, in a leading study, Khodadadi and Hosseinizadeh [20] analysed the potential of NePCMs to improve TES performance. Water PCM, enhanced with copper nanoparticles, demonstrated a higher heat release rate than conventional PCM due to increasing the thermal conductivity and decreasing the latent heat of fusion. Recently, Yazdanifard et al. [26] innovatively combined NePCMs with nanofluids to filter radiation and store solar energy in a photovoltaic thermal system. A wide range of nano additives, such as Al_2O_3, ZnO, CuO, Cu, SiO_2, and MWCNT, have been analysed in previous studies [27–30].

3. Methodology

Due to the significance of energy supply in the world, a large number of publications have focused on using nanoparticles in LHTES units to improve the performance of thermal storage. Bibliometrics is classified under information science to manage knowledge and monitor information. While the number of review papers on the content analysis of NePCM articles is considerable, there is no bibliometric analysis in this field. However, this method has been widely used in different fields. For instance, Calderón et al. [31] adopted a bibliometric analysis to identify the trend of TES research over the last twenty years. Later, Saikia et al. [32] monitored and planned the following research trend in solar cooling using the bibliometric approach. In addition, a bibliometric analysis was conducted to understand the perspective of thermal management in lithium-ion batteries [33]. Yataganbaba et al. [34] analysed, through bibliometric methods, the trend of encapsulation of PCMs between 1990 and 2015. The present study is a preliminary bibliometric analysis of LHTES, revealing that NePCM is an emerging topic of interest. This bibliometric study will lead researchers and predict a future trend in the field of NePCMs.

Bibliometrics is a statistical method that illustrates an overview of a specific research field's trend and introduces leading publications, publishers, countries, institutions, and researchers in such research areas [35,36]. Web of Science (WoS) Core Collection is a proper source of high-quality data for bibliometric analyses [37]. In engineering research fields, the Scopus database includes a higher number of publications than WoS [37,38]. The present methodology aims to combine the collected data extracted from both rich databases and remove all duplicates to include all publications about the integration of NePCMs in TESs.

To conduct the research process, different keywords have been selected to search relevant publications on the two websites, such as "phase change material" AND "nanoparticle", or "phase change material" AND "nano-particle", or "PCM" AND "particle", or "nano enhanced phase change material", or "nano-enhanced phase change material", or "nepcm", or "ne-pcm", AND NOT "nano encapsulated", AND (LIMIT-TO (language,"English") in titles, abstracts, keywords, topics. The search is limited to publications published in English, while there is no specific time span. With the queries, 926 and 807 publications are found in Scopus and WoS, respectively. Looking closely at the titles and abstracts of the collected data results in further refinement. Eventually, exported files are merged, and duplications are removed. The total number of relevant publications in this field is equal to 864. The present bibliometric analysis is carried out using VOSviewer [39] and R package Bibliometrix [40]. The following sections present bibliometric results about relevant authors, sources, institutions, worldwide networks, distribution of publications per year, and other information.

4. Results and Discussion

4.1. Main Information

Due to the improvement of LHTES by NePCMs, most of the publications have focused on this field. As shown in Figure 3, a bulk of publications (79%) collected from the WoS and Scopus databases have been published as journal papers. Furthermore, 12% and 6% of publications are distributed as conference papers and review papers, respectively. Proceedings papers and book chapters have slight portions of 2% and 1%, respectively.

The starting year of the NePCM papers is 1977, and the total document number is 864. However, this research field has become popular since 2005. Hence, Figure 4 shows the annual growth rate of publications between 2005 and 2021, and a few contributions are not shown only in this figure. While an overall increasing trend of 20.34% can be seen in this time span, the number of publications has significantly increased since 2015. It shows that NePCM is a hot topic that has been able to attract more interest due to its effectiveness. For instance, the number of publications reached its peak at 181 in 2021.

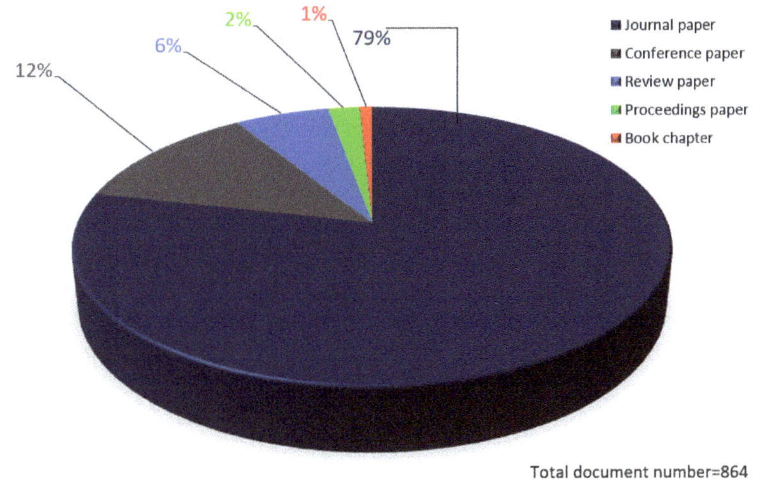

Figure 3. Distribution of publications.

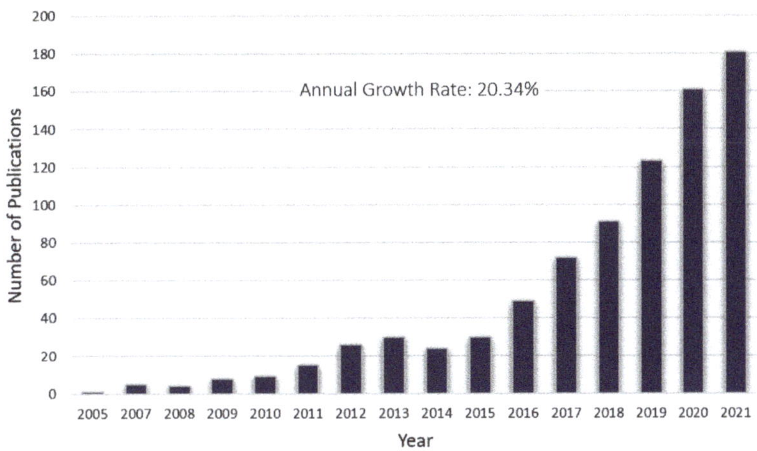

Figure 4. Number of all publications per year.

In this analysis, the collaboration through joint publications between countries is identified by the authors' affiliations. Bibliographic networks are created to provide the reader with a visual analysis of bibliometrics pertaining to international collaboration. Figure 5 illustrates the relationship between countries publishing more than 15 relevant publications, excluding other countries. In addition, the size of the circles and country names is weighted based on the number of published articles in the NePCM research field. As can be seen, China is the most productive country in this research field (168), followed by India (130), Iran (113), and the United States (67). Saudi Arabia and Iran published more remarkable multiple-country publications (18 and 15, respectively) in this research field. Moreover, the thickness of the lines between countries represents interaction strength, and the colours show different clusters. As the first country with the most significant interactions, Saudi Arabia collaborated more with institutions in Vietnam, Egypt, India, the United States, and Malaysia. Therefore, these countries are shown in one cluster in blue. Moreover, the second cluster in red is formed by Iran, Canada, Turkey, Vietnam, Spain, and France. The third group in green includes China, the United Kingdom, Pakistan, UAE, Morocco, Taiwan, and Singapore. Therefore, these countries are shown in one cluster in blue. Similarly, the second cluster in red is formed by Iran, Canada, Turkey, Vietnam, Spain, and France. The third group in green includes China, the United Kingdom, Pakistan, UAE, Morocco, Taiwan, and Singapore.

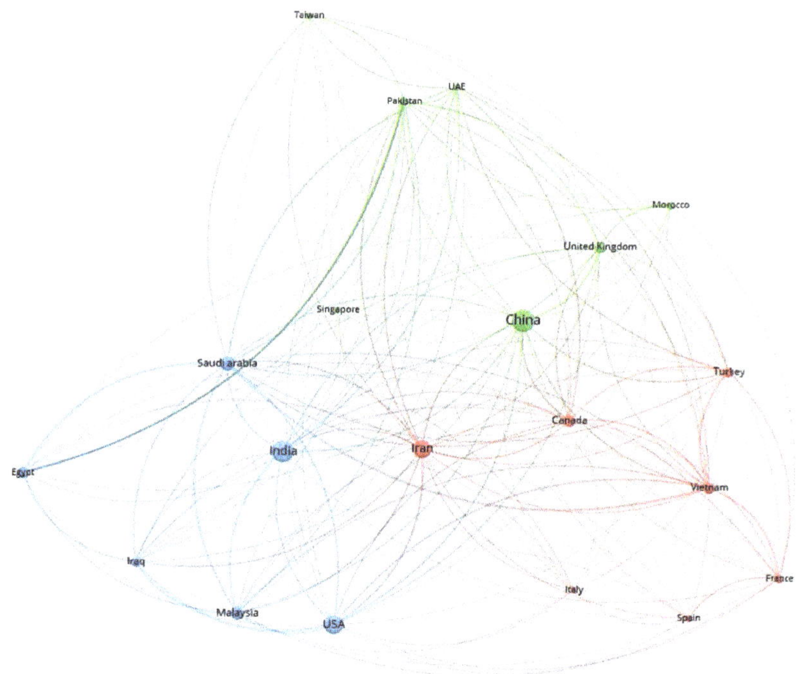

Figure 5. Collaboration between countries.

4.2. Analysis of Keywords

To shed more light on the main idea and research methodology, Table 1 shows 10 authors' keywords with an occurrence over 35. "Phase change material", with 296 occurrences, is the most frequent keyword chosen by authors, followed by "nanoparticles" and "thermal energy storage" with the frequency of 160 and 146, respectively. It is consistent with the main purpose of the current bibliometric study (i.e., the application of NePCMs in TES systems). All collected publications have focused on nanotechnology to improve the performance of LHTES units.

Table 1. Wordcloud of author's keywords with over 35 occurrences.

Terms	Frequency
Phase change material	296
Nanoparticles	160
Thermal energy storage	146
Thermal conductivity	119
PCM	90
Solidification	83
Melting	78
Nanoparticle	78
Nanofluid	41
NEPCM	36

For further evaluation, Figure 6 shows a treemap of the 30 most occurred words in research abstracts with the aim of introducing research gaps and trends to researchers. Treemaps consist of a series of rectangles with different sizes proportional to the frequencies of the words used in abstracts. As shown, "thermal", "PCM", and "heat" have been widely used in the abstracts of this research field with the frequency of 2759, 2184, and 1964, respectively. As shown in the figure, every three words with similar frequencies are shown vertically or horizontally next to each other. The second criterion is colour adopted to distinguish the difference in each group of words. As evident, "Thermal" is in blue, while "PCM" and "Heat" are in red and grey, respectively. Then, "nanoparticles", "phase", and "material" are the following widespread words in scientific research on NePCMs, whereas "volume" has the minimum occurrence (380) in the figure. It also shows that some influential parameters in NePCM, such as volume fraction, thermal conductivity, and solar application, are popular in previous studies.

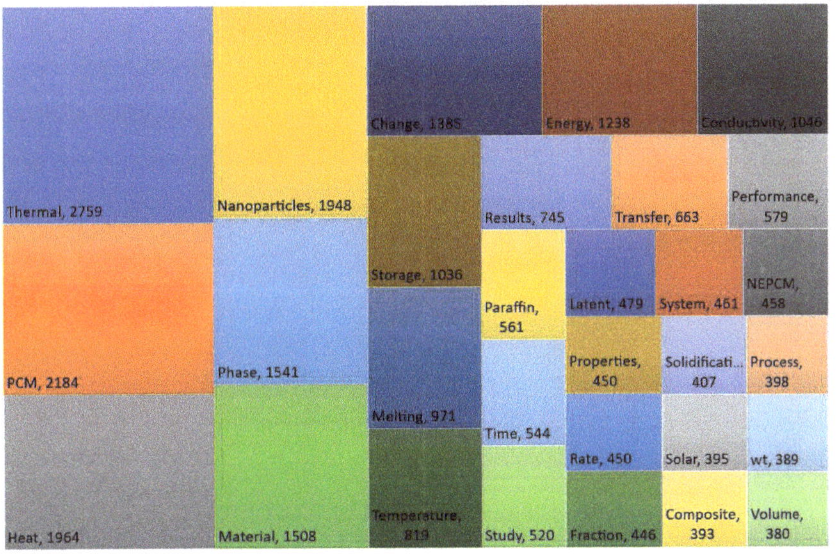

Figure 6. Treemap of 30 most frequent words in abstracts with the minimum occurrence of 380.

4.3. Productivity and Growth Rate

This section discusses the productivity and growth rate of countries and sources. Figure 7 illustrates the productivity of 10 countries with over 17 articles in the NePCM research field. China holds the highest number of articles (168), followed by India and Iran with 130 and 113, respectively. Later, the United States has shared 67 articles, and the other countries' publications are significantly lower than these four productive countries, which is consistent with Figure 5.

The scatter of publications in a given field can be defined by Bradford's law. In the present study, it is adopted to identify the leading publishers in the field of NePCM in TES units. Based on Bradford's law [41], journals published in a research field can be divided into three zones, while each zone includes approximately one-third of all publications. Zone 1 has a few journals, whereas Zone 2 includes more sources, and Zone 3 has the bulk of sources. In the present bibliometric analysis, 248 sources collected from WoS and Scopus are divided into three clusters based on Bradford's law, including Cluster 1 with 8, Cluster 2 with 33, and Cluster 3 with 207 sources. One-third of the 864 publications in NePCMs have been published by 8 sources, as shown in Figure 8. *"Journal of Energy Storage"* holds the highest contribution with 78 articles, and then two journals of, *"International*

Journal of Heat and Mass Transfer" and *"Applied Thermal Engineering"* have published 55 and 43 publications, respectively.

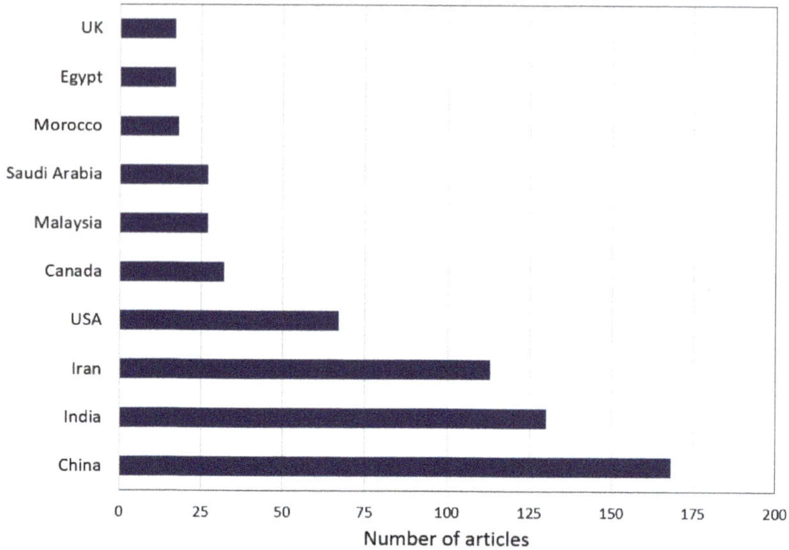

Figure 7. Productive countries.

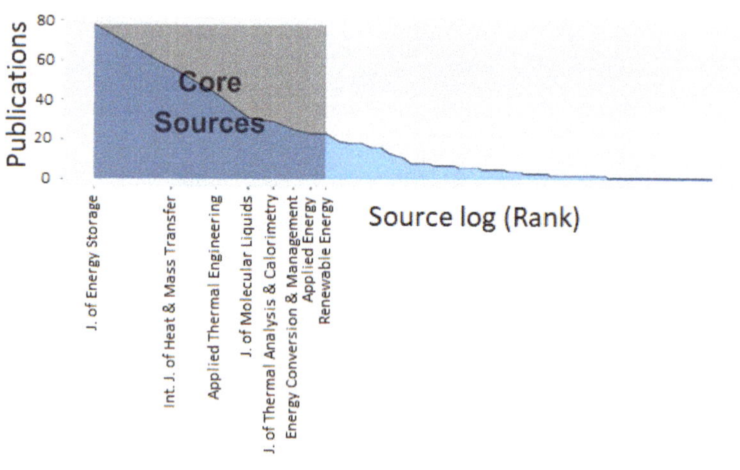

Figure 8. Bradford's Law of Scattering.

From another perspective, Figure 9 compares the cumulative evolution of ten productive sources per year since 2012. It is highly beneficial for researchers since the figure shows which sources have attracted a higher number of publications in the last 10 years. In recent years, *"Journal of Energy Storage"* has seen the largest publication in NePCM papers, followed by *"International Journal of Heat and Mass Transfer"* and *"Applied Thermal Engineering"*. However, *"International Journal of Heat and Mass Transfer"* published more articles over the years until 2020. The number of published articles in *"Journal of Energy Storage"* has steeply increased since 2019. In addition, a closer look indicates that NePCM has become a hot topic, and all journals have experienced significant growth.

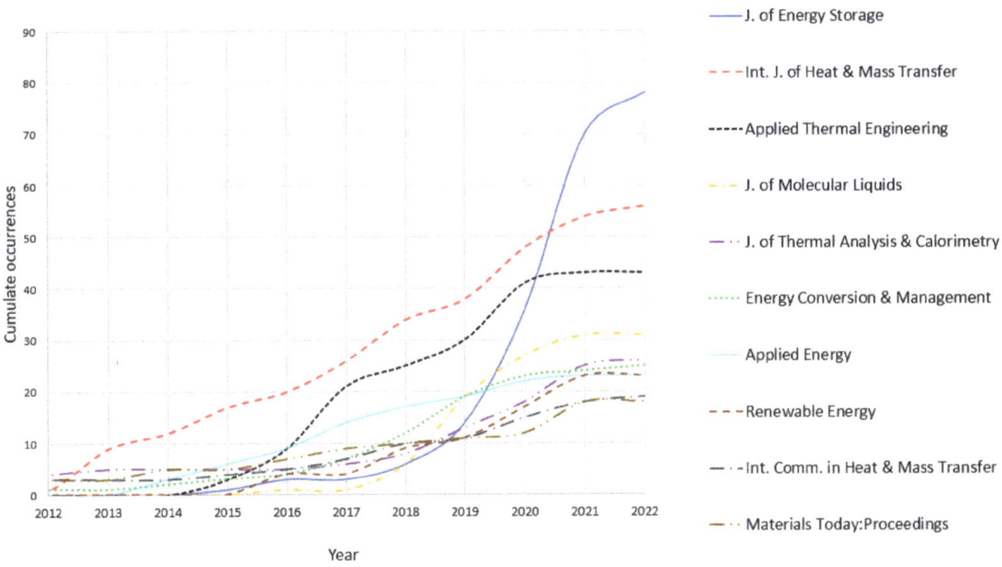

Figure 9. Cumulative evolution of sources in the last 10 years.

4.4. Authorship and Countries Evolution

In this study, the scientific authorship output is analysed based on the collected database from WoS and Scopus using several measurements, including H-index, G-index, M-index, the total citations, and the number of publications. H-index measures the number of publications with a citation number greater than or equal to H [42]. However, G-index applies a higher weight to publications, attracting more citations [43]. Unlike H-index, M-index considers the starting year of authors and is expressed as a ratio of the H-index to years publishing [42].

Table 2 compares the research output of the 10 top authors in NePCM with some measurements. The top-one author is Mohsen Sheikholeslami from Babol Noshirvani University of Technology, who records outstanding progress with 33 NePCM articles and holds the highest bibliometric measurements (H-index = 19, G-index = 33, and M-index = 2.714). Shafee A., Ganji D.D., and Ghalambaz M. are the other productive authors with a high number of papers and indexes. The publications of Sheikholeslami M. and Khodadadi J.M. in the field of NePCM have attracted 2039 and 1404 total citations, respectively. The total citations of the other authors are less than 1000 in this field. Moreover, the H-index of all the top 10 authors is higher than 6, and the minimum citation is 189. China and Iran have 3 authors individually among the 10 top authors in this research field.

Furthermore, Figure 10 presents the most highly-cited relevant publications with a minimum of 249 total citations. The main significance of a publication's citations is to guarantee its popularity on the research topic and indicate the scholarly impact of an article. The purpose of this figure is to identify the most cited publications and introduce these valuable references to those who are keen to study in this field. Khodadadi and Hosseinizadeh's article [20] attracted the highest citation of 474. In this numerical and experimental study, they revealed early findings confirming the superiority of NEPCM for TES applications. The second most-cited publication belongs to Kim and Drzal [44]. They experimentally studied increasing the thermal conductivity of paraffin wax by exfoliated graphite nanoplatelets.

Table 2. Top 10 authors in the NePCM field.

Author	Affiliation	N. of Papers	Index			Citations
			H	G	M	
Sheikholeslami M.	Babol Noshirvani University of Technology, Iran	33	19	33	2.7	2039
Shafee A.	Duy Tan University, Vietnam	20	13	20	2.6	776
Ganji D.D.	Babol Noshirvani University of Technology, Iran	19	17	19	2.4	809
Ghalambaz M.	Ton Duc Thang University, Vietnam	19	10	19	1.7	623
Khodadadi J.M.	Auburn University, USA	13	12	13	0.7	1404
Wang J.	Hebei University of Technology, China	13	9	13	0.6	929
Fan L.W.	Zhejiang University, China	13	8	13	0.8	281
Mehryan S.A.M.	Islamic Azad University, Iran	13	6	13	2	189
Kalaiselvam S.	Anna University, India	12	12	12	1. 1	586
Zhang X.	University of Science and Technology Beijing, China	12	6	12	0.5	390

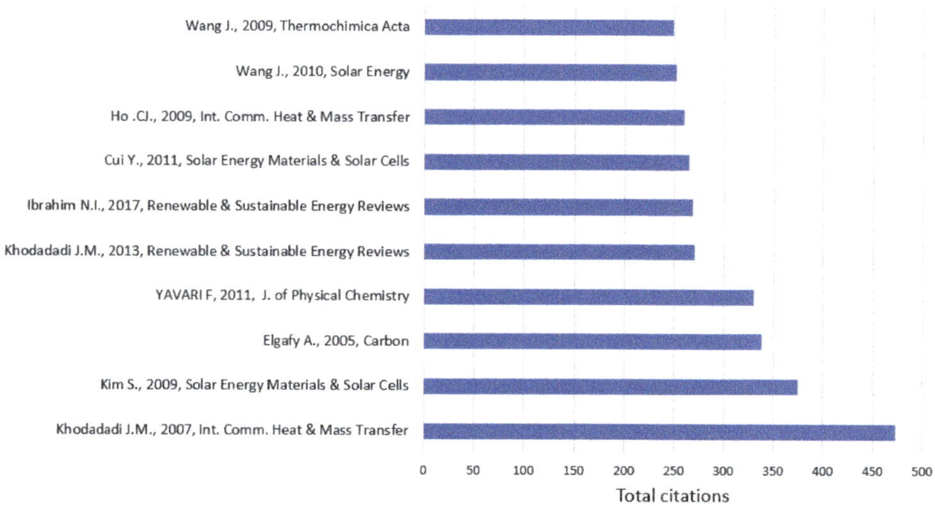

Figure 10. Ten high-cited publications in the NePCM field are Khodadadi and Hosseinizadeh [20], Kim and Drzal [44], Elgafy and Lafdi [25], Yavari et al. [45], Khodadadi et al. [22], Ibrahim et al. [46], Cui et al. [47], Ho and Gao [48], Wang et al. [49], and Wang et al. [50].

Besides the country's productivity shown in Figure 7, Figure 11 provides evidence and information about the number of citations attracted by each country. These two factors reveal the leading countries with impactful roles in a specific research field in the world. From the perspective of the most cited countries, China holds the highest cited rank (6184) in the NePCM field in the world. Then, the United States and Iran have attracted the second and third ranks of global citations, with 4158 and 3184, respectively. India accounts for more than 2859 citations, while the citation number of other countries is below 808. Hence, the total number of citations from the top four countries is 2.3 times greater than all the citations from other countries.

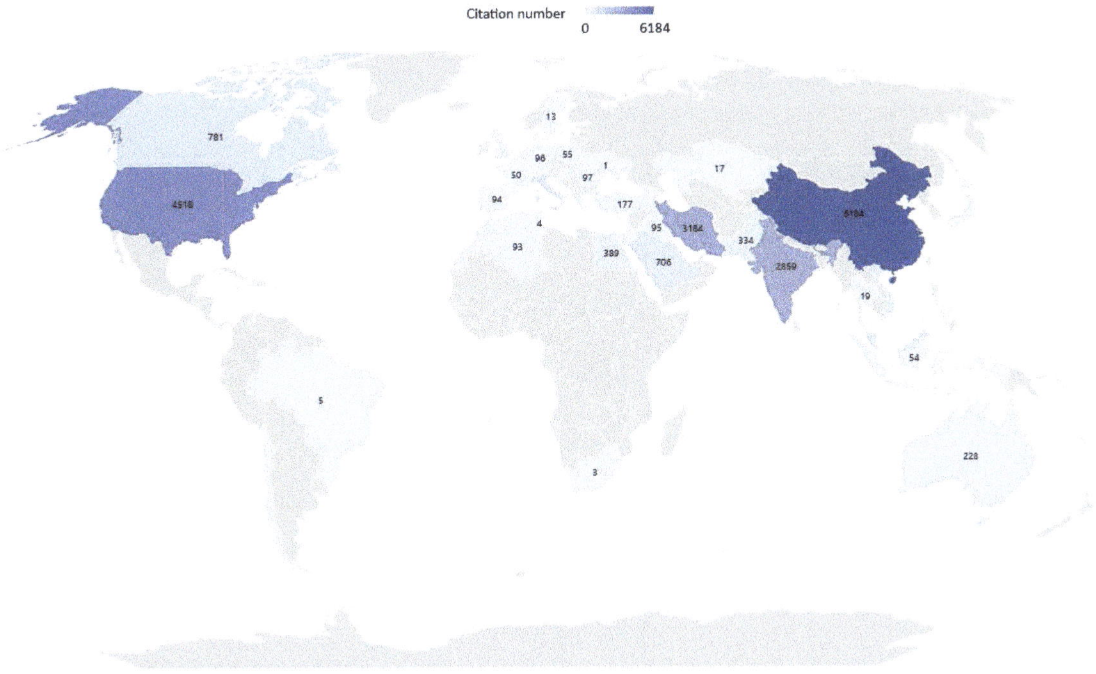

Figure 11. Most cited countries for research in TES using NePCM.

5. Conclusions

This paper presents the evolution of the LHTES study with the application of NePCM through the statistical approach for the first time. An overview of the NePCM literature review in LHTES in TES units was initially provided. Then, a bibliometric analysis was adopted to characterise the application of NePCM in thermal energy storage based on the Scopus and WoS databases.

Despite different techniques to compensate for the inherently low thermal conductivity of PCMS, the effective method is the dispersion of nanoparticles in PCMs. However, it can be further improved by some parameters, such as the size and shape of particles, particle concentration, and particle type. The bibliometric study revealed that the literature on NePCM in LHTES has exponentially grown over the past 16 years. Journal articles were the most frequently used document type, consisting of 79% of the records. Moreover, 12% and 6.1% of publications were distributed as conference papers and review papers, respectively. The analysis showed that the coupling of NePCM with LHTES is a hot topic that has been able to attract more and more interest due to its proven heat transfer enhancement. Exponential growth in the publication output is observed, and a peak record of 181 publications was attained for 2021. The annual growth rate of production was 20.34%.

China was the most important contributor to the research in NePCM in thermal energy storage with the most publications (168), followed by India (130) and Iran (113), respectively. The United States recorded 67 publications in this research field. In terms of multiple-country publications, Saudi Arabia and Iran published 18 and 15 publications, respectively, in this research field. Moreover, as the first country with the most significant interactions, Saudi Arabia collaborated more with institutions in Vietnam, Egypt, India, the United States, and Malaysia.

Based on the bibliometric analysis, 248 sources collected from Scopus and WoS are divided into three clusters. Clusters 1, with 8 sources, included 305 publications in NePCM, where *"Journal of Energy Storage"* held the highest contribution with 78 articles, followed by *"International Journal of Heat and Mass Transfer"* which accounted for 55, and then *"Applied Thermal Engineering"* with a record of 43 publications. Therefore, in recent years, NePCM has become a hot topic in the aforementioned journals based on the significant growth statistics.

In terms of the occurrence of words in the literature, "Phase change material", "Nanoparticles", and "Thermal energy storage" were the most frequent keywords. However, "Thermal", "PCM", and "Heat" were frequently used in the previous studies with occurrences of 2759, 2184, and 1964, respectively.

"Journal of Energy Storage", *"International Journal of Heat and Mass Transfer"*, and *"Applied Thermal Engineering"* were the leading players in the field of NePCM. While the *"International Journal of Heat and Mass Transfer"* published more articles over the years until 2020, the number of published articles in the *"Journal of Energy Storage"* steeply increased in the last four years. The rapid advances on the scientific frontiers of NePCM and LHTES resulted in ever-increasing citations. From the perspective of the most cited countries, China held the highly-cited rank (6184). Then, the United States and Iran have attracted the second and third ranks of global citations, with 4158 and 3184, respectively. The top four countries attracted 2.3 times more citations than the other countries.

The leading publications, publishers, countries, and researchers were successfully identified. Mohsen Sheikholeslami recorded outstanding progress with 33 NePCM articles and the highest bibliometric measurements (H-index = 19, G-index = 33, and M-index = 2.714). Most importantly, the lack of research in specific regions and areas was recognised to allow for possible collaboration opportunities between the research communities. In addition, it was observed that researchers' interest has rapidly increased in this field due to the importance of LHTES in using renewable energy. It is expected that China will remain the most productive country, and *"Journal of Energy Storage"* will have the greatest production growth in the near future. Mohsen Sheikholeslami's publications, which attracted the highest number of total citations, can be suggested as valuable references to those who are keen to study in this field. Moreover, it can be concluded that the maturity of this technology is moderate as there are still drawbacks associated with problems, such as corrosion, phase separation, and supercooling. These drawbacks require further attention, and therefore there is still a place to continue performing research in the TES field using NePCM.

Author Contributions: Conceptualization, J.M. and A.L.; methodology, J.M.; software, J.M.; bibliometric analysis, J.M.; investigation, J.M.; resources, J.M.; data curation, J.M.; writing—original draft preparation, J.M. and A.L.; writing—review and editing, J.M., A.L., V.T. and R.T.; supervision, A.L., V.T. and R.T. All authors have read and agreed to the published version of the manuscript.

Funding: This research received no external funding.

Data Availability Statement: Data are contained within the article.

Conflicts of Interest: The authors declare no conflict of interest.

Nomenclature

Acronyms

FEM	Finite element method
FVM	Finite volume method
HTF	Heat transfer fluid
LHTES	Latent heat thermal energy storage
NePCM	Nano-enhanced phase change material
NEPCM	Nano encapsulated phase change material
PCM	Phase change material
TES	Thermal energy storage

References

1. Khodadadi, M.; Sheikholeslami, M. Numerical simulation on the efficiency of PVT system integrated with PCM under the influence of using fins. *Sol. Energy Mater. Sol. Cells* **2021**, *233*, 111402. [CrossRef]
2. Tofani, K.; Tiari, S. Nano-Enhanced Phase Change Materials in Latent Heat Thermal Energy Storage Systems: A Review. *Energies* **2021**, *14*, 3821. [CrossRef]
3. Nazir, H.; Batool, M.; Osorio, F.J.B.; Isaza-Ruiz, M.; Xu, X.; Vignarooban, K.; Phelan, P.; Inamuddin; Kannan, A.M. Recent developments in phase change materials for energy storage applications: A review. *Int. J. Heat Mass Transf.* **2019**, *129*, 491–523. [CrossRef]
4. Barnes, F.; Levine, J. *Large Energy Storage Systems*; Taylor & Francis Group: New York, NY, USA, 2011.
5. Huang, Z.; Xie, N.; Luo, Z.; Gao, X.; Fang, X.; Fang, Y.; Zhang, Z. Characterization of medium-temperature phase change materials for solar thermal energy storage using temperature history method. *Sol. Energy Mater. Sol. Cells* **2018**, *179*, 152–160. [CrossRef]
6. Chaturvedi, R.; Islam, A.; Sharma, K. A review on the applications of PCM in thermal storage of solar energy. *Mater. Today Proc.* **2021**, *43*, 293–297. [CrossRef]
7. Qiu, L.; Ouyang, Y.; Feng, Y.; Zhang, X. Review on micro/nano phase change materials for solar thermal applications. *Renew. Energy* **2019**, *140*, 513–538. [CrossRef]
8. Saffari, M.; de Gracia, A.; Fernández, C.; Cabeza, L.F. Simulation-based optimization of PCM melting temperature to improve the energy performance in buildings. *Appl. Energy* **2017**, *202*, 420–434. [CrossRef]
9. Jia, J.; de Gracia, A.; Fernández, C.; Cabeza, L.F. Energy saving performance optimization and regional adaptability of prefabricated buildings with PCM in different climates. *Case Stud. Therm. Eng.* **2021**, *26*, 101164. [CrossRef]
10. Park, J.H.; Berardi, U.; Chang, S.J.; Wi, S.; Kang, Y.; Kim, S. Energy retrofit of PCM-applied apartment buildings considering building orientation and height. *Energy* **2021**, *222*, 119877. [CrossRef]
11. Mozafari, M.; Lee, A.; Mohammadpour, J. Thermal management of single and multiple PCMs based heat sinks for electronics cooling. *Therm. Sci. Eng. Prog.* **2021**, *23*, 100919. [CrossRef]
12. Sarrafha, H.; Kasaeian, A.; Jahangir, M.H.; Taylor, R.A. Transient thermal response of multi-walled carbon nanotube phase change materials in building walls. *Energy* **2021**, *224*, 120120. [CrossRef]
13. Senthil, R. Effect of position of heat transfer fluid tube on the melting of phase change material in cylindrical thermal energy storage. *Energy Sources Part A Recovery Util. Environ. Eff.* **2019**, *44*, 2374–2384. [CrossRef]
14. Gong, Z.-X.; Mujumdar, A.S. A New Solar Receiver Thermal Store for Space-Based Activities Using Multiple Composite Phase-Change Materials. *J. Sol. Energy Eng.* **1995**, *117*, 215–220. [CrossRef]
15. Al Siyabi, I.; Khanna, S.; Mallick, T.; Sundaram, S. Multiple Phase Change Material (PCM) Configuration for PCM-Based Heat Sinks—An Experimental Study. *Energies* **2018**, *11*, 1629. [CrossRef]
16. Tiari, S.; Qiu, S. Three-dimensional simulation of high temperature latent heat thermal energy storage system assisted by finned heat pipes. *Energy Convers. Manag.* **2015**, *105*, 260–271. [CrossRef]
17. Tiari, S.; Qiu, S.; Mahdavi, M. Numerical study of finned heat pipe-assisted thermal energy storage system with high temperature phase change material. *Energy Convers. Manag.* **2015**, *89*, 833–842. [CrossRef]
18. Mohammadpour, J.; Salehi, F.; Lee, A. Performance of nano encapsulated phase change material slurry heat transfer in a microchannel heat sink with dual-circular synthetic jets. *Int. J. Heat Mass Transf.* **2022**, *184*, 122265. [CrossRef]
19. Mohammadpour, J.; Lee, A. Investigation of nanoparticle effects on jet impingement heat transfer: A review. *J. Mol. Liq.* **2020**, *316*, 113819. [CrossRef]
20. Khodadadi, J.M.; Hosseinizadeh, S.F. Nanoparticle-enhanced phase change materials (NEPCM) with great potential for improved thermal energy storage. *Int. Commun. Heat Mass Transf.* **2007**, *34*, 534–543. [CrossRef]
21. Leong, K.Y.; Rahman, M.R.A.; Gurunathan, B.A. Nano-enhanced phase change materials: A review of thermo-physical properties, applications and challenges. *J. Energy Storage* **2019**, *21*, 18–31. [CrossRef]
22. Khodadadi, J.M.; Fan, L.; Babaei, H. Thermal conductivity enhancement of nanostructure-based colloidal suspensions utilized as phase change materials for thermal energy storage: A review. *Renew. Sustain. Energy Rev.* **2013**, *24*, 418–444. [CrossRef]
23. Siegel, R. Solidification of low conductivity material containing dispersed high conductivity particles. *Int. J. Heat Mass Transf.* **1977**, *20*, 1087–1089. [CrossRef]
24. Seeniraj, R.; Velraj, R.; Narasimhan, N.L. Heat transfer enhancement study of a LHTS unit containing dispersed high conductivity particles. *J. Sol. Energy Eng.* **2002**, *124*, 243–249. [CrossRef]
25. Elgafy, A.; Lafdi, K. Effect of carbon nanofiber additives on thermal behavior of phase change materials. *Carbon* **2005**, *43*, 3067–3074. [CrossRef]
26. Yazdanifard, F.; Ameri, M.; Taylor, R. Parametric investigation of a nanofluid-NEPCM based spectral splitting photovoltaic/thermal system. *Energy Convers. Manag.* **2021**, *240*, 114232. [CrossRef]
27. Hosseini, S.M.J.; Ranjbar, A.A.; Sedighi, K.; Rahimi, M. Melting of nanoprticle-enhanced phase change material inside shell and tube heat exchanger. *J. Eng.* **2013**, *2013*, 784681. [CrossRef]
28. Aqib, M.; Hussain, A.; Ali, H.M.; Naseer, A.; Jamil, F. Experimental case studies of the effect of Al2O3 and MWCNTs nanoparticles on heating and cooling of PCM. *Case Stud. Therm. Eng.* **2020**, *22*, 100753. [CrossRef]
29. TEMEL, Ü.N.; ÇİFTÇİ, B.Y. Determination of thermal properties of A82 organic phase change material embedded with different type nanoparticles. *Isı Bilimi Ve Tek. Derg.* **2018**, *38*, 75–85.

30. Javadi, H.; Urchueguia, J.F.; Ajarostaghi, S.S.M.; Badenes, B. Numerical study on the thermal performance of a single U-tube borehole heat exchanger using nano-enhanced phase change materials. *Energies* **2020**, *13*, 5156. [CrossRef]
31. Calderón, A.; Barreneche, C.; Hernández-Valle, K.; Galindo, E.; Segarra, M.; Fernández, A.I. Where is Thermal Energy Storage (TES) research going—A bibliometric analysis. *Sol. Energy* **2020**, *200*, 37–50. [CrossRef]
32. Saikia, K.; Vallès, M.; Fabregat, A.; Saez, R.; Boer, D. A bibliometric analysis of trends in solar cooling technology. *Sol. Energy* **2020**, *199*, 100–114. [CrossRef]
33. Cabeza, L.F.; Frazzica, A.; Chàfer, M.; Vérez, D.; Palomba, V. Research trends and perspectives of thermal management of electric batteries: Bibliometric analysis. *J. Energy Storage* **2020**, *32*, 101976. [CrossRef]
34. Yataganbaba, A.; Ozkahraman, B.; Kurtbas, I. Worldwide trends on encapsulation of phase change materials: A bibliometric analysis (1990–2015). *Appl. Energy* **2017**, *185*, 720–731. [CrossRef]
35. Björk, B.-C.; Hedlund, T. Emerging new methods of peer review in scholarly journals. *Learn. Publ.* **2015**, *28*, 85–91. [CrossRef]
36. Glänzel, W.; Moed, H.F. Journal impact measures in bibliometric research. *Scientometrics* **2002**, *53*, 171–193. [CrossRef]
37. de Paulo, A.F.; Porto, G.S. Solar energy technologies and open innovation: A study based on bibliometric and social network analysis. *Energy Policy* **2017**, *108*, 228–238. [CrossRef]
38. Cabeza, L.F.; Chàfer, M.; Mata, É. Comparative analysis of web of science and scopus on the energy efficiency and climate impact of buildings. *Energies* **2020**, *13*, 409. [CrossRef]
39. Van Eck, N.J.; Waltman, L. Software survey: VOSviewer, a computer program for bibliometric mapping. *Scientometrics* **2010**, *84*, 523–538. [CrossRef]
40. Aria, M.; Cuccurullo, C. Bibliometrix: An R-tool for comprehensive science mapping analysis. *J. Informetr.* **2017**, *11*, 959–975. [CrossRef]
41. Black, P.E. *Bradford's Law, in Dictionary of Algorithms and Data Structures*; US National Institute of Standards and Technology: Gaithersburg, MD, USA, 2004; p. 24.
42. Hirsch, J.E. An index to quantify an individual's scientific research output. *Proc. Natl. Acad. Sci. USA* **2005**, *102*, 16569–16572. [CrossRef]
43. Egghe, L. Theory and practise of the g-index. *Scientometrics* **2006**, *69*, 131–152. [CrossRef]
44. Kim, S.; Drzal, L.T. High latent heat storage and high thermal conductive phase change materials using exfoliated graphite nanoplatelets. *Sol. Energy Mater. Sol. Cells* **2009**, *93*, 136–142. [CrossRef]
45. Yavari, F.; Raeisi Fard, H.; Pashayi, K.; Rafiee, M.A.; Zamiri, A.; Yu, Z.; Ozisik, R.; Borca-Tasciuc, T.; Koratkar, N. Enhanced thermal conductivity in a nanostructured phase change composite due to low concentration graphene additives. *Am. Chem. Soc.* **2011**, *115*, 8753–87583.
46. Ibrahim, N.I.; Al-Sulaiman F., A.; Rahman, S.; Yilbasb, B.S.; Sahin, A.Z. Heat transfer enhancement of phase change materials for thermal energy storage applications: A critical review. *Renew. Sustain. Energy Rev.* **2017**, *74*, 26–50. [CrossRef]
47. Cui, Y.; Liu, C.; Yu, X. The experimental exploration of carbon nanofiber and carbon nanotube additives on thermal behavior of phase change materials. *Sol. Energy Mater. Sol. Cells* **2011**, *95*, 1208–1212. [CrossRef]
48. Ho, C.J.; Gao, J.Y. Preparation and thermophysical properties of nanoparticle-in-paraffin emulsion as phase change material. *Int. Commun. Heat Mass Transf.* **2009**, *36*, 467–470. [CrossRef]
49. Wang, J.; Xie, H.; Xin, Z.; Li, Y.; Chen, L. Enhancing thermal conductivity of palmitic acid based phase change materials with carbon nanotubes as fillers. *Sol. Energy* **2010**, *84*, 339–344. [CrossRef]
50. Wang, J.; Xie, H.; Xin, Z. Thermal properties of paraffin based composites containing multi-walled carbon nanotubes. *Thermochim. Acta* **2009**, *488*, 39–42. [CrossRef]

energies

MDPI

Review

Modern Thermal Energy Storage Systems Dedicated to Autonomous Buildings

Michał Musiał [1,*], Lech Lichołai [1] and Dušan Katunský [2]

[1] The Faculty of Civil and Environmental Engineering and Architecture, Rzeszow University of Technology, 35959 Rzeszow, Poland; lech.licholai@prz.edu.pl
[2] Faculty of Civil Engineering, Technical University of Košice, 04001 Košice, Slovakia; dusan.katunsky@tuke.sk
* Correspondence: mmusial@prz.edu.pl; Tel.: +48-17-865-1005

Abstract: This paper presents a detailed analysis of the research into modern thermal energy storage systems dedicated to autonomous buildings. The paper systematises the current state of knowledge concerning thermal energy storage systems and their use of either phase change materials or sorption systems; it notes their benefits, drawbacks, application options, and potential directions for future development. The rapid proliferation of studies on installation systems, new composites, and phase change materials requires a systematisation of the subject related to short- and long-term thermal energy storage in building structures. This paper focuses on assessing the validity of the current improved thermal energy storage solutions for buildings with very high energy efficiency standards and buildings that are energy-independent. The paper presents the current results of the energy and economic analyses of the use of heat storage systems in buildings. This paper shows the optimal heat storage systems for autonomous buildings. Moreover, it also shows other potential ways to develop systems and composites capable of storing heat in autonomous buildings.

Keywords: autonomous buildings; composite thermal energy storage; sorption storage; low-emission engineering; phase change material

check for updates

Citation: Musiał, M.; Lichołai, L.; Katunský, D. Modern Thermal Energy Storage Systems Dedicated to Autonomous Buildings. *Energies* **2023**, *16*, 4442. https://doi.org/10.3390/en16114442

Academic Editor: Ann Lee

Received: 28 April 2023
Revised: 26 May 2023
Accepted: 29 May 2023
Published: 31 May 2023

1. Introduction

The development of modern energy-efficient construction engineering has been determined by changes occurring during building design. A logical choice is to increase the use of thermal energy from renewable sources for building heating, to reduce the heat losses, and, ultimately, to reduce the building's demand for heat or cold. In this context, a satisfactory solution in terms of architectural and construction design of a building is to achieve a state where the direct and indirect solar radiation gains and internal gains from the building's operation make up for the heat losses due to penetration and building ventilation. Nevertheless, when striving to achieve low-emission buildings, their functioning across their entire life needs to be considered.

In addition to their obvious energy efficiency, autonomous buildings are powered with energy from renewable sources and, characteristically, are as independent as possible from an outside energy and matter supply throughout their lives. Such buildings are also analysed from the perspective of the construction materials used and the way that they relate to closed-loop systems; these analyses use the life cycle cost (LCC), life cycle assessment (LCA), and life sustainability cost analysis (LSCA) methods [1–6]. In this context, autonomous buildings fulfil the criteria of sustainable growth [7–11] with regard to reductions in the consumption of non-renewable resources and their replacement with renewable resources, reductions in waste production and pollution, and the consumption of renewable resources no faster than they can be replenished.

An intuitive response to the criteria of sustainable growth in construction is to increase the share of renewable energy generation, primarily solar energy, in passive systems which require no additional power supply. Such systems can be successful if highly efficient heat and cold short- and medium-term storage systems are applied. Heat and cold storage systems used within building components are a solution frequently described in scientific papers; these studies enable autonomous construction objectives to be achieved in accordance with sustainable growth principles. Due to the dynamic development of energy-efficient construction, as well as the numerous scientific studies on new materials, composites, and heat and cold storage technologies for building structures, along with their installation systems, the authors of this paper see a need to systematise the current state of knowledge on this subject. To facilitate the use of the information included in this paper, a graphical diagram of the subject is shown in Figure 1.

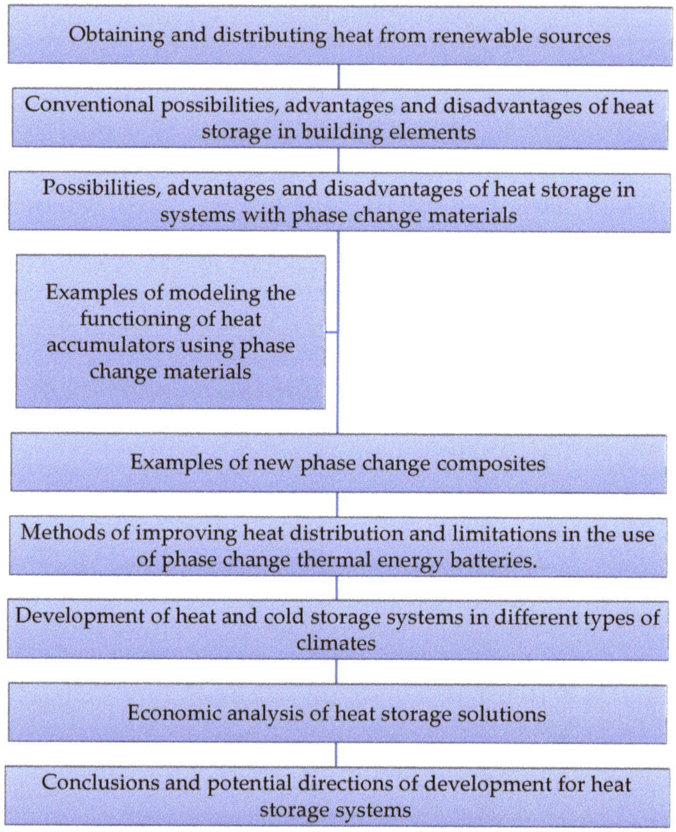

Figure 1. Graphical diagram of the literature available on phase change thermal energy storage systems.

1.1. *Thermal Energy Acquisition and Distribution Methods from Renewable Sources*

Renewable energy sources are naturally renewed during their use. These are supplies that are replenished as quickly as they are consumed. By definition, renewable energy (electricity or heat) is produced from solar, wind, water, geothermal energy, biomass, biofuels, biogas, or hydrogen obtained from renewable sources.

- Solar energy. Most renewable sources are directly or indirectly dependent on the sun. Most of the direct gains are absorbed at latitudes around the equator, but this energy is then dispersed across the planet in the form of winds and ocean currents.
- Wind energy. Air currents can be captured and used to drive wind turbines. Wind energy shows the fastest growth among all renewable sources.

- Hydropower. We can also obtain energy from water, based on either its movement or its temperature differences.
- Geothermal energy. This is obtained by capturing the heat of the earth itself, usually from depths of up to several kilometres below its surface. It is an expensive source of renewable energy.
- Biomass. We know different forms of solid biomass: wood fuel, organic components of municipal waste, or unused parts of agricultural crops. Most types of biomass contain usable energy.
- Biofuels. Liquid biofuels are generally bioalcohols (e.g., bioethanol) or biooils (biodiesel or pure vegetable oils). Their biggest advantage is the lower emissions.
- Biogas. This can easily be produced from biologically active waste substances that arise, for example, from the production of paper or sugar and from sewage, animal waste, and other substances. These various wastes must be allowed to settle together and to undergo natural fermentation to produce methane.

1.2. ZeroEnergy and Autonomous Buildings

Zeroenergy buildings (ZBs) are designed and built to use as little energy as possible. When renewable energy is added to these buildings, they are able to produce enough energy to meet or exceed their operating requirements. The idea of zeroenergy buildings is based on ensuring that the heat gains are equal to the heat losses generated during their operation. This is achieved by reducing heat losses through the thermal enclosure and ventilation. This should be accompanied by maximum heat gains from solar radiation and the profits from building equipment and users.

Autonomous buildings (ABs) are designed to be energy self-sufficient. They require the unique modelling and engineering of a forward-looking building with renewable technology integration, clean energy storage, and demand reduction as focal points. Moreover, the AB is not only a zero carbon and zero energy building, but also has zero grid connection and zero energy bills (ZZZZ)—engineering zero. These are original and innovative approaches, leading to practical solutions to national environmental challenges, such as climate change, air quality, green energy production, and local management. The study found that ABs are more expensive than traditional buildings because they embody the integration of renewable technologies and highly energy-efficient materials; yet, they offer the best engineering services and products, which can raise the bar for Kuwaiti villas and provide multiple solutions: increased housing, increases in green energy production, increases in air quality, and significant reductions in CO_2 emissions, along with the saving of money and the built environment.

An extension of the idea of zeroenergy buildings is the concept of netzeroenergy buildings (NZEB). This is based on the aim of increasing the energy self-sufficiency of a building. The NZEB concept includes the achievement of a zero-heat balance for the building as well as the full renewability of the energy sources required to power the auxiliary equipment for the heating system and everyday appliances. The construction of such buildings relies on an extensivephotovoltaic (PV) system and small wind turbines. Legal acts specifying the energy performance building design (EPBD) take into consideration only the energy demand for the heating, cooling, and, sometimes, lighting of buildings. Such acts omit the energy needed to power home appliances and the aspects related to the temporary thermal comfort of building users. Unfortunately, the temporary sense of satisfaction with the thermal conditions inside the building determines the actual demand for electricity, heat, and cold, such as by opening the window when the ventilation system is temporarily inefficient. Furthermore, an important factor that is currently not considered when establishing the building energy performance is the formingof building heat balances on an average monthly, not hourly, basis, which neglects the effects of the demand for heat and cooling that vary over the day. The above deficiencies in the procedures for assessing the energy efficiency of buildings lead to major discrepancies between the theoretical and the actual performance. Therefore, heat storage in the building structure can reduce the temporary demand for heat and cooling by the building users. Such actions significantly contribute to reducing the actual energy demand of the building by improving the thermal comfort of its users.

2. Conventional Possibilities, Advantages, and Disadvantages of Heat Storage in Building Elements

The main factors determining the efficiency of heat storage systems are their cost, the efficiency of their heat storage and distribution, and, indirectly, their environmental impact. The important factors determining the efficiency of the heat accumulator are enthalpy of the phase change, heat transferability, chemical neutrality, low flammability, and low toxicity.

The most important property is the heat capacity of the substance, which directly affects the amount of accumulated heat in the substance in relation to the volume. The higher the heat capacity, the smaller the volume of the substance needed to accumulate a certain amount of heat. Thermal conductivity is also an important property; it affects the transfer of heat at the interface between the substance that accumulates heat and the heat-carrying substance that distributes the heat from the accumulation system to the point of consumption. Higher thermal conductivity ensures better heat transfer and increased efficiency. Last, but not least, the reversibility of the substance is also important, i.e., the ability to heat/cool repeatedly without degrading the material. This property is extremely important for substances that change states in the heating or cooling process. In the accumulation of sensible heat, thermal energy is stored during the heating of a substance that has suitable properties for these purposes. Most often, water, which has a high heat capacity (4.18 kJ/(kg·K)), is used to accumulate thermal energy in the form of sensible heat.

The advantages of this system mainly include low investment costs, the non-toxicity of the heat storage substance, and the available range of hot water storage tanks. The fact that the thermal capacity of the accumulated sensible heat is limited can be considered a negative. In order to store a large amount of energy, a large storage volume is theoretically required, which reduces the efficiency in terms of heat loss. It is also true that increasing the temperature of the substance in order to store more heat increases the heat loss at the interface of the material with the surrounding environment, which reduces the efficiency of the process.

Heat storage using phase changes is a method frequently discussed in the scientific literature. The advantages of this method are high heat storage density, almost isothermal storage, and the capability to enable wide application in the building structure. Nevertheless, the most commonly used groups of phase change materials are often characterised by a limited heat transfer capacity when they are in the solid state, with a thermal conductivity of 0.2–0.7 W/m·K. This problem has been addressed by adding thermal conductors, such as copper and aluminium alloys or carbon fibres with a thermal conductivity of 200, 370, or 470 W/m·K, respectively. There are various heat storage materials. With each method of heat accumulation, it is advisable to use different materials and to consider their physical and chemical properties. Water, oils (with the ability to accumulate higher temperatures), and solid substances (aggregates and concrete) are mainly used to accumulate sensible heat. The advantages of using solid substances are the elimination of the risk of liquid substance leakage and their affordability, although their heat capacity is significantly lower compared to that of liquids.

The accumulation of latent heat mainly involves the use of organic (paraffins, fatty acids) and inorganic substances (inorganic salts). The melting point of paraffins is in the range of 12–71 °C, while inorganic salts are in the range of 30–120 °C. The disadvantages of organic substances are their flammability and low thermal conductivity, which limits their use. For the accumulation of thermal energy bound in a chemical reaction, many organic and inorganic substances can be used which primarily meet the perfect reversibility of the chemical reaction. Currently, there are many systems that use accumulated solar energy. Its use finds application in heating applications (hot water heating, radiant heating), but also for technological purposes and electricity production. In practice, the most widespread is the heating of drinking water and the accumulation of energy in the form of sensible heat in hot water tanks. However, for the needs of heating and seasonal accumulation, the systems are still limited in terms of efficiency and financial return. Inorganic salts, which are able to accumulate thermal energy, appear to be an advantageous solution, but their use is limited due to their low thermal conductivity. To optimize this method, it is necessary to use measures that increase the thermal conductivity and thus the overall efficiency of the heat accumulation system. Currently, the accumulation of heat in the form of group heat can mainly be solved experimentally, although putting it into practice still requires further research. Similarly, the method of heat accumulation in the form of a chemical reaction is currently the subject of experiments, and so far, it appears to be a highly advantageous process in terms of the efficient storage of a large amount of heat in a small volume with minimal heat loss.

The literature describes methods to better systematise the options for thermal energy storage in the structures of different materials; Figure 2 shows a diagram of examples.

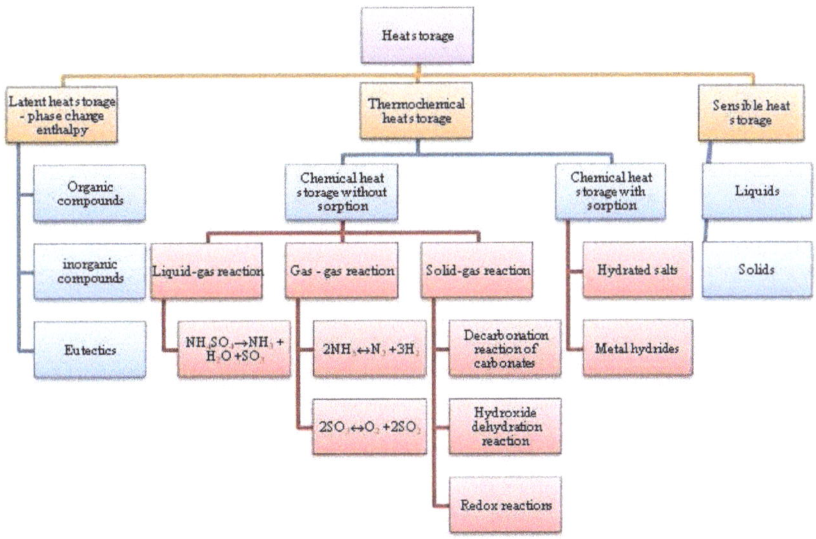

Figure 2. Summary of thermal energy storage systems in terms of material structures and chemical compounds.

Another group of systems that enable thermal energy to be stored comprises sorption systems. Sorption-based heat and cold storage systems utilise the bond energy of the molecules that form hydrated salts or metal hydrates. In construction, they are used as bothheat and cold storage systems, often on a large scale, and as substances that improve heat pump performance. Fan et al. [12] presents a study on a three-phase sorption thermal energy storage system that improves heat pump performance. The results the study presents show that a coefficient of performance (COP) of 7.53 was achieved with a top and bottom heat source temperature difference of 20K. Similar methods of using sorbents to improve heat pump performance are discussed elsewhere [13–18]. Zhang et al. [19] indicate that solid–gas sorption thermal energy storage systems are currently the most effective for obtaining thermal energy from solar plants. The advantages of such systems include high thermal energy storage density, low heat losses during storage, the ability to store thermal energy for prolonged periods of time, and flexible operation modes. As an example of such compounds, $SrCl_2$, with an evaporation/condensation heat of 1630 J/g at −5 °C, was studied. Similarly, Yan et al. [20] noted that salt sorbents were a leading solution that enabled waste heat or renewable energy to be reclaimed. It was also demonstrated that the primary limitations in using this technology, as with PCM, are the ensuring of the necessary kinetics for the thermal energy storage process and the selection of the sorbent's parameters to suit its application.

The drawbacks of using sorption systems for storing thermal energy include a frequently encountered non-linear relation between the temperature increase and the actual amount of heat released. This issue was investigated by Fumey et al. [21], who demonstrated that the thermal energy storage potential of the sorbent was not fully utilised. A sample solution for this issue was presented by Lin et al. [22], where an Al_2O_3 and LiCl composite absorbent was used. The resulting sorbent with a 16.44% salt content showed an absorption of 0.39 g/g and a thermal energy storage density of 345.88 kWh/m^3.

Another solution with which to improve sorption cold storage system performance is to use a sorption thermal battery (STB), which contains a zeolite and $MgCl_2$, as described by Choa et al. [23]. The battery is dedicated to high-cooling power storage systems, providing a 15.1% increased thermal storage density, compared to a pure zeolite sorbent, and allows storage of up to 686.86 kJ/kg. On the other hand, Nguyen et al. [24] discuss the results of the experimental testing of an activated carbon and $MgSO_4$ sorption composite. Thanks to the high specific area of activated carbon, the test composite showed an improved water sorption efficiency and a hydration heat of 920 J/g.

Another use for sorption systems is the new, efficient method of generating hydrogen from biomass gasification and decarbonisation, as described by Zhang et al. [25]. The method improves the hydrogen generation efficiency by 17.4% compared to conventional methods.

Heat and cold storage capabilities by salt or metal sorption have a high potential in theory, but due to the lower actual ability to fix water, their applicability is limited. The papers [20,23–26] discuss methods for reducing this issue under certain specific conditions, such as a particular reaction environment, the addition of a new component, or a changed place where they are applied. Sorption systems provide an important potential for heat and cold storage at high volumes and high power, which has found practical application in industry.

3. Possibilities, Advantages, and Disadvantages of Heat Storage in Systems with Phase Change Materials

Due to the possibility of stabilising daily temperature changes in building components through the use of substances and materials in their structures which are capable of isothermal heat storage, the thermal comfort for the users of buildings modified in this manner has begun to improve. Scientific studies have begun testing phase change materials in different forms, such as microgranulate, or directly as mortar additives and elements of walls or ceilings and as autonomously applied micro- or mini-batteries of thermal energy in transparent and opaque building partitions, as well as in ventilation, air conditioning, and heating systems. Using the same phase change material, but in different construction partition types, can result in diametrically different effects, which is why this paper separately analyses the use of PCMs in transparent and opaque partitions as well as complex thermal energy storage systems.

3.1. *Phase Change Materials in Transparent Partitions*

Numerous examples of using phase change materials combined with transparent partitions were presented by Soares et al. [27]. The main objectives of using PCMs in this partition class was to substantially increase their thermal inertia and to better adapt the absorbed solar radiation energy to the heat gains and demand profile. In this context, it is justified to use phase change materials for modifying windows, the shading elements working in conjunction with windows, or the window woodwork. To this end, PCMs were used as filling for inter-pane spaces [28,29] in window packages or as light-permeable coatings [30]. Furthermore, phase change materials are used as the filling in blinds and roller shades [31–35] or as the filling for thermal energy batteries installed within double facades [36]. Additionally, PCM thermal energy storage performance has also been tested by applying it inside glass bricks [37]. In the above examples, phase change materials were used in the form of granulate or mini-packages by coating pure PCMs [38–41]. An example of an organic phase change mixture in the form of microcapsules and stable capsules is shown in Figure 3.

(**a**) (**b**)

Figure 3. (**a**) Image of Micronal—a coated PCM microgranulate; (**b**) stable capsule with an organic phase change material.

Another PCM form frequently described in scientific papers is shape-stabilised composites, SSPCM [42–45]. They take the shape of capsules or packages covered with a coating that keeps the PCM in place and prevent its controlled unsealing [42,46]. Examples of using transparent materials directly in windows or window blinds are described elsewhere [31,32,47]. An image of the use of phase change materials in the form of stable stalled polycarbonate composite capsules, as well as directly, is shown in Figure 4.

(**a**) (**b**)

Figure 4. (**a**) Organic phase change material during melting of installed polycarbonate; (**b**) melting of phase change eutectic mixture.

3.2. Phase Change Materials in Opaque Partitions

As with opaque partitions, increasing the thermal accumulation capacity of opaque construction partitions is the object of numerous scientific studies. The daily temperature fluctuation reduction effect of PCMs will be particularly visible in buildings with a lightweight frame construction. Even if provided with sufficient thermal insulating power for the partitions forming the external envelope, the thermal conditions in these buildings are sensitive to changing conditions in the external environment.

Studies on solving this issue are described by Soares et al. [48], who investigated the use of granulated PCM-containing drywall. The tests and their analyses were conducted for the Mediterranean climate in Coimbra, Portugal, and concerned the functioning of lightweight steel-framed (LSF) buildings. The test results demonstrated that building cooling and heating performance improved by 62%. Similar analyses and results in using PCM granulate are described by Saffarin et al. [49]. Similar conclusions concerning the reduction in the cold demand in buildings located in a hot climate were made by Zahir et al. [50]. Significant potential was shown for solar radiation energy absorption using thermal energy storage (TES) systems, as well as for solar radiation energy use for the needs of the building's users. However, due to the limited ability to regulate heat accumulation and distribution, these solutions are not widespread. There is an example [51] of a large-scale study on buildings with PCM-containing partitions; the study holistically discusses their impact on thermal load levelling reduction (TLLR), the reduction in CO_2 emissions (CO_2 ES), the average indoor temperature reduction (AITR), the average heat gain reduction (AHGR), and the energy cost savings (ECS). The experiment was conducted in the hot climate of Egypt, and the results demonstrated that head load was reduced by 8.71%, supplied electric power by 56 W, and CO_2 ES by 1.35 kg per day. It was also demonstrated that the PCM performance was the highest when used as part of the roof structure.

On the other hand, Al-Absi et al. [52] discussed an external wall envelope made of foamed concrete (FC) and a PCM. The tests were performed under real climate conditions in Malaysia, achieving a 6.75 °C temperature reduction inside the test chamber and leading to a reduced temperature amplitude attenuation decrement and a 32.1% reduced structure heating rate.

Another example of using PCMs in opaque partitions is the use of foamed cement with a PCM, as described by Li et al. [53]. The effects of the test composite were verified in five climate types in China, and the results demonstrated that PCMs performed the most effectively in a temperate, warm climate.

Phase change materials are also used as additions to brick ceramic elements. Agarwal et al. [54] present the results of testing a PCM composite made of n-Eicosane and OM35, applied to ceramic bricks. The analysis was performed in the Indian climate, where a room was tested made of bricks modified in the above manner, equipped with a gravity ventilation system. The results showed a 32% reduction in cold demand, but the solution was deemed economically non-viable due to the estimated payback period of 181 years. An example of using PCMs in ceramic hollow bricks is shown in Figure 5.

Figure 5. Example of using PCM packages in ceramic hollow bricks [55].

An interesting solution to the problem of temperature regulation limitations in buildings using PCMs is described by Leitzke et al. [56]. Based on the example of a building in the climate of Brazil, a system was proposed involving separate areas designed with different internal air temperatures. This enabled the streamlining of the heat distribution system and the improvement of the efficiency of the heat and cold management within the building. A system designed in this manner showed a 65% reduction in cold demand. Another solution to the issue of heat distribution in rooms with PCMs was described by Wang et al. [57], where an adaptive dynamic building envelope integrated with PCMs (AD-BEIPCM) was proposed. An analysis of multiple previously collected cases demonstrated a significant improvement in heat distribution within the modified building components.

An awareness of the performance variability of PCM solutions when applied in different climate conditions is fundamental. Imafidion et al. [58] discusses the results of numerical simulations of the effects of using retrofitted phase change prefabricated walls for the thermal modernisation of rooms in the Canadian climate and in the tropical climate. It was shown that the payback period in the Canadian climate was 35 years, while in the tropical climate it was 7.6 years.

The use of phase change materials in the form of passive composites and heat and cold storage systems faces major limitations in the distribution of the stored heat and cold, as shown in [48–57]. For this reason, the solution for some of these limitations is ready-made thermal energy batteries, which often form part of active heat and cold distribution systems. Some examples of PCM use in combination with opaque partitions, as described by Lichołai et al. [59], are shown in Figure 6.

(**a**) (**b**)

Figure 6. (**a**) Image of a wall with hollow bricks containing PCM packages [59]; (**b**) aluminium PCM battery [59].

3.3. Phase Change Materials in the Form of Thermal Energy Batteries/Storage Systems

Thermal energy battery storage systems are used both as elements working passively with building elements, such as a thermal envelope [48–55], and as active systems supporting the processed heat and cold distribution and storage installations [60].

Zhu et al. [60] present a high-temperature phase change material (H-PCM) system, which uses the phenomenon of carbon dioxide evaporation and condensation. The solution is characterised by a high thermal energy storage and heat distribution efficiency due to the fact that it maintains the properties of a liquid and convective heat transfer by CO_2 in either state of matter.

Another example of using phase change materials in systems utilising heat exchanges between liquid and gas is the system described by Momeni et al. [61]; this system stores waste and residual heat from electric and hybrid car ventilation systems. The researchers demonstrated that using a PCM with a latent heat of 323 J/g extends the time to heat a space by 6 min after the power is turned off, and greatly accelerates its heating after the power is turned on.

Examples of a passive thermal energy battery dedicated to the transparent partitions used in moderate climates are shown elsewhere [62,63]. It is a ready-made product containing a new phase change material in the form of a eutectic mixture of propyl palmitate and butyl stearate, with a melting/freezing enthalpy of 186 J/g at 18–28 °C [64–66]. The external surface of the battery is made of aluminium covered with an absorption coating. The results of numerical analyses and empirical tests show an increased thermal performance of the transparent partitions with thermal energy batteries on sunny days. At the same time, increased heat losses by transparent partitions were noted on cloudy days when compared to an identical reference window. An example of a PCM thermal energy battery applied directly beside the building windows is shown in Figure 7.

Figure 7. Thermal energy battery containing a new PCM in the form of a eutectic mixture.

A different application for phase change materials in solar plants or electronic systems is described by Li et al. [67]. The ability to isothermally store large amounts of thermal energy was used to reduce the temperature of electronic systems working as thermoelectric energy generators (TEG). The results showed that the peak temperature values for the test electric systems were reduced by 33.5 °C, and the demand for the current to power the test system was lowered by 1/10.

One example of natural, environmentallyfriendly thermal energy batteries involves the salt gradient solar ponds described by Rghif et al. [68]. These are salted natural or artificial water bodies of various sizes which, due to the dissolved salts, have an increased heat storage capacity compared to pure water. These reservoirs can be used as natural stabilisers for the local microclimate or as a source of low-temperature heat. A similar solution for phase change systems was discussed by Gu et al. [69], who demonstrated that combining low-temperature heat pump (HP) systems with PCMs that improve the thermal capacity of a low-temperature source reduces the power needed to supply the heat pump. It is particularly important today, when reducing building energy demand and heat emission levels is called for.

In a critical approach to modern applications, when considering the results so far achieved and the realities of using phase change materials in buildings and components, one needs to compare the limitations of using the PCMs themselves and the possibilities of applying them effectively.

It is obvious that increasing a partition or building element's thermal capacity alone, without first investigating the thermal conditions at the location where a phase change material is to be applied, will not only not bring the assumed results but, as shown by Musiał et al. [62], may in fact degrade the thermal parameters of the partition. This is a consequence of, for example, improperly selecting the location to apply a PCM with a particular phase change enthalpy and temperature range [38,64]. Additionally, a frequently encountered problem is the selection of the form and geometry of the PCM capsules or mini-packages, which limit heat distribution. A consequence is incomplete PCM melting and an additional reduction in stored heat removal during, for example, its solidification. The above issue is discussed elsewhere [38,64].

Another phenomenon affecting phase change performance is the material's behaviour under overheating and overcooling conditions. Excessive heating or cooling of the PCM may lead to a situation where, despite increasing its temperature above the melting point or reducing it below the solidification point, the phase change does not occur or is noticeably delayed. At best it leads to the incomplete utilisation of the heat and cold storage capacity by the PCM, while at worst it causes irreversible changes in its structure, preventing it from functioning as designed. A comprehensive description of this phenomenon is given elsewhere [38].

Due to PCM phase change reversibility within the melting and solidification range and the ensured immutability of its physicochemical parameters, even after many thousands or tens of thousands of phase change cycles, the most commonly used PCMs are organic or derivatives. According to different papers [70,71], the stable capabilities and thermal energy storage performance are necessary conditions to make PCM application economically viable in composites with resin-, cement-, gypsum-, or epoxy-based matrices, etc. In this respect, a significant proportion of the inorganic phase change materials, including light metal salt hydrates, are inferior to organic PCMs. As described by Vogel [71], this is a consequence of the phenomenon of congruence, which is an irreversible separation of crystallisation water from the solid part of the salt as a result of its sedimentation. There are ways, as described by Smolec [72], to counteract this phenomenon by adding emulsifying agents. Nevertheless, despite the available methods of preventing the congruence of inorganic salt hydrates and their much higher phase change enthalpy values than organic PCMs, they are infrequently used in construction and scientific research.

Another limitation in using organic phase change materials is their relatively low self-ignition temperature (about 150 °C), which limits their use in installation systems, structurally important building elements, or electric cars. Example comparisons of the physical properties of organic and inorganic PCMs can be found elsewhere [38,64,71]. Furthermore, an extremely important issue that applies to all groups of phase change materials is the ensuring of the sealing of the capsule, microcapsule, and larger PCM package coatings. In addition to the obvious need to ensure the chemical inertness of the PCM and matrix, it is important to ensure similar thermal expansion coefficient values and the ability of the coating to withstand significant plastic deformation. This is necessary due to the phenomenon of PCM volume change by up to 20% when they change their state of matter during melting or solidification [38,40,41,71]. The aspect of ensuring PCM capsule sealing is also important due to the undesirable PCM reactions that may occur with conventional construction materials should their protective coating lose sealing.

The above considerations summarise the limitations and scientific and technical problems most frequently described in the scientific papers which investigate the use of phase change materials in construction. For better clarity on this, a detailed summary of the places and methods of phase change material application in construction materials and building components is provided in Table 1.

Table 1. Methods of PCM use in construction and methods of combining them with traditional construction materials.

		Place of Application of Phase Change Materials in Buildings (Bibliography Numbers)				
		Walls	Floors	Windows and blinds	Heat storage	Active solar system
Method of combining PCM with conventional building materials	Combining capsules, microcapsules containing PCM with cement or gypsum	53–55, 59	53	27–29, 38–41	67	
	Impregnation of porous materials, e.g., aerated concrete, ceramic bricks	55, 58–59	73		87	
	Direct mixing of PCM with cement or gypsum	38	38		86	
	Production of stable composites containing up to 80% of pure PCM with a polymer matrix (HDPE—shape-stabilized PCM)		64–66	42–66	20	94
	Boards laminated with an inner layer of PCM	57	56		42–45	78, 94
	Heat accumulators in the form of cylinders or cuboids with dimensions of a few to several centimetres covered with a polymer coating, placed in the free spaces of hollow elements	12		31–31, 47	60–61	67, 72

4. Examples of Modelling the Functioning of Heat Accumulators Using Phase Change Materials

Modelling the thermal functioning of phase change materials is the subject of numerous deliberations and scientific studies. Attempts to represent the physical and chemical processes of the state of matter changes in substances, mixtures, or endo- and exergonic chemical reactions as a function of heat flux density over time have utilised several principal methods:

- Finite elements method, using splines and specific boundary conditions.
- Finite differences method, using splines and specific boundary conditions and meeting the thermal diffusivity condition.
- Finite volumes method, using splines and specific boundary conditions.
- Statistical methods linked to the experiment plan and multi-variable function response planes.
- Methods using fuzzy sets, e.g., Mamdani–Assilian models, used when a large pool of empirical results for verifying the model is not available.
- Methods using artificial neural networks.

The methods listed above, except the methods using fuzzy sets [73,74] and the artificial neural network methods [75–77], are modified solutions of the J. Stefan problem [78]. The problem involves determining the position of a variable relative to time and a coordinate system of a discrete boundary grid between a liquid and solid substance.

The majority of scientific studies use finite element methods to model the prediction of the thermal functioning of phase change materials or buildings and composites containing PCMs [78–81]; some use the finite differences method [82–85] and, more rarely, the finite volumes method [86,87].

The finite elements method and finite volumes method are characterised by a high calculation accuracy, although preparing the model is often time-consuming and requires knowledge of the boundary condition characteristics at the model construction stage. On the other hand, the finite differences method is more intuitive and allows the obtaining of explicit analysis results and variables over time for the entire time section studied. Unfortunately, it is severely limited by the need to meet the condition of thermal diffusiveness and the need to properly select the time step value, which requires validating the model and the discrete grid.

Currently, the starting point in modelling non-stationary, complex heat flow in explicit systems is to determine the temperature or energy level values at consecutive points of the assumed discrete grid for moment t+1 (the next time step). Examples of a general equation in the finite differences method for a two-dimensional discrete grid [62,70] are expressed by Formulas (1) and (2).

$$T_{i,j}^{t+1} = \frac{\Delta t}{C_{wi,j} \cdot \rho_{i,j}} \cdot \left(\begin{array}{c} \frac{T_{i,j-1}^{t} - T_{i,j}^{t}}{R_{i,j-1 \leftrightarrow i,j}} + \frac{T_{i,j+1}^{t} - T_{i,j}^{t}}{R_{i,j \leftrightarrow i,j+1}} + \\ + \frac{T_{i-1,j}^{t} - T_{i,j}^{t}}{R_{i-1,j \leftrightarrow i,j}} + \frac{T_{i+1,j}^{t} - T_{i,j+1}^{t}}{R_{i+1,j \leftrightarrow i,j}} \end{array} \right) + T_{i,j}^{t} \quad (1)$$

$$\Delta t \le \min \left[\frac{\Delta x_{i,j}^2}{2 \cdot \frac{\lambda_{i,j}}{\rho_{i,j} \cdot C_{w,i,j}}} \right] \quad (2)$$

where $R_{(i-1,j)}$—heat resistance between points i − 1 and j; $\rho_{i,j}$—density of points i,j; $C_{wi,j}$—specific heat of the material at points i,j; $T_{ti,j-t}$—temperature of points i,j at time t; Δt—time step; $\Delta x_{i,j}^2$—square of the thickness of the elements i,j; $\lambda_{i,j}$—thermal conductivity coefficient i,j.

On the other hand, representing the proper thermal functioning of a PCM during its phase change is performed by modifying Formula (1) and changing the scalar specific heat with a spline. This function describes the rate of PCM heating over time and over a discrete grid; the function is used separately when the PCM is solid orliquid or during the phase change. In theory [63], the relationship in question can be described using Formula (3).

$$C_{wi,j.PCM} = \begin{cases} m_s \cdot C_{W.S} \cdot (T_T - T_0) & \text{if } T_{PCM} > T_T \\ m_T \cdot \Delta H_T & \text{if } T_{PCM} = T_T \\ m_l \cdot C_{W.l} \cdot (T_l - T_T) & \text{if } T_{PCM} < T_T \end{cases} \quad (3)$$

where T_T—phase change temperature of the PCM; C_w—specific heat of the PCM; ΔH_T—PCM phase change enthalpy.

In this case, changing the PCM enthalpy value during its phase change is determined as the difference between the enthalpy of a specific point in the discrete grid at the next and current time step, as in Formula (4).

$$\Delta H_T = H_{i,j}^{t+1} - H_{i,j}^{t} \quad (4)$$

The enthalpy values of any points in the discrete grid, for consecutive time steps, is determined using Formula (5) [62], by using the temperatures at these points to determine whether or not the phase change temperature has been reached.

$$\Delta H_T^{t+1} = \left\{ \begin{array}{ll} \int_{T=0}^{T^{t+1}} h(T)_1 dT - \int_{T=0}^{T^{t}} h(T)_1 dT & \left| \begin{array}{c} h(T)_1 \rightarrow \text{for melting} \\ \text{when } T^{t+1} > T^{t} \end{array} \right. \\ \int_{T=0}^{T^{t+1}} h(T)_2 dT - \int_{T=0}^{T^{t}} h(T)_2 dT & \left| \begin{array}{c} h(T)_2 \rightarrow \text{for clotting} \\ \text{when } T^{t+1} \le T^{t} \end{array} \right. \end{array} \right\} \quad (5)$$

The above equations are dedicated to pure phase change materials or their isotropic composites. On the other hand, the approach described by Musiał et al. [87] allows the determination of the heat flux flowing through an anisotropic composite using Formula (6).

$$Q_{A.E} = \begin{cases} \text{if } T_p < T_l \rightarrow A_O \int_{t=1}^{t=O} q_r dt \\ \text{if } T_p = T_l \rightarrow m_L \int_{t=O}^{t=n} \Delta H_{E.r} dt \\ \text{if } T_p > T_l \rightarrow A_O \int_{t=1}^{t=O} q_r dt \end{cases} \quad (6)$$

where T_l—phase change temperature; T_p—sample temperature; A_O—external surface of the PCM sample; q_r—heat flux density flowing through the sample from the PCM; m_L—mass of the PCM being melted; $\Delta H_{E.r}$—PCM melting/solidification heat.

When modelling more complex problems concerning PCM systems, where it is required to consider the heat transfer between a thermal energy battery and the environment inside a building or another building component by convection and radiation, then the finite elements method equations are used. Such calculations are often performed using MATLAB [63], Adina [59] Energy Plus [86], or Ansys fluent [88] software. A different approach is presented for using the fuzzy inference method and artificial neural networks [73–77]. In this case, it is enough to have a pool of empirical results from experiments to obtain a result at an acceptable error level.

Selecting the method with which to model the functioning of thermal energy battering with phase change materials is dependent on the available empirical data and knowledge of the experimental boundary conditions. An additional factor determining whether numerical method simplification can be used is the need to meet additional solution stability conditions.

5. Examples of New Phase Change Composites

Attempts to solve the technical and scientific issues with the use, application, and functioning of phase change materials, as described in the previous section, have resulted in the design of new phase change composites. A large portion of the scientific articles published on the subject concern composite phase change materials with improved heat distribution properties.

In this context, a logical and scientifically valid method of improving the thermal conductivity of solid phase change materials is to ensure they work together with metals that conduct heat well, as described by Xu et al. [89]. The metals applied directly with PCMs may be used in various forms, as the steel fibres presented by Cui et al. [90] or as the distributed nanoparticles discussed by Liu et al. [91]. One needs to be aware that improving PCM thermal conductivity using highly processed materials with a large carbon footprint, i.e., metal alloys, may not be the most effective solution. Another method of applying metals to intensify heat distribution in PCMs is to use them in a foamed form [92,93] and to investigate the relationship between a foamed copper content of 0.43 to 2.15% and the PCM melting time. The results showed that the most beneficial copper content in the composite was 0.86%, which allowed the conductive heat exchange to be intensified, while not restricting the heat exchange by the free convection of the liquid PCM.

Improving heat permeation by using spatial structure conductors is described by Zhao et al. [94]. The article presents a method of producing metal foams as a framework for PCM composites obtained using Kelvin cells.

An unusual and interesting method of intensifying PCM heat exchange is the use of hypergravity. Zhang et al. [95] demonstrated that by using a high centrifugal force to change gravity from 1 g to 9 g the total melting time of the test PCM could be reduced by 60.24%. An additional factor affecting PCM heat transfer performance was its method of application and its arrangement. Liu et al. [96] investigated changes in the melting time of a mixture of aliphatic RT27 alkanes as affected by their position relative to the vertical and the heat-conducting rods inserted in them. An increase in the intensity of PCM heat transfer can be obtained by increasing the speed of its convection. Ho et al. [97] proved that by increasing the flow rate of liquid PCM to $18.4 \cdot 10^{-3}$ m/s, its melting time could be reduced 2.83 times, down to 6 min.

Another option to improve heat distribution within a PCM, as described in the scientific articles, is forced convection. Often, as in studies [94,98–102], it is performed by applying the PCM in a twin pipe system for the convective transfer of heat. In this context, the use of nano-liquids as a medium that intensifies heat and liquid flow in a unit of time is viable. Some studies [103] used a nano-liquid (1% Al_2O_3 solution) as the heat transfer medium in a twin pipe system containing a PCM. The effect of the above changes was the achievement of a heat transfer that increased by 32% compared to pure water.

Another category of phase change material modifications intended to create new composites is the inclusion of conductor materials. It is a compromise between increasing the conductivity of the composite and maintaining its thermal capacity. A good example is the new, metallic, wood-derivative phase change material presented by Lin et al. [102]. Coated microcapsules with a PCM, covered with a copper layer, were used to improve the thermal capacity of porous wooden elements. The results confirmed an improvement in the composite's thermal conductivity of 362%, compared to pure wood, while maintaining an average melting/solidification enthalpy of 92 J/g.

Ryms et al. [103] discuss the use of a free PCM with an addition of post-pyrolytic carbon obtained from tyre recycling. The resulting composite was used as a concrete additive. The composite contained the PCM RT21; free concrete and the recycled material enabled a PCM content of up to 32% to be maintained within the structure of the concrete elements. Other experimental results on a composite of PCM and expanded glass aggregate are described by Yousefi et al. [104].

These tests resulted in a composite of PCM and glass fibres with a saturation capacity of up to 80% and a thermal conductivity of concrete that was reduced by 47%.

Another example of the improvement of PCM thermal conductivity is the PCM and biocarbon composite, as discussed elsewhere [105]. The biocarbon was obtained through a pyrolysis of invasive aquatic plants. The testing of the newly obtained composite showed a 13.82-fold increase in thermal conductivity compared to a pure PCM. An important technical issue is soaking, with the application of the phase change materials to fibrous, porous structures. Zhang et al. [106] show an interesting way to solve this problem. The soaking of a pipe matrix made of carbon fibres with a liquid PCM was performed using centrifugal force while rotating the matrix. The test results of the organic PCM and carbon fibre composite produced in this manner showed that a yield point of 19 MPa and a phase change enthalpy of 77.83 J/g were achieved. An example of the versatility of phase change materials is provided by Vennapusa et al. [107], with a PCM and nanographene composite used as a component for clothing materials. The study showed an increased thermal accumulation capacity and improved user thermal comfort. Example images of PCM and conductor material thermal energy mini-batteries and their components are shown in Figure 8.

Another group of phase change composites consists of those where the direct application of a pure PCM is possible thanks to the use of a cement, silica, or polymer matrix. While the matrices improve heat distribution within the composite, they also provide the composite with a framework and a suitable strength and keep the PCM where it was applied even when it melts. Examples of PCM composites with a cement matrix that are used in construction and have a sufficient load-bearing capacity are described [108,109]. The papers present the strength test results for composites of an organic PCM with a slag aggregate and silicon carbide matrix, which confirm their higher compressive and bending strength compared to a PCM composite containing a natural aggregate. Zhao et al. [94] present a wood and plastic phase change composite. The tests showed that the initial strength properties of the material were maintained after the PCM was applied, while its thermal conductivity was increased by 26.7%, which improved the heat exchange efficiency between the environment and the PCM.

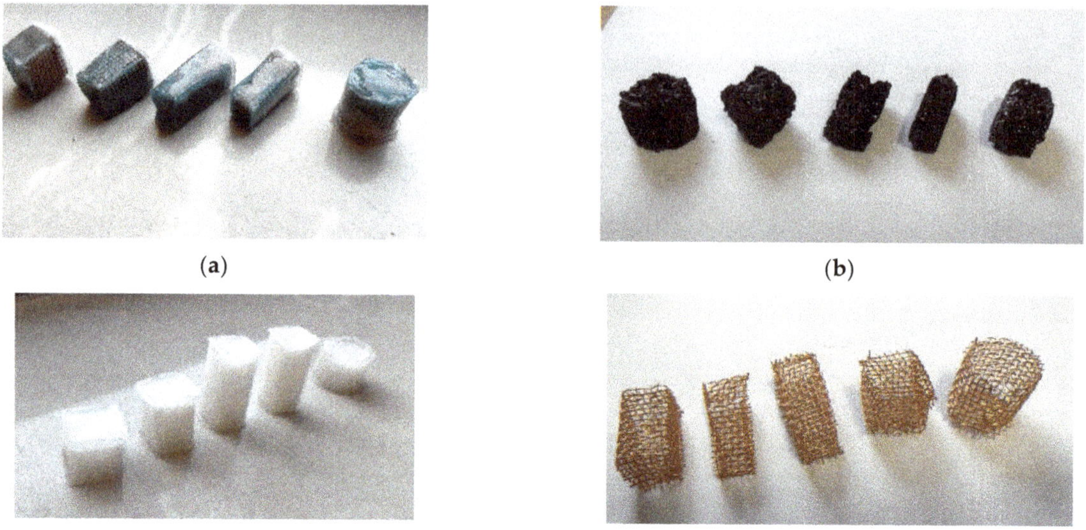

(a) (b) (c) (d)

Figure 8. (**a**) Compact thermal energy batteries with a PCM and copper conductor; (**b**) conductor coke recyclate; (**c**) compact thermal energy batteries with a pure PCM; (**d**) thermal energy battery conductor framework made of copper mesh.

The temperature stabilisation effect of PCMs is also used when designing food packaging materials. One study [110] examined several composites of porous materials and PCMs, investigating how they affected the heat and cold storage capabilities of the test packaging.

The tests have proven the formation of stable composites of organic PCM (caprylic acid) with porous materials, such as molecular sieves, clay soils, or perlite.

6. Methods of Improving Heat Distribution and the Limitations in the Use of Phase Change Thermal Energy Batteries

Improvements in heat distribution within a phase change material and between the material and the element of a building or installation are principally achieved in one of two ways. One is to increase the thermal conductivity of the PCM itself by modifying it with high heat conductivity materials, such as copper or aluminium. The other method involves intensifying the heat transfer through free convection or forced convection, and it depends on the character of the PCM used and on the location where the material is applied. Between the two main solutions to limited PCM heat distribution, active systems (which require additional power supply) are used to improve heat transfer through convection and apply mainly to modifying PCMs in transparent partitions, building installation systems, and larger thermal energy storage systems [111–115]. On the other hand, the intensification of PCM heat conduction is usually used in passive solutions (which do not require an additional power supply) and dedicated modifications for opaque building partitions, as well as for small, compact thermal energy batteries. [52,59,116–120]. An example of an active system for a mobile indoor blind with a PCM is shown in Figure 9.

Phase change materials in the form of coated microcapsules are frequently used in many scientific studies [38,41], and they are an element of several finished products in the form of additives for cement matrix-based mortars or drywall systems. PCM microcapsules are characterised by a satisfactory thermal stability, but unfortunately, they are also characterised by an insufficient effectiveness in transferring the heat they absorb. Zhao et al. [121] analysed microcapsule compositions in the context of a polymer matrix content to phase change core content ratio. Their results showed that a 2:1 weight ratio of the matrix (styrene-divinylbenzene) to the PCM core (n-octadecane) enabled themaintenance of the necessary thermal stability while increasing the microcapsule mass loss temperature from 115.5 °C to 160.5 °C when compared to conventional matrices.

(a) (b) (c)

Figure 9. (**a**) Large-scale test chamber; (**b**) mobile indoor blinds with a PCM; (**c**) phase change thermal energy battery during testing in an isothermal large-scale test chamber [62].

Kant et al. [122] present an overview of the fundamental research on the possibilities to improve heat distribution. The article notes the necessity to achieve long-term functioning for thermal energy storage systems in the context of their widespread use in construction.

Chao et al. [123] present a system for the passive absorption of solar radiation and the accumulation of this energy by phase change polyethylene glycol (PEG) packages placed in a concrete composite. The analysis results demonstrated an improvement in the test building's thermal energy accumulation system by 23.4% with a payback period estimated at 4.2 years.

Another group of studies showing improvements in PCM heat conduction are those that used allotropic forms of carbon, such as graphene, fullerenes, or recycled waste materials from metalworking or refining processes [116].

The use of phase change materials to improve the effectiveness of heat distribution in ventilation installation systems or in convective heat exchangers is the subject of numerous scientific papers [73–75,116]. One of the more frequently referenced technical issues in scientific studies on PCM applications in these systems concerns the type, geometry, and shape of the conductor exchanger (usually made of copper of an aluminium alloy). Wang et al. [124] demonstrated that among the flow-based heat exchangers, the most effective in terms of distributing the heat stored in the PCM are conductor exchangers with a concentric design and radially arranged heat conducting elements.

Wang et al. [57] presented options for adapting thermally dynamic building envelopes containing phase change materials in sealed packages. Empirical and simulation calculation results were analysed, showing frequent limitations in the PCM applications in construction, such as their thermal capacity not being fully utilised, limited heat distribution ability within the PCM, or limited ability to control the PCM.

An interesting solution is to use the cascade heat storage system (CLHS), as described by Shen et al. [125], in ventilation systems with phase change materials. This partially solves the problem of the thermal overload of conventional thermal energy storage systems containing a single PCM type. The test system allowed the risk of overloading the thermal energy batteries to be reduced by 73.21%.

7. Development of Heat and Cold Storage Systems in Different Types of Climates

The intended purposes of using phase change materials, such as by adapting the heat or cold demand curves to fit their reservoirs in thermal energy storage systems, are factors that decide the effectiveness of the thermal energy storage systems. Depending on the outdoor and indoor climate conditions or the need to acquire and store energy from renewable sources, as in moderate climate conditions, or the need to prevent room overheating in a subtropical climate, PCMs can be placed either on the indoor or the outdoor side of the partition. The above relations are discussed by Lu et al. [126].

Due to reasons related to technology, the PCM ability to isothermally store heat is usually used for short-term thermal energy storage systems, which limits the viability of their use. The following papers provide examples of solutions to the issues of overcooled or overheated rooms in different climate types. A method for preventing window overheating is to place vertical grids of window packages between the glass panes, as described by Lai et al. [127]. The empirical results obtained in China and the USA showed a reduction in the cooling load of air-conditioned buildings by 37.8% and 24.8%, respectively.

Samandi et al. [128] propose the use of horizontal mobile roller blinds, which simultaneously provide better natural lighting in rooms and reduce costs. In addition to the ability to store heat, the tested roller blinds can reflect part of the solar radiation to the interior thanks to the PCM they contain.

Improved building window shading performance in warm climates has been achieved with the use of controlled shading. The system described in [129,130] is based on changing the blind position depending on the time of day, season, latitude, and light demand. Chi et al. [131] show an example from India, demonstrating that the use of such a system leads to a reduction in room cooling costs by 7.57%. A similar fixed outdoor blind solution was tested in cold Chinese climate conditions [132]. The test results showed that the most effective angle for outdoor window blinds was 90°. The result of the tests was an energy saving of 21.77% compared to the reference rooms.

The proper use of transparent partition shading systems with PCMs [31,32,133] can reduce the costs of room heating or cooling. There are also modifications and modernisation options that improve the thermal performance of transparent partitions with PCMs that use electrochromic installations [134], solutions supported with air flow systems [135], and greenery systems that stabilise the temperature [136].Vinga et al. [137] considered a modification of the vertical greenery surfaces with a mobile window shutter system. The results showed that windows combined with greenery systems enabled an 11.5% reduction in the need for room cooling during the summer.

The essence of window and roller blind modernisations in buildings is the viability of the modernisation across the entire building life cycle. Li et al. [108] provide an empirical and numerical analysis of the viability of window modernisation with phase change materials under Brazilian climate conditions. The analysis was performed using the Energy Plus software, and the results demonstrated a significant reduction in the air conditioning costs of the adjoining rooms when using external wooden shading systems.

Due to their purpose of providing transparency and light permeability, transparent partitions are characterised by low thermal accumulation capacity. For this reason, their temperature is sensitive to momentary changes in atmospheric conditions, external climate, and room microclimate. Due to the solar radiation gains, they also exhibit high instantaneous efficiency, suggesting broad opportunities for successful modifications with phase change materials. Furthermore, for the above reasons, PCM modifications of transparent partitions may bring more palpable effects than the modification of opaque partitions.

Consequently, another study [138] analysed nine different methods of applying PCMs in transparent partitions. This was conducted in the inter-pane spaces using aerogels and transparent thermal insulation. When discussing the advantages and drawbacks of each solution, the authors noted the benefits of increasing the thermal inertia of windows and showed the limited efficiency of these solutions during the winter season. On the other hand, Qui et al. [139] present the technology and application limitations as well as the detrimental visual aspects of applying different forms of PCMs to window structures.

When modifying transparent partitions with PCMs, their sensitivity to changing outdoor climatic conditions restricts the viable usefulness of PCMs with specific phase change temperature ranges. In this context, PCM usefulness is decided by the phase change enthalpy value and the temperature range in which the change occurs. According to some papers [63,64,134,139–143], it is viable to use PCMs with melting/solidification temperatures within the 16–28 °C range. Ultimately, they depend on the properties of the building element to be modified, the climate type, and the location of the PCM application.

Ensuring the right heat transfer rate between the PCM and the adjacent building elements is one of the main issues in using them. In one study [144], the authors reduced this problem by using honeycombed aluminium panels. This allowed them to increase the thermal energy adsorption, removal, and storage potential in a PCM structure.

Another option for reducing the costs of thermal functioning of building rooms is to use PCMs in the window blind structure. According to one paper [60], the year-long use of blinds with a phase change material in the South Korean climate extended the thermal comfort period by 34% and reduced the cooling load by 44%.

An intensified heat flow in a ventilated window design with a phase change material was achieved in another study [143]. The tests were performed in the Copenhagen climate, and they concerned an analysis of the permanent thermal energy capacity of 3.19 MJ of a PCM applied as 5 mm, 10 mm, or 20 mm thick boards. The results showed that the most thermally effective were the thinnest PCM boards.

Forced air flow as a mechanism to improve PCM heat distribution is important, especially in warm climates. Under such conditions, transparent partitions with external walls can cause building interior overheating. Li et al. [145], on the other hand, analysed the calculated simulations of liquid flow in the inter-pane space of a window package with a surface area of 9 m^2. Feeding this system with the heat or cold obtained from renewable sources allowed the stabilisation of the window temperatures throughout the day. The deliberations concerned the renewable energy acquisition from multiple different sources, such as low-temperature geothermal springs and active solar systems. The results confirmed that a zero-heat growth in the test room was achieved. A similar solution is described by Lyu et al. [146], and the results show a reduction of 43% in the amount of heat transferred through the test window during the summer period.

Modifying conventional transparent partitions with phase change materials [63,66,147] may result in the degrading of their thermal insulation properties during the night and on cloudy days. An example described by Musiał [66] is a phase change battery applied in the inter-pane space of double windows, enabling direct and indirect solar radiation gains to be absorbed [66]; see Figure 10.

For this reason, another study [148] investigated a new composite of a PCM and silica aerogel. The results showed that the aerogel and PCM packages were effective only in a cold climate.

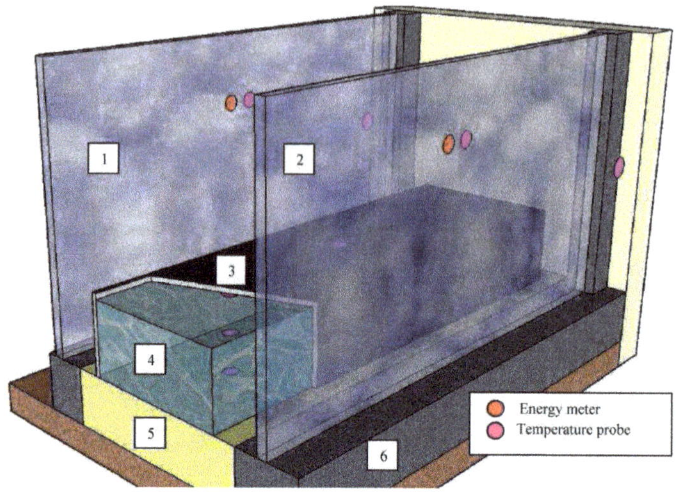

Figure 10. Diagram of inter-pane PCM thermal energy battery functioning.

8. Economic Assessment of the Use of Heat Storage in the Building

One of the main criteria justifying the use of heat storage in buildings is the reduction in financial costs related to the heating and cooling of the building. In addition to the desire to reduce the costs of thermal functioning, thermal self-sufficiency is also important in autonomous buildings. The thermal self-sufficiency is also important for autonomous buildings. Faraj et al. [149] studied the economic benefits of using PCM in the walls, floor, and roof of a test room. The results showed an economic benefit of USD 225 per year; so, it was possible to determine the payback period of 6.82 years for a PCM room upgrade. This result was obtained by using PCM simultaneously in all three components of the building and applying it to active and passive systems.

Hou et al. [150] presented the test results of test rooms with PCM-applied walls. The results of the financial analysis for the most effective PCM parameters and its location show savings of 21.07%. Due to the cost of the PCM and its passive functioning, the investment may only pay off after 25 years.

The thickness of the layer of PCM applied in the building structure has a major impact on the efficiency of heat storage and the investment payback period. Elefky et al. [151] presented an economic and energy analysis of the use of different thicknesses of PCM. The results demonstrated that the most effective combination of PCM in a building component was 70% PCM and 30% secondary component, which meant savings of 15%. Similar conclusions were presented by Tafuni et al. [152], who showed the possible storage of cold and heat in the stone deposits of greenhouses and glass extensions; the result was a potential savings of 153 USD/MWh. On the other hand, Li et al. [153] presented the results of the analysis of waste heat savings through the installation of a buffer tank in the building. This allowed a 31% reduction in the thermal load and annual savings of 5%, and the investment was paid off after 15 years. An alternative to the examples described above may be the use of a hybrid, multi-source heat recovery and storage system [154–157]. According to Fu et al. [156], hybrid systems can store heat for up to 3–4 months, resulting in savings of 16.9%. On the other hand, according to Taheri et al. [155], it is not the use of phase change materials but a multi-source system of heat pumps with a pressure buffer storage that allows energy cost reductions of up to 78.96%.

An important economic aspect is the supply of electric energy to active heat storage systems. Shayan et al. [158] proved that the use of flexible photovoltaic panels on various oval-shaped structures increased the return on the investment from 34.8% to 40.5%.

The above papers show great opportunities to reduce the costs of heating and cooling in buildings when using either phase change materials or hybrid, complex systems. The least efficient method is the passive operation of pure phase change batteries, while the most efficient solution is complex hybrid systems, which are not used frequently in single-family residential buildings due to their high initial cost.

9. Discussion

This review of the current state of knowledge concerning thermal energy storage systems dedicated to autonomous buildings highlights the current advantages, drawbacks, and limitations of their applications. It must be noted that current heat and cold storage solutions and technologies exhibit similar limitations in terms of the use of their potential. The primary and not fully solved issue is the insufficient heat and cold removal system, which limits their potential.

In the context of improving the thermal performance of autonomous buildings, there is no sufficiently effective and easily available thermal energy storage and distribution system that would use PCMs without a phase change or sorption. However, the goal of the thermal autonomy of buildings can be achieved more easily by using a combination of sorption and phase change thermal energy storage systems. To this end, high-performance composite sorption systems modified with high specific surface area compounds, such as activated carbon, need to be used in heat pump systems and heat and cold recuperation and distribution systems. Additionally, sorption-based installation systems enable effective, isothermal, medium- and long-term heat and cold storage over a broad range of temperatures.

The use of phase change materials within their melting/solidification temperature range can enable the stabilisation of the daily temperature fluctuations in transparent partitions and can extend the time to consume the solar radiation energy absorbed. Additionally, composite phase change materials with high heat conductivity values successfully increase the thermal capacity of lightweight frame construction partitions.

It needs to be noted that an additional factor affecting the viability of using heat and cold storage stems is the character and type of the outdoor climate and the designed temperature inside the building rooms. In this context, PCMs applied in passive systems achieve better performance in warm climates, noticeably reducing peak temperature values and reducing the demand for cooling power.

However, in economic terms, the simple sorption systems and PCM result in savings of around 20–25% of the heating and cooling costs in buildings.

To summarise the above conclusions, the most effective solution is currently a hybrid combination of sorption and phase change thermal energy storage systems applied in those locations and installations where they exhibit the greatest energy benefits.

At the same time, it needs to be noted that further development and improvement is needed in storage systems for waste heat and the heat obtained from renewable sources; this would enable them to be widely applied in construction.

According to the authors, the solutions that should be developed are multi-source hybrid systems for supplying buildings with heat and cold. Other promising solutions are systems for storing heat and electricity using buffer pressure tanks, as well as energy and cooling storage systems for hydrogen turbines.

Author Contributions: Conceptualisation, M.M.; methodology, M.M.; formal analysis, M.M., L.L. and D.K.; investigation, M.M.; resources, M.M., L.L. and D.K.; writing—original draft preparation, M.M., L.L. and D.K.; writing—review and editing, M.M., L.L. and D.K.; supervision, M.M., L.L. and D.K.; funding acquisition, M.M. and L.L. All authors have read and agreed to the published version of the manuscript.

Funding: This research received no external funding.

Conflicts of Interest: The authors declare no conflict of interest.

Abbreviation

Symbol	Name
ADBEIPCM	Adaptive Dynamic Building Envelope Integrated with PCM
AHGR	Average Heat Gain Reduction
AITR	Average Indoor Temperature Reduction
AB	Autonomous Buildings
CLHS	Cascade Latent Heat Storage
COP	Coefficient Of Performers
CO_2 ES	Associated CO_2 Emissions Saving
ECS	Energy Cost Saving
EPBD	Energy Performance Building Design
FC	Foamed Concrete
HDPE	Shape-stabilized PCM
HP	Heat pump
H-PCM	High-temperature Phase Change Material
LCA	Live Cycle Assessment
LCC	Live Cycle Cost
LSCA	Live Sustainability Cost Analysis
LSF	Lightweight Steel-Framed
NZEB	Net Zero Energy Buildings
PCM	Phase Change Materials
PCM-FC	Phase Change Material—Foamed Concrete
PGE	Polyethylene Glycol
PV	Photovoltaic panel
SGSP	Salt Gradient Solar Pond
SSPCM	Shape-Stable Phase Change Materials
STB	Sorption Thermal Battery
TEG	Thermoelectric Power Generator
TES	Thermal Energy Storage
TLLR	Thermal Load Levelling Reduction
ZB	Zero Energy Buildings
ZZZZ	Zero Carbon, Zero Energy, Zero Grid Connection, and Zero Energy Bills

References

1. Sun, S.; Liu, Z.; Hou, Y.; Li, X.; Xie, Y.; Zhai, H.; Wei, X. Analysis of harmonic characteristics based on improved double-LCC compensation network structure. *Energy Rep.* **2022**, *8*, 891. [CrossRef]
2. Larsen, V.G.; Tollin, N.; Sattrup, P.A.; Birkved, M.; Holmboe, T. What are the challenges in assessing circular economy for the built environment? A literature review on integrating LCA, LCC and S-LCA in life cycle sustainability assessment, LCSA. *J. Build. Eng.* **2022**, *50*, 104203. [CrossRef]
3. Cardoso, V.E.M.; Simões, L.M.; Ramos, N.M.M.; Almeida, R.M.S.F.; Almeida, M.; Mateus, R. Impact of air barriers application in LCA and LCC of naturally ventilated dwellings in mild climate regions. *Energy Build.* **2023**, *279*, 112667. [CrossRef]
4. Ropo, M.; Mustonen, H.; Knuutila, M.; Luoranen, M.; Kosonen, A. Considering embodied CO_2 emissions and carbon compensation cost in life cycle cost optimization of carbon-neutral building energy systems. *Environ. Impact Assess. Rev.* **2023**, *101*, 107100. [CrossRef]
5. Vasishta, T.; Mehany, M.H.; Killingsworth, J. Comparative life cycle assesment (LCA) and life cycle cost analysis (LCCA) of precast and cast–in–place buildings in United States. *J. Build. Eng.* **2023**, *67*, 105921. [CrossRef]
6. Harter, H.; Willenborg, B.; Lang, W.; Kolbe, T.H. Life Cycle Assessment of building energy systems on neighbourhood level based on semantic 3D city models. *J. Cleaner Prod.* **2023**, *407*, 137164. [CrossRef]
7. Byrne, E.P. The evolving engineer; professional accreditation sustainability criteria and societal imperatives and norms. *Educ. Chem. Eng.* **2023**, *43*, 23. [CrossRef]
8. Wijesooriya, N.; Brambilla, A.; Markauskaite, L. Biophilic design frameworks: A review of structure, development techniques and their compatibility with LEED sustainable design criteria. *Clean. Prod. Lett.* **2023**, *4*, 100033. [CrossRef]
9. Bhyan, P.; Shrivastava, B.; Kumar, N. Allocating weightage to sustainability criteria's for performance assessment of group housing developments: Using fuzzy analytic hierarchy process. *J. Build. Eng.* **2023**, *65*, 105684. [CrossRef]
10. Ranjbar, N.; Balali, A.; Valipour, A.; Yunusa-Kaltungo, A.; Edwards, R.; Pignatta, G.; Moehler, R.; Shen, W. Investigating the environmental impact of reinforced-concrete and structural-steel frames on sustainability criteria in green buildings. *J. Build. Eng.* **2021**, *43*, 103184. [CrossRef]

11. Gilani, G.; Hosseini, S.M.A.; Pons-Valladares, O.; de la Fuente, A. An enhanced multi-criteria decision-making approach oriented to sustainability analysis of building facades: A case study of Barcelona. *J. Build. Eng.* **2022**, *54*, 104630. [CrossRef]
12. Fan, Y.B.; Jiang, L.; Zhang, X.J.; Xu, X.G.; Frazzica, A. Heat pump assisted open three-phase sorption thermal battery for efficient heat storage. *Energy Convers. Manag.* **2022**, *277*, 116630. [CrossRef]
13. Tzinnis, E.; Baldini, L. Combining sorption storage and electric heat pumps to foster integration of solar in buildings. *Appl. Energy* **2021**, *301*, 117455. [CrossRef]
14. Aydin, D.; Casey, S.P.; Chen, X.; Riffat, S. Numerical and experimental analysis of a novel heat pump driven sorption storage heater. *Appl. Energy* **2018**, *211*, 954. [CrossRef]
15. Yu, Y.; Wang, L. Solid sorption heat pipe coupled with direct air cooling technology for thermal control of rack level in internet data centers: Design and numerical simulation. *Int. J. Heat Mass Transf.* **2019**, *145*, 118714. [CrossRef]
16. Yu, Y.; Wang, L.W.; An, G.L. Experimental study on sorption and heat transfer performance of NaBr-NH3 for solid sorption heat pipe. *Int. J. Heat Mass Transf.* **2018**, *117*, 125. [CrossRef]
17. Zhang, X.; Tian, Y.; Wang, L.; Zhang, B. Performance regulation characteristics of bi-halide sorbents in the thermochemical sorption system for cold/heat storage. *Appl. Therm. Eng.* **2023**, *219 Pt A*, 119453. [CrossRef]
18. Li, W.; Klemeš, J.J.; Wang, Q.; Zeng, M. Efficient thermal management strategy of Li-ion battery pack based on sorption heat storage. *Energy Convers. Manag.* **2022**, *256*, 115383. [CrossRef]
19. Zhang, H.; Yan, T.; Yu, N.; Wang, C.Y. Experimental investigation on a strontium chloride/ammoniasorption heat storage system. *Appl. Therm. Eng.* **2023**, *219 Pt B*, 119478. [CrossRef]
20. Yan, T.; Zhang, H. A critical review of salt hydrates as thermochemical sorption heat storage materials: Thermophysical properties and reaction kinetics. *Sol. Energy* **2022**, *242*, 157. [CrossRef]
21. Fumey, B.; Weber, R.; Baldini, L. Heat transfer constraints and performance mapping of a closed liquid sorption heat storage process. *Appl. Energy* **2023**, *335*, 120755. [CrossRef]
22. Lin, Y.C.; Liu, W.; Zhang, X.J.; Jiang, L. Performance analysis on open thermochemical sorption heat storage from a real mass transfer perspective. *J. Energy Storage* **2022**, *54*, 105267. [CrossRef]
23. Chao, J.; Xu, J.; Bai, Z.; Wang, P.; Wang, R.; Li, T. Integrated heat and cold storage enabled by high-energy-density sorption thermal battery based on zeolite/$MgCl_2$ composite sorbent. *J. Energy Storage* **2023**, *64*, 107155. [CrossRef]
24. Nguyen, M.H.; Zbair, M.; Dutournié, P.; Bennici, S. Thermochemical sorption heat storage: Investigate the heat released from activated carbon beads used as porous host matrix for $MgSO_4$ salt. *J. Energy Storage* **2023**, *59*, 106452. [CrossRef]
25. Zhang, C.; Li, Y.; Yang, L.; Fan, X.; Chu, L. Analysis on H_2 production process integrated $CaO/Ca(OH)_2$ heat storage and sorption enhanced staged gasification using calcium looping. *Energy Convers. Manag.* **2022**, *253*, 115169. [CrossRef]
26. Tawalbeh, M.; Khan, H.A.; Al-Othman, A.; Almomani, F.; Ajith, S. A comprehensive review on the recent advances in materials for thermal energy storage applications. *Int. J. Thermofluids* **2023**, *18*, 100326. [CrossRef]
27. Soares, N.; Costa, J.J.; Gaspar, A.R.; Santos, P. Review of passive PCM latent heat thermal energy storage systems towards building's energy efficiency. *Energy Build.* **2013**, *59*, 82. [CrossRef]
28. Weinläder, H.; Beck, A.; Fricke, J. PCM-facade-panel for daylighting and room heating. *Sol. Energy* **2005**, *78*, 177. [CrossRef]
29. Li, S.; Zou, K.; Sun, G.; Zhang, X. Simulation research on the dynamic thermal performance of a novel tripleglazed window filled with PCM. *Sustain. Cities Soc.* **2018**, *40*, 266. [CrossRef]
30. Li, D.; Wu, Y.; Liu, C.; Zhang, G.; Arıcı, M. Numerical investigation of thermal and optical performance of window units filled with nanoparticle enhanced PCM. *Int. J. Heat Mass Transf.* **2018**, *125*, 1321. [CrossRef]
31. Musiał, M. Analysis of the impact of selected factors on the effectiveness of using PCM in mobile window insulation. *E3S Web Conf.* **2018**, *49*, 00073. [CrossRef]
32. Musiał, M. Evaluation of the energy efficiency of an internal blind containing PCM. *E3S Web Conf.* **2018**, *49*, 00074. [CrossRef]
33. Lichołai, L.; Musiał, M.; Szyszka, J. International Patent Application EP.15461528.0, 2015.
34. Mehling, H. Strategic project 'Innovative PCM-Technology'—Results and future perspectives. In Proceedings of the 8th Expert Meeting and Workshop, Kizkalesi, Turkey, 18–20 April 2005.
35. Weinläder, H.; Koerner, W.; Heidenfelder, M. Monitoring results of an interior sun protection system with integrated latent heat storage. *Energy Build.* **2011**, *43*, 2468. [CrossRef]
36. Li, Y.; Darkwa, J.; Kokogiannakis, G.; Su, W. Phase change material blind system for double skin façade integration: System development and thermal performance evaluation. *Appl. Energy* **2019**, *252*, 113376. [CrossRef]
37. Bontemps, A.; Ahmad, M.; Johannčs, K.; Sallée, H. Experimental and modelling study of twin cells with latent heat storage walls. *Energy Build.* **2011**, *43*, 2456. [CrossRef]
38. Jaworski, M.; Abeid, S. Thermal conductivity of gypsum with incorporated phase change material (PCM) for building applications. *J. Power Technol.* **2011**, *91*, 49.
39. Wnuk, R.; Jaworski, M. Investigation of thermal characteristics of heat accumulating building elements containing PCM (phase change materials). *Pol. Sol. Energy* **2010**, *2*, 5. (In Polish)
40. Cabeza, L.F.; Castellón, C.; Nogués, M.; Medrano, M.; Leppers, R.; Zubillaga, O. Use of microencapsulated PCM in concrete walls for energy savings. *Energy Build.* **2007**, *39*, 113. [CrossRef]
41. Cabeza, L.F.; Medrano, M.; Castellón, C.; Castell, A.; Solé, C.; Roca, J.; Nogués, M. Thermal energy storage with phase change materials in building envelopes. *Contrib. Sci.* **2007**, *3*, 501. [CrossRef]

42. Zhu, N.; Wang, S.; Xu, X.; Ma, Z. A simplified dynamic model of building structures integrated with shaped-stabilized phase change materials. *Int. J. Therm. Sci.* **2010**, *49*, 1722. [CrossRef]
43. Xiao, M.; Feng, B.; Gong, K. Preparation and performance of shape stabilized phase change thermal storage materials with high thermal conductivity 2002. *Energy Convers. Manag.* **2002**, *43*, 103. [CrossRef]
44. Sar, A. Form-stable paraffin/high density polyethylene composites as solid-liquid phase change material for thermal energy storage: Preparation and thermal properties. *Energy Convers. Manag.* **2004**, *45*, 2033. [CrossRef]
45. Xiao, M.; Feng, B.; Gong, K. Thermal performance of a high conductive shape stabilized thermal storage material. *Sol. Energy Mater. Sol. Cells* **2001**, *69*, 293. [CrossRef]
46. Zhou, G.; Yang, Y.; Xu, H. Energy performance of a hybrid space-cooling system in an office building using SSPCM thermal storage and night ventilation. *Sol. Energy* **2011**, *85*, 477. [CrossRef]
47. Musiał, M. Analysis of the impact of PCM elements' geometry on their heat storage efficiency. *Bauphysik* **2019**, *41*, 324–330. [CrossRef]
48. Soares, N.; Gaspar, A.R.; Santos, P.; Costa, J.J. Multi-dimensional optimization of the incorporation of PCM-drywalls in lightweight steel-framed residential buildings in different climates. *Energy Build.* **2014**, *70*, 411. [CrossRef]
49. Saffari, M.; Roe, C.; Finn, D.P. Improving the building energy flexibility using PCM-enhanced envelopes. *Appl. Therm. Eng.* **2022**, *217*, 119092. [CrossRef]
50. Zahir, M.H.; Irshad, K.; Shafiullah, M.; Ibrahim, N.I.; Islam, A.K.M.K.; Mohaisen, K.O.; Sulaiman, F.A.A. Challenges of the application of PCMs to achieve zero energy buildings under hot weather conditions: A review. *J. Energy Storage* **2023**, *64*, 107156. [CrossRef]
51. Al-Yasiri, Q.; Szabó, M. Experimental study of PCM-enhanced building envelope towards energy-saving and decarbonisation in a severe hot climate. *Energy Build.* **2023**, *279*, 112680. [CrossRef]
52. Al-Absi, Z.A.; Hafizal, M.I.M.; Ismail, M. Innovative PCM-incorporated foamed concrete panels for walls' exterior cladding: An experimental assessment in real-weather conditions. *Energy Build.* **2023**, *288*, 113003. [CrossRef]
53. Li, Q.; Ju, Z.; Wang, Z.; Ma, L.; Jiang, W.; Li, D.; Jia, J. Thermal performance and economy of PCM foamed cement walls for buildings in different climate zones. *Energy Build.* **2022**, *277*, 112470. [CrossRef]
54. Agarwal, P.; Prabhakar, A. Energy and thermo-economic analysis of PCM integrated brick in composite climatic condition of Jaipur—A numerical study. *Sustain. Cities Soc.* **2023**, *88*, 104294. [CrossRef]
55. Krasoń, J.; Miąsik, P.; Lichołai, L.; Dębska, B.; Starakiewicz, A. Analysis of the Thermal Characteristics of a Composite Ceramic Product Filled with Phase Change Material. *Buildings* **2019**, *9*, 217. [CrossRef]
56. Leitzke, R.K.; Grala da Cunha, E.; Ferrugem, A.P.; Leal da Silva, T. The use of multi-objective evolutionary algorithms to assess phase change materials in a residence with light framings in bioclimatic zones 1, 2 and 3. *Energy Build.* **2023**, *284*, 112847. [CrossRef]
57. Wang, P.; Liu, Z.; Zhang, X.; Hu, M.; Zhang, L.; Fan, J. Adaptive dynamic building envelope integrated with phase change material to enhance the heat storage and release efficiency: A state-of-the-art review. *Energy Build.* **2023**, *286*, 112928. [CrossRef]
58. Imafidon, O.J.; Ting, D.S.-K. Retrofitting buildings with phase change materials (PCM)—The effects of PCM location and climatic condition. *Build. Environ.* **2023**, *236*, 110224. [CrossRef]
59. Lichołai, L.; Starakiewicz, A.; Krasoń, J.; Miąsik, P. The Influence of Glazing on the Functioning of a Trombe Wall Containing a Phase Change Material. *Energies* **2021**, *14*, 5243. [CrossRef]
60. Zhu, M.; Nan, W.; Wang, Y. Analysis on the thermal behaviour of the latent heat storage system using S-CO_2 and H-PCM. *Renew. Energy* **2023**, *208*, 240. [CrossRef]
61. Momeni, M.; Fartaj, A. Numerical thermal performance analysis of a PCM-to-air and liquid heat exchanger implementing latent heat thermal energy storage. *J. Energy Storage* **2023**, *58*, 106363. [CrossRef]
62. Musiał, M.; Lichołai, L. The Impact of a Mobile Shading System and a Phase-Change Heat Store on the Thermal Functioning of a Transparent Building Partition. *Materials* **2021**, *14*, 2512. [CrossRef]
63. Lichołai, L.; Musiał, M. Experimental analysis of the functioning of a window with a phase change heat accumulator. *Materials* **2020**, *13*, 3647. [CrossRef]
64. Musiał, M. Use of organic aliphatic esters to obtain an energy-efficient eutectic mixture. *Przem. Chem.* **2018**, *97*, 1855. [CrossRef]
65. Lichołai, L.; Musiał, M. Use of copolymers of vinyl acetate and alkyl acrylates for coating organic phase change materials. *Przem. Chem.* **2018**, *97*, 1852. [CrossRef]
66. Musiał, M. Experimental and Numerical Analysis of the Energy Efficiency of Transparent Partitions with a Thermal Storage Unit. *J. Ecol. Eng.* **2020**, *21*, 201. [CrossRef]
67. Li, W.Q.; Li, Y.X.; Yang, T.H.; Zhang, T.Y.; Qin, F. Experimental investigation on passive cooling, thermal storage and thermoelectric harvest with heat pipe-assisted PCM-embedded metal foam. *Int. J. Heat Mass Transf.* **2023**, *201*, 123651. [CrossRef]
68. Rghif, Y.; Colarossi, D.; Principi, P. Salt gradient solar pond as a thermal energy storage system: A review from current gaps to future prospects. *J. Energy Storage* **2023**, *61*, 106776. [CrossRef]
69. Gu, H.; Chen, Y.; Yao, X.; Huang, L.; Zou, D. Review on heat pump (HP) coupled with phase change material (PCM) for thermal energy storage. *Chem. Eng. J.* **2023**, *455*, 140701. [CrossRef]
70. Pogorzelski, J.A. *Building Thermal Physics*; Polish Scientific Publisher: Warsaw, Poland, 1976. (In Polish)
71. Vogel, A.I. *Organic Preparation*; Polish Scientific Publisher: Warsaw, Poland, 2006. (In Polish)

72. Smolec, W. *Photothermal Solar Energy Conversion*; Polish Scientific Publisher: Warsaw, Poland, 2000. (In Polish)
73. Tak, N. Type-1 possibilistic fuzzy forecasting functions. *J. Comput. Appl. Math.* **2020**, *370*, 112653. [CrossRef]
74. Askari, S. A novel and fast MIMO fuzzy inference system based on a class of fuzzy clustering algorithms with interpretability and complexity analysis. *Expert Syst. Appl.* **2017**, *84*, 301. [CrossRef]
75. Liu, J.; Tavakoli, F.; Sajadi, S.M.; Mahmoud, M.Z.; Heidarshenas, B.; Aybar, H.S. Numerical evaluation and artificial neural network modeling of the effect of oval PCM compartment dimensions around a triple lithium-ion battery pack despite forced airflow. *Eng. Anal. Bound. Elem.* **2022**, *142*, 71. [CrossRef]
76. Alnaqi, A.A.; Alsarraf, J.; Al-Rashed, A.A.A.A. Transient numerical study on injecting PCM in buildings along with extra comfort ventilation: Use of artificial neural network to decline energy utilization. *Eng. Anal. Bound. Elem.* **2022**, *143*, 559. [CrossRef]
77. Eldokaishi, A.O.; Abdelsalam, M.Y.; Kamal, M.M.; Abotaleb, H.A. Modeling of water-PCM solar thermal storage system for domestic hot water application using Artificial neural networks. *Appl. Therm. Eng.* **2022**, *204*, 118009. [CrossRef]
78. Stefan, J. On the theory of ice formation, in particular on ice formation in the polar sea. *Wiedemann Ann. Phys. Chem.* **1891**, *42*, 269–286. (In Germany) [CrossRef]
79. Baskar, I.; Chellapandian, M. Experimental and finite element analysis on the developed real-time form stable PCM based roof system for thermal energy storage applications. *Energy Build.* **2022**, *276*, 112514. [CrossRef]
80. Cueto, O.; Trabelsi, A.; Cagli, C.; Cyrille, M.C. Coupling a phase field model with an electro-thermal solver to simulate PCM intermediate resistance states for neuromorphic computing. *Solid-State Electron.* **2023**, *200*, 108542. [CrossRef]
81. Feng, Y.; Qin, Y. Nanoparticles effect on freezing of PCM utilizing finite element approach. *J. Mol. Liq.* **2022**, *366*, 120168. [CrossRef]
82. Benhamou, J.; Lahmer, E.B.; Jami, M. Three-dimensional simulation of conjugate heat transfer using the hybrid lattice Boltzmann-finite difference method. *Int. Commun. Heat Mass Transf.* **2022**, *139*, 106486. [CrossRef]
83. Ullah, H.; Hayat, T.; Ahmad, S.; Alhodaly, M.S. Entropy generation and heat transfer analysis in power-law fluid flow: Finite difference method. *Int. Commun. Heat Mass Transf.* **2021**, *122*, 105111. [CrossRef]
84. Patil, H.M.; Maniyeri, R. Finite difference method based analysis of bio-heat transfer in human breast cyst. *Therm. Sci. Eng.Prog.* **2019**, *10*, 42. [CrossRef]
85. Deng, D.; Jiang, Y.; Liang, D. High-order finite difference methods for a second order dual-phase-lagging models of microscale heat transfer. *Appl. Math. Comput.* **2017**, *309*, 31–48. [CrossRef]
86. Li, M.-J.; Chen, J.; Lian, Y.; Xiong, F.; Fang, D. An efficient and high-fidelity local multi-mesh finite volume method for heat transfer and fluid flow problems in metal additive manufacturing. *Comput. Methods Appl. Mech. Eng.* **2023**, *404*, 115828. [CrossRef]
87. Musiał, M.; Licholai, L.; Pękala, A. Analysis of the thermal performance of isothermal composite heat accumulators containing organic phase change material. *Energies* **2023**, *16*, 1409. [CrossRef]
88. Hedayat, A.; Davari, A. Feasibility study to increase the reactor power at natural convection mode in Tehran Research Reactor (TRR) through a hybrid thermal-hydraulic simulation and analysis using the RELAP5 code and Computational Fluid Dynamic (CFD) modeling by ANSYS-FLUENT. *Prog. Nucl. Energy* **2022**, *150*, 104285. [CrossRef]
89. Xu, C.; Zhang, H.; Fang, G. Review on thermal conductivity improvement of phase change materials with enhanced additives for thermal energy storage. *J. Energy Storage* **2022**, *51*, 104568. [CrossRef]
90. Cui, H.; Zou, J.; Gong, Z.; Zheng, D.; Bao, X.; Chen, X. Study on the thermal and mechanical properties of steel fibre reinforced PCM-HSB concrete for high performance in energy piles. *Constr. Build. Mater.* **2022**, *350*, 128822. [CrossRef]
91. Liu, Y.; Zheng, R.; Li, J. High latent heat phase change materials (PCMs) with low melting temperature for thermal management and storage of electronic devices and power batteries: Critical review. *Renew. Sustain. Energy Rev.* **2022**, *168*, 112783. [CrossRef]
92. Zhu, M.; Wang, Z.; Zhang, H.; Sun, X.; Dou, B.; Wu, W.; Zhang, G.; Jiang, L. Experimental investigation of the comprehensive heat transfer performance of PCMs filled with CMF in a heat storage device. *Int. J. Heat Mass Transf.* **2022**, *188*, 122582. [CrossRef]
93. Wang, L.; Kan, A.; Yu, W. Melting behavior and heat transfer performance in a modified PCM-filled enclosure with fins under hypergravity conditions. *Int. Commun. Heat Mass Transf.* **2022**, *138*, 106415. [CrossRef]
94. Zhao, J.; Li, Y.; Fang, X.; Sun, J.; Zhang, W.; Wang, B.; Xu, J.; Liu, Y.; Guo, H. High interface compatibility and phase change enthalpy of heat storage wood plastic composites as bio-based building materials for energy saving. *J. Energy Storage* **2022**, *51*, 104293. [CrossRef]
95. Zhang, S.; Pu, L.; Mancin, S.; Ma, Z.; Xu, L. Experimental study on heat transfer characteristics of metal foam/paraffin composite PCMs in large cavities: Effects of material types and heating configurations. *Appl. Energy* **2022**, *325*, 119790. [CrossRef]
96. Liu, D.; Xie, K.; Zhang, H.; Qiang, Y.; Yang, D.; Wang, Z.; Zhu, L.; Akkurt, N.; Du, Y.; Shen, M.; et al. Numerical evaluation of convective heat transfer properties of two-dimensional rotating PCM melt in the unilaterally heated rectangular container. *Renew. Energy* **2022**, *193*, 920. [CrossRef]
97. Ho, C.J.; Huang, S.H.; Lai, C.M. Enhancing laminar forced convection heat transfer by using Al2O3/PCM nanofluids in a concentric double-tube duct. *Case Stud. Therm. Eng.* **2022**, *35*, 102147. [CrossRef]
98. Mozafari, M.; Lee, A.; Cheng, S. A novel dual-PCM configuration to improve simultaneous energy storage and recovery in triplex-tube heat exchanger. *Int. J. Heat Mass Transf.* **2022**, *186*, 122420. [CrossRef]
99. Ho, C.J.; Jang, C.; Lai, C.M. Natural convection heat transfer in PCM suspensions in a square enclosure with bottom heating and top cooling. *Therm. Sci. Eng. Prog.* **2021**, *255*, 101037. [CrossRef]

100. Kouravand, A.; Kasaeian, A.; Pourfayaz, F.; Amin, M.; Rad, V. Evaluation of a nanofluid-based concentrating photovoltaic thermal system integrated with finned PCM heatsink: An experimental study. *Renew. Energy* **2022**, *11*, 1010–1025. [CrossRef]
101. Abdelrazik, A.S.; Saidur, R.; Al-Sulaiman, F.A.; Al-Ahmed, A.; Ben-Mansour, R. Multiwalled CNT and graphene nanoplatelets based nano-enhanced PCMs: Evaluation for the thermal performance and its implications on the performance of hybrid PV/thermal systems. *Mater. Do Day Commun.* **2022**, *31*, 103618. [CrossRef]
102. Lin, X.; Chen, X.Y.; Weng, L.; Hu, D.; Qiu, C.; Liu, P.; Zhang, Y.; Fan, M.; Sun, W.; Guo, X. In-situ copper ion reduction and micro encapsulation of wood-based composite PCM with effective anisotropic thermal conductivity and energy storage. *Sol. Energy Mater. Sol. Cells* **2022**, *242*, 111762. [CrossRef]
103. Ryms, M.; Januszewicz, K.; Haustein, E.; Kazimierski, P.; Lewandowski, W.M. Thermal properties of a cement composite containing phase change materials (PCMs) with post-pyrolytic char obtained from spent tyres as a carrier. *Energy* **2022**, *239*, 121936. [CrossRef]
104. Yousefi, A.; Tang, W.; Khavarian, M.; Fang, C. Development of novel form-stable phase change material (PCM) composite using recycled expanded glass for thermal energy storage in cementitious composite. *Renew. Energy* **2021**, *175*, 14. [CrossRef]
105. Das, D.; Bordoloi, U.; Muigaia, H.H.; Kalita, P. A novel form stable PCM based bio composite material for solar thermal energy storage applications. *J. Energy Storage* **2020**, *30*, 101403. [CrossRef]
106. Zhang, X.; Li, Y.; Luo, C.; Pana, C. Fabrication and properties of novel tubular carbon fiber-ionic liquids/stearic acid composite PCMs. *Renew. Energy* **2021**, *177*, 411. [CrossRef]
107. Vennapusa, J.R.; Konala, A.; Dixit, P.; Chattopadhyay, S. Caprylic acid based PCM composite with potential for thermal buffering and packaging applications. *Mater. Chem. Phys.* **2020**, *253*, 123453. [CrossRef]
108. Kim, H.G.; Qudoos, A.; Jeon, I.K.; Ryou, J.S. Assessment of PCM/SiC-based composite aggregate in concrete: Mechanical, physical and microstructural properties. *Constr. Build. Mater.* **2020**, *262*, 120088. [CrossRef]
109. Kim, H.G.; Qudoosb, A.; Jeon, I.K.; Woo, B.H.; Ryou, J.S. Assessment of PCM/SiC-based composite aggregate in concrete: Energy storage performance. *Constr. Build. Mater.* **2020**, *258*, 119637. [CrossRef]
110. Ali, M.A.; Fayaza; Viegas, R.F.; Kumar, M.B.S.; Kannapiran, R.K.; Feroskhana, M. Enhancement of heat transfer in paraffin wax PCM using nano graphene composite for industrial helmets. *J. Energy Storage* **2019**, *26*, 100982. [CrossRef]
111. Wang, G.; Bai, L.; Lin, J. Liquid-structure coupled simulation study on impacts of HTF temperature on discharge and mechanics performances of PCM heat storage system. *Case Stud. Therm. Eng.* **2023**, *44*, 102885. [CrossRef]
112. Abhinand, S.; Sharma, A.; Kothadia, H.B. Performance analysis of PCM melting in a fin-assisted thermal energy storage system—A numerical study. *Int. Commun. Heat Mass Transf.* **2023**, *144*, 106747. [CrossRef]
113. Afaynou, I.; Faraji, H.; Choukairy, K.; Arshad, A.; Arıcı, M. Heat transfer enhancement of phase-change materials (PCMs) based thermal management systems for electronic components: A review of recent advances. *Int. Commun. Heat Mass Transf.* **2023**, *143*, 106690. [CrossRef]
114. Bentivoglio, F.; Rouge, S.; Soriano, O.; de Sousa, A.T. Design and operation of a 180 kWh PCM heat storage at the Flaubert substation of the Grenoble urban heating network. *Appl. Therm. Eng.* **2021**, *185*, 116402. [CrossRef]
115. Guan, Y.; Meng, Q.; Ji, T.; Hu, W.; Li, W.; Liu, T. Experimental study of the thermal characteristics of a heat storage wall with micro-heat pipe array (MHPA) and PCM in solar greenhouse. *Energy* **2023**, *264*, 126183. [CrossRef]
116. Musiał, M.; Pękala, A. Functioning of Heat Accumulating Composites of Carbon Recyclate and Phase Change Material. *Materials* **2022**, *15*, 2331. [CrossRef] [PubMed]
117. Yu, J.; Wang, Y.; Qi, C.; Zhang, W. Reverse latent heat power generation properties of PCM/nanoparticles/metal foam at night. *Colloids Surf. A* **2023**, *665*, 131240. [CrossRef]
118. Rathod, M.K.; Joshi, V.; Rakshit, D. Chapter 4-Heat transfer in metal foam infused PCM: A case study. In *Phase Change Materials for Heat Transfer*; Elsevier: Amsterdam, The Netherlands, 2023; Volume 103. [CrossRef]
119. Baruah, J.S.; Athawale, V.; Rath, P.; Bhattacharya, A. Melting and energy storage characteristics of macro-encapsulated PCM-metal foam system. *Int. J. Heat Mass Transf.* **2022**, *182*, 121993. [CrossRef]
120. Kong, X.; Li, J.; Fan, M.; Li, W.; Li, H. Study on the thermal performance of a new double layer PCM trombe wall with multiple phase change points. *Sol. Energy Mater. Sol. Cells* **2022**, *240*, 111685. [CrossRef]
121. Zhao, K.; Wang, J.; Xie, H.; Guo, Z. Microencapsulated phase change n-Octadecane with high heat storage for application in building energy conservation. *Appl. Energy* **2023**, *329*, 120284. [CrossRef]
122. Kant, K.; Pitchumani, R. Advances and opportunities in thermochemical heat storage systems for buildings applications. *Appl. Energy* **2022**, *321*, 119299. [CrossRef]
123. Chao, L.; Sun, C.; Peng, L.; Li, J.; Sun, M.; Bao, L.; Liu, J.; Ma, Y. Passive energy-saving buildings realized by the combination of transparent heat-shielding glass and energy storage cement. *Constr. Build. Mater.* **2023**, *365*, 130023. [CrossRef]
124. Wang, Z.; Wang, M.; Qin, H. The energy storage characteristic analysis and optimization of latent heat energy storage component with finned tubes in building envelope. *Case Stud. Therm. Eng.* **2022**, *33*, 101946. [CrossRef]
125. Shen, Y.; Liu, S.; Mazhar, A.R.; Wang, J.; Li, Y. Phase change materials embedded with tuned porous media to alleviate overcharging problem of cascaded latent heat storage system for building heating. *Energy Build.* **2023**, *281*, 112746. [CrossRef]
126. Lu, S.; Lin, Q.; Liu, Y.; Yue, L.; Wang, R. Study on thermal performance improvement technology of latent heat thermal energy storage for building heating. *Appl. Energy* **2022**, *323*, 119594. [CrossRef]

127. Lai, K.; Wang, W.; Giles, H. Solar shading performance of window with constant and dynamic shading function in different climate zones. *Sol. Energy* **2017**, *147*, 113. [CrossRef]
128. Samadi, S.; Noorzai, E.; OBeltrán, L.; Abbasi, S. A computational approach for achieving optimum daylight inside buildings through automated kinetic shading systems. *Front. Archit. Res.* **2020**, *9*, 335. [CrossRef]
129. Ghosh, A.; Neogi, S. Effect of fenestration geometrical factors on building energy consumption and performance evaluation of a new external solar shading device in warm and humid climatic condition. *Sol. Energy* **2018**, *169*, 94. [CrossRef]
130. Huo, H.; Xua, W.; Li, A.; Cui, G.; Wu, Y.; Liu, C. Field comparison test study of external shading effect on thermal-optical performance of ultralow-energy buildings in cold regions of China. *Build. Environ.* **2020**, *180*, 106926. [CrossRef]
131. Chi, F.; Zhu, Z.; Jin, L.; Bart, D. Calculation method of shading area covering southerly orientated windows for the Tangwu building in Sizhai village. *Sol. Energy* **2019**, *180*, 39–56. [CrossRef]
132. Bhatia, A.; Sangireddy, S.A.R.; Garg, V. An approach to calculate the equivalent solar heat gain coefficient of glass windows with fixed and dynamic shading in tropical climates. *J. Build. Eng.* **2019**, *22*, 90. [CrossRef]
133. Sherif, A.; Sabry, H.; Arafa, R.; Wagdy, A. The Impact of Alternative Window Glazing Types and a Shading System on the Daylighting of Hospital Patient Rooms: Simulation Analysis under a Desert Clear Sky. *Energy Procedia* **2015**, *78*, 1805. [CrossRef]
134. Abazari, T.; Mahdavinejad, M. Integrated Model for Shading and Airflow Window in BSk. *Energy Procedia* **2017**, *122*, 571. [CrossRef]
135. Zheng, X.; Dai, T.; Tang, M. An experimental study of vertical greenery systems for window shading for energy saving in summer. *J. Clean. Prod.* **2020**, *259*, 120708. [CrossRef]
136. Invidiata, A.; Ghisi, E. Life-cycle energy and cost analyses of window shading used to improve the thermal performance of houses. *J. Clean. Prod.* **2016**, *133*, 1371. [CrossRef]
137. Vigna, I.; Bianco, L.; Goia, F.; Serra, V. Phase Change Materials in Transparent Building Envelopes: A Strengths, Weakness, Opportunities and Threats (SWOT) Analysis. *Energies* **2018**, *11*, 111. [CrossRef]
138. Nyoumbi, C.D.; Tambue, A. Convergence of a fitted finite volume method for pricing two dimensional assets with stochastic volatilities. *Math. Comput. Simul.* **2023**, *207*, 388. [CrossRef]
139. Qiu, S.; Solomon, L.; Fang, M. Study of Material Compatibility for a Thermal Energy Storage System with Phase Change Material. *Energies* **2018**, *11*, 572. [CrossRef]
140. Park, H.J.; Yun, Y.B.; Chang, J.S.; Wi, S.; Jeon, J.; Kim, S. Impact of a passive retrofit shading system on educational building to improve thermal comfort and energy consumption. *Energy Build.* **2020**, *216*, 109930. [CrossRef]
141. Sá, A.V.; Azenha, M.; Guimarães, A.S.; Delgado, J.M.P.Q. FEM Applied to Building Physics: Modeling Solar Radiation and Heat Transfer of PCM Enhanced Test Cells. *Energies* **2020**, *13*, 2200. [CrossRef]
142. Zhou, Q.; Liu, P.-F.; Tzeng, C.-T.; Lai, C.-M. Thermal Performance of Microencapsulated Phase Change Material (mPCM) in Roof Modules during Daily Operation. *Energies* **2018**, *11*, 679. [CrossRef]
143. Hu, Y.; Heiselberg, P.K. A new ventilated window with PCM heat exchanger-Performance analysis and design optimization. *Energy Build.* **2018**, *169*, 185. [CrossRef]
144. Lundström, L.; Akander, J.; Zambrano, J. Development of a Space Heating Model Suitable for the Automated Model Generation of Existing Multifamily Buildings—A Case Study in Nordic Climate. *Energies* **2019**, *12*, 485. [CrossRef]
145. Li, C.; Lyu, Y.; Li, C.; Qiu, Z. Energy performance of water flow window as solar collector and cooling terminal under adaptive control. *Sustain. Cities Soc.* **2020**, *59*, 102152. [CrossRef]
146. Lyu, Y.-L.; Liu, W.-J.; Sua, H.; Wuc, X. Numerical analysis on the advantages of evacuated gap insulation of vacuum-water flow window in building energy saving under various climates. *Energy* **2019**, *175*, 353. [CrossRef]
147. Szyszka, J.; Starakiewicz, A. A quasi-box window concept to improve the thermal-insulation property of old windows—Case study. *E3S Web Conf. SOLINA* **2018**, *49*, 00115. [CrossRef]
148. Li, D.; Zhang, C.; Li, Q.; Liu, C.; Arıcı, M.; Wu, Y. Thermal performance evaluation of glass window combining silica aerogels and phase change materials for cold climate of China. *Appl. Therm. Eng.* **2020**, *165*, 114547. [CrossRef]
149. Faraj, K.; Khaled, M.; Faraj, J.; Hachem, F.; Chahine, K.; Castelain, C. Energetic and economic analyses of integrating enhanced macro-encapsulated PCM's with active underfloor hydronic heating system. *Energy Rep.* **2022**, *8*, 848. [CrossRef]
150. Hou, J.; Liu, Z.-A.; Zhang, L.; Zhang, T.; Hou, C.; Fukuda, H. Parametric and economic analysis of incorporating phase change material (PCM) into exterior walls to reduce energy demand for traditional dwellings in northeast of Sichuan hills, China. *Appl. Therm. Eng.* **2023**, *223*, 119982. [CrossRef]
151. Elfeky, K.E.; Mohammed, A.G.; Wang, Q. Thermo-economic evaluation of PCM layer thickness change on the performance of the hybrid heat storage tank for concentrating solar power plants. *Energy* **2022**, *253*, 124128. [CrossRef]
152. Tafuni, A.; Giannotta, A.; Mersch, M.; Pantaleo, A.M.; Amirante, R.; Markides, C.N.; De Palma, P. Thermo-economic analysis of a low-cost greenhouse thermal solar plant with seasonal energy storage. *Energy Convers. Manag.* **2023**, *288*, 117123. [CrossRef]
153. Li, H.; Hou, J.; Hong, T.; Ding, Y.; Nord, N. Energy, economic, and environmental analysis of integration of thermal energy storage into district heating systems using waste heat from data centres. *Energy* **2021**, *219*, 119582. [CrossRef]
154. Pelella, F.; Zsembinszki, G.; Viscito, L.; Mauro, A.W.; Cabeza, L.F. Thermo-economic optimization of a multi-source (air/sun/ground) residential heat pump with a water/PCM thermal storage. *Appl. Energy* **2023**, *331*, 120398. [CrossRef]
155. Taheri, B.; Jabari, F.; Foroud, A.A. A green cogeneration microgrid composed of water-source heat pumps, a gravity energy storage, and a bio-fueled gas turbine: Design and techno-economic optimization. *Sustain. Cities Soc.* **2023**, *95*, 104594. [CrossRef]

156. Fu, H.; Fang, L.; Yu, M.; Cui, P.; Zhang, W.; Mao, Y.; Zhuang, Z.; Fang, Z. Influence and economic analysis of heat storage in the non-heating season on the heat extraction capacity of mid-deep borehole heat exchangers. *Energy Build.* **2023**, *278*, 112619. [CrossRef]
157. Shayan, M.E.; Najafi, G.; Ghobadian, B.; Gorjian, S.; Mamat, R.; Ghazali, M.F. Multi-microgrid optimization and energy management under boost voltage converter with Markov prediction chain and dynamic decision algorithm. *Renew. Energy* **2022**, *201*, 179. [CrossRef]
158. Shayan, M.E.; Najafi, G.; Ghobadian, B.; Gorjian, S.; Mazlan, M.; Samami, M.; Shabanzadeh, A. Flexible Photovoltaic System on Non-Conventional Surfaces: A Techno-Economic Analysis. *Sustainability* **2022**, *14*, 3566. [CrossRef]

energies

MDPI

Article

Solar Salt above 600 °C: Impact of Experimental Design on Thermodynamic Stability Results

Julian Steinbrecher [1,*], Markus Braun [1], Thomas Bauer [2], Sebastian Kunkel [1] and Alexander Bonk [1]

[1] German Aerospace Center (DLR), Institute of Engineering Thermodynamics, D-70569 Stuttgart, Germany; alexander.bonk@dlr.de (A.B.)
[2] German Aerospace Center (DLR), Institute of Engineering Thermodynamics, D-51147 Cologne, Germany
* Correspondence: author: julian.steinbrecher94@gmail.com

Abstract: Thermal energy storage (TES) based on molten salts has been identified as a key player in the transition from fossil fuels to renewable energy sources. Solar Salt, a mixture of $NaNO_3$ (60 wt%) and KNO_3 (40 wt%), is currently the most advanced heat transfer and storage material used in concentrating solar power (CSP) plants. Here, it is utilized to produce electricity via a Rankine cycle, with steam temperatures reaching 550 °C. The goal of this study is to increase the operating temperature of solar salt to over 600 °C, allowing it to be adapted for use in high-temperature Rankine cycles with steam temperatures greater than 600 °C. Yet, this goal is impaired by the lack of available thermodynamic data given the salt's complex high-temperature decomposition and corrosion chemistry. The study explores the thermodynamics of the decomposition reactions in solar salt, with a focus on suppressing decomposition into corrosive oxide ions up to a temperature of 620 °C. The results provide a new understanding of the stabilization of solar salt at previously unexplored temperatures with effective utilization of gas management techniques.

Keywords: molten salt; nitrite formation; oxide formation; high-temperature chemistry; thermal energy storage; concentrated solar power (CSP)

Citation: Steinbrecher, J.; Braun, M.; Bauer, T.; Kunkel, S.; Bonk, A. Solar Salt above 600 °C: Impact of Experimental Design on Thermodynamic Stability Results. *Energies* **2023**, *16*, 5241. https://doi.org/10.3390/en16145241

Academic Editor: Fabio Montagnaro

Received: 31 May 2023
Revised: 2 July 2023
Accepted: 6 July 2023
Published: 8 July 2023

1. Introduction

Providing dispatchable renewable energy at a GWh scale has become increasingly significant in the last decade. Thermal energy storage (TES) based on molten nitrate salts is currently the standard solution in several concentrated solar power (CSP) plants to provide dispatchable electricity during day- and night-times. Solar salt, a mixture of 60 wt% $NaNO_3$ and 40 wt% KNO_3, is a commonly used TES material for temperatures up to 565 °C [1–3]. The overall efficiency of the power block and storage capacity could still be enhanced if the stability limit could be extended well above 600 °C. Yet, temperatures above 600 °C lead to the formation of toxic gases (e.g., NO_x) and corrosive decomposition products in the salt, which are potentially harmful to, e.g., the container materials [4,5]. Additionally, it has been demonstrated that contact of molten salt with alloy materials (e.g., Inconel 625, 45# carbon steel) is related to the evolution of NO and NO_2 [6,7]. These toxic gases cannot be released into the environment without further gas treatment [8]. The chemical nature of solar salt has been intensively studied, and there is general agreement on a decomposition mechanism with two reaction steps. Initially, the nitrate anion (NO_3^-) will decompose to form a nitrite ion (NO_2^-) under the evolution of oxygen (O_2) (Equation (1)).

$$NO_3^- \rightleftharpoons NO_2^- + 0.5O_2 \quad (1)$$

At temperatures beyond 565 °C, the NO_2^- -ion will decompose to form oxide ion species (e.g., O_2^{2-}, O_2^-, O^{2-}) under the evolution of nitrous gases (e.g., NO, NO_2, N_2O).

The following equilibrium reactions (Equations (2) and (3)) represent the most likely possibility for the second degradation step of solar salt at elevated temperatures [9,10].

$$2NO_2^- \rightleftharpoons O^{2-} + 2NO + 0.5O_2 \tag{2}$$

$$2NO_2^- \rightleftharpoons O^{2-} + N_2 + 1.5O_2 \tag{3}$$

So far, the first decomposition step of NO_3^--anions (Equation (1)) has been proven valid [11,12]. For the second decomposition step, several reaction mechanisms were suggested, and, for solar salt above 600 °C, the majority of them can be described by a combination of Equations (2) and (3) [13–16]. Equation (2) is considered to include the temperature-dependent gas–gas equilibrium reaction (Equation (4)), which lies on the side of NO for temperatures above 300 °C [17].

$$2NO + 0.5O_2 \rightleftharpoons 2NO_2 \tag{4}$$

Equation (3) is understood as a subordinated decomposition reaction related to Equation (2), although Yang et al. [6] reported that it is favorable from a thermodynamic perspective. However, in the studies of Wright et al. [9] and Olivares et al. [18], the N_2 emission was measured to be an order of magnitude less than the emission of NO. Owing to this fact, Equation (2) is understood as the primary reaction path for nitrite decomposition in the presented work. This reaction generally leads to the formation of corrosive oxide ions coupled with the release of nitrous gases. The exact stoichiometry and mechanism, however, still need to be confirmed experimentally.

The authors believe that the whole potential of solar salt as a TES medium, in terms of its maximum operating temperature, has not yet been exploited. The main reason for this limitation is the usage of a system configuration where the hot and cold tanks of the TES system are openly connected to ambient air. In this case, NO_3^- and NO_2^- concentrations equilibrate as a function of the present oxygen partial pressure, according to Equation (1) [2,19]. However, nitrite decomposition (Equation (2)) is not stabilized since nitrous gases are constantly flushed out of the tanks [8,20,21]. Consequently, oxide ion formation can thermodynamically never reach equilibrium and should continue to increase during the lifetime of the plant to an unacceptable level. This can potentially result in aggravated corrosion, which is a crucial factor hindering TES with nitrate salts by moving to higher operating temperatures [22–24].

Retaining reactive gases such as O_2 and NO or NO_2 in the system would effectively prevent the decomposition of the nitrate salt, which has been partly demonstrated previously [25].

In this study, the presence of nitric oxide (NO) in the purge gas was proven to reduce the rate of O^{2-}-formation significantly [26,27]. Yet, a fully stabilized solar salt system with a constant oxide ion concentration at 620 °C has not been presented in the literature. It has been suggested that transport limitations and corrosive side reactions may have affected the solar salt chemistry, leading to misinterpretation of chemical equilibria, especially on small-scale experiments (e.g., sub-kg scale). In their kinetic study, Nissen and Meeker [11] have already revealed that it is necessary to eliminate oxygen (and nitrite) mass transport effects in order to achieve reproducible data, but they assumed that there were no errors in the thermodynamic equilibrium data. However, in a more recent study [28], it was found that above 600 °C, all of the present thermodynamic datasets for solar salt are most likely afflicted by decomposition reactions or mass transport limitations and do not reflect a thermodynamic equilibrium. In consequence, the interpretation of those results leads to false thermodynamic predictions, especially on the high-temperature properties of solar salt. However, the authors suggested that the data of Sötz et al. [12] generated at temperatures below 550 °C are in fact applicable for elevated temperatures if side reactions are reduced to a minimum. With this study, by more careful experimental design improvements, the requirements for generating data without side reactions were met.

These include the replacement of metal components of test rigs previously utilized for the generation of high-temperature thermodynamic data with a fully ceramic experimental setup. Additionally, experimental design improvements involved the appropriate selection of process parameters (e.g., purge gas). A new experimental series with an improved setup up to 620 °C and a thermodynamic evaluation is presented. For the first time, the experiments demonstrate the impact of several nitrous gas species (NO, NO_2, or N_2O) as well as N_2 on the oxide ion-producing reaction up to 620 °C to identify the most relevant reactions for the stabilization of solar salt.

2. Material and Methods

Solar salt (60 wt% $NaNO_3$, 40 wt% KNO_3) was produced by stoichiometric mixing of $NaNO_3$ (Merck, purity > 99%) and KNO_3 (Merck, purity > 99%). For each experiment, an autoclave test rig (described elsewhere [25]) was loaded with 150 g of solar salt contained in an Al_2O_3 crucible. The in-house setup (see Figure 1a) was designed in such a way that only the lower part of the autoclave is heated. Salt creeping up the crucible walls will solidify in the cold part and fall back into molten salt. The reported setup was optimized by substituting stainless steel components with full ceramic (Al_2O_3) components (see Figure 1b). The autoclave test rig is temperature controlled with a maximum deviation of ±4 °C. To fix the atmospheric composition, the salt is purged with a constant gas flow (100 mL/min) adjusted by calibrated flow meters (EL-Flow; Bronkhorst high-tech B.V., AK Ruurlo, Netherlands). To avoid concentration gradients in the molten salt, every experiment is stirred at 60 rpm if not mentioned otherwise. After melting the salt mixture at 300 °C under a flow of synthetic air (5.0 grade, Linde Gas), the auto sampling unit (detailed picture in [25]) is mounted onto the autoclave test rig. Subsequently, the desired purge gas flow is adjusted by mixing O_2, N_2, and a reference gas containing 2000 ppm NO (NO_2, N_2O) in N_2 (all 5.0-grade Linde gas). The concentrations of O_2 (vol%), CO_2 (ppm), NO_2 (ppm), and NO (ppm) in the exhaust gas are continuously measured with the Emerson X-SREAM analyzer (Model XEGP, Emerson Electric Co., St. Louis, USA). The salt is heated to the target temperature (e.g., 600 °C, 620 °C). Upon reaching the target temperature, the first salt sample is collected (t = 0 h). Over the course of the isothermal experiments (approx. 1200 h, ±2 °C), salt samples are extracted and analyzed by wet chemistry to monitor salt composition. To investigate whether oxide ion formation is limited by mass transport, the agitator speed is adjusted between 0, 60 (default speed), and 120 rpm.

To reveal the ion content of the salt samples, ion chromatography (IC; Metrohm model 883 Basic IC plus) was utilized. About 125 mg of the salt sample was dissolved in ultrapure water (conductivity < 0.055 μScm^{-1}) and analyzed. As a result, the content of nitrate and nitrite is given in absolute values based on a standardized calibration of the IC method. A detailed description of the experimental procedure and calibration is given elsewhere [28]. The oxide ion contents were determined by automated acid base titration (Metrohm Titrando 800, Herisau, Switzerland) under an inert purge gas of N_2 (5.0-grade Linde gas). This technique is based on the reaction of basic oxide ions (e.g., O^{2-}) in water to form two hydroxide ions (HO^-). By titration of HCl solution (0.01 M) until neutrality, the amount of hydroxide ion can be determined. A detailed description is given in our previous work [27]. The limit of detection (LOD) for oxide ions, represented as O^{2-} in this work, was below 0.01 mol% (20 ppm) with regard to salt composition. The measuring errors of nitrate and nitrite content were ±0.3% and ±0.9%, respectively. For the detection of oxide ions, the relative error at the LOD was ±14%, which is the maximum deviation and is applied below the limit of quantification (LOQ > 0.07 mol%). Above the LOQ, the error was as low as 2% and is shown as error bars for the oxide data points in the plots of this work. For nitrate and nitrite, the error bars lay within the symbol size. Cation concentrations of all experiments were monitored and remained unchanged over time and hence are not part of the discussion. Anion concentrations are given and discussed in relative molar percentage (mol%), where the sum of all anions detected (NO_3^-, NO_2^-, O^{2-}) is equal to 100%.

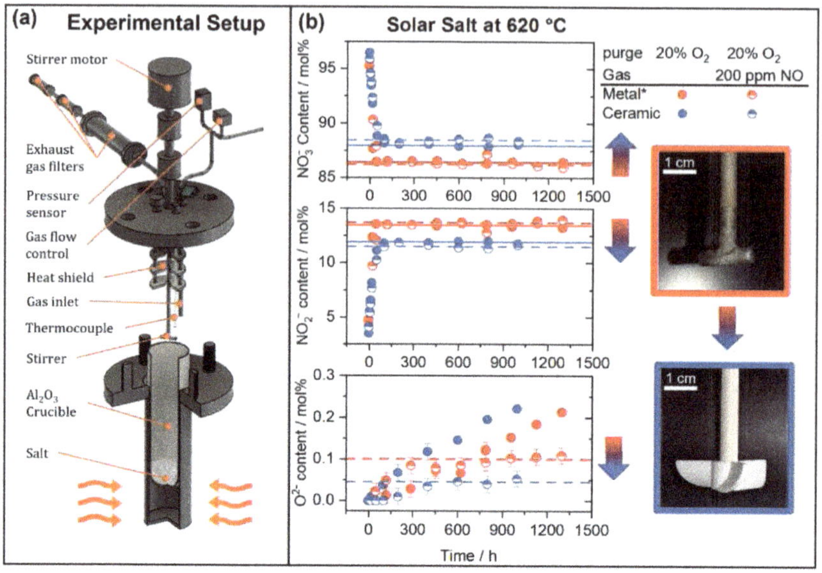

Figure 1. (**a**) Schematic image of the autoclave test rig with inside view of different components. Hot section indicated by orange arrows. (**b**) Anion content over time of solar salt stored at 620 °C, purged with synthetic air (closed symbols), and purged with synthetic air and 200 ppm NO (half open symbols). Equilibrium composition is marked as horizontal lines. Picture of metal stirrer (red) and ceramic stirrer (blue). Arrow with color transition from red to blue indicate the direction of salt composition change with the optimization of the setup. * Red data points are republished (Reprinted/adapted with permission from Ref. [26], 2020, International Solar Energy Society).

Equilibrium levels were calculated from anion content after the reaction quotient (ratio between educt and product) obtained a stable value. For NO_3^- and NO_2^- (Equation (1)), this is the case after approximately 300 h. Oxide formation rates are often afflicted with a strong error, especially if a corrosion reaction cannot be excluded [28]. It has been shown in a different study [27] that nitrous gases stabilize the molten salt, and an established oxide equilibrium allows comparison between different experiments. The O^{2-} equilibrium (Equation (2)) takes approximately 500 h to be attained. After the respective time, the median of the remaining data points is calculated and displayed as horizontal lines in the graphs, which are defined as equilibrium composition.

3. Results and Discussion

The principle of oxide ion formation has been the subject of many studies using molten nitrate salts for thermal energy storage. Unfortunately, despite decades of research, there is no general agreement about one single reaction to describe the oxide formation in molten nitrate salt. Missing data on the equilibrium composition of solar salt at elevated temperatures prevents sophisticated predictions of the long-term salt composition development, which is necessary for next-generation nitrate salt TES. In subsequent sections, the optimized setup is introduced and its effect on the salt composition is evaluated. Further, the role of N_2 gas in the stabilization of solar salt is discussed. To generate insight into the role of nitrous gases, for the first time, equilibrium data for solar salt purged with NO_2 and N_2O are presented. At 620 °C, the potential boundaries for the stabilization of molten salt with NO purge gas are introduced. Finally, the generated data are evaluated from a thermodynamic perspective, and process implications are made.

3.1. Setup Optimization

In the above-mentioned studies, with increasing operating temperatures, the authors experienced a significant alteration of the test rig, in particular the corrosion of the metal agitator unit. Corrosion reactions involve redox reactions between nitrate/nitrite ions and the alloying elements Fe, Cr, and Ni, leading to an increase in the impurity levels (e.g., chromate (CrO_4^{2-}) ions) in the molten salt but also changes to the decomposition mechanism [27,29]. Consequently, the metal agitator (Figure 1) was exchanged for a full ceramic (γ-Al_2O_3) 3D-printed propeller to prevent side reactions and, at the same time, increase mixing efficiency by geometry optimization. To identify any changes due to the optimized setup, this study comprises two experiments with identical parameter sets to two experiments already published [26]. Figure 1 displays the results of salt composition over time of two experiments with metallic (red, previous work [25]) and inert ceramic (blue, this work) propeller setups.

Closed symbols in Figure 1 depict the data of experiments under a synthetic air (20% O_2) atmosphere at 620 °C. The nitrate and nitrite content of both experiments attain a steady state after about 300 h, but for the experiment with the ceramic stirrer (blue), the NO_2^- content (11.9 mol%) is significantly lower compared to the data in red (13.5 mol%) gained with the metallic stirrer. The oxide ion concentrations increased in both experiments, as expected because the purge gas did not contain the gases (e.g., NO) necessary to obtain an oxide ion equilibrium.

A similar result was found in the experiment with a purge gas of 20% O_2 and 200 ppm of NO (rest N_2). The concentration of NO_3^- and NO_2^- (88.5 mol%, 11.5 mol%) for the optimized setup (blue) reveals a lower amount of the decomposition product compared to the red dataset (86.2 mol%, 13.7 mol%). In the case of the oxide ion concentration (in red), the equilibrium concentration was measured to be 0.1 mol%; albeit, a slow increase is still visible. Using the ceramic setup, the oxide ion concentration (in blue) stabilizes at 0.045 mol% under identical conditions.

This outcome is noteworthy in various aspects. The optimized setup not only stabilized the conditions for the molten salt to reach equilibrium in all reaction steps but also significantly decreased the fraction of decomposed salt due to side reactions with metallic components, specifically a reduction of up to 50% in the concentration of oxide ions. If the degree of salt decomposition is defined by the fraction of decomposition products (NO_2^-, O^{2-}) versus pure nitrate salt, our results become clearer. With this definition, it can be stated that the improvements to the experimental setup led to a lowering of the degree of salt decomposition by 16%. This result can be explained by the reduction in side reactions, which in the former publications led to higher decomposition product formation. Without corrosion reactions, the concentrations of NO_2^- and O^{2-} (in steady state) present the pure chemistry of solar salt at 620 °C and are now applicable for thermodynamic calculations. Consequently, solar salt appears to be more stable under these conditions than suggested by previous investigations. The next section includes results on the potential impact of nitrogen gas on the oxide ion producing a reaction in molten solar salt.

3.2. Salt Stabilization without N_2

In the before-mentioned study of Sötz et al. [26], an equilibrium description was developed based on experimental data and an intensive literature review. It was found that the entirety of the published reaction mechanisms diverges in the fraction of nitrogen produced in the course of the oxide ion formation. Based on comparing the listed thermodynamic enthalpies and experimental results, a closed agreement for the following reaction equation (Equation (5)) was found.

$$2NO_2^- \rightleftharpoons O^{2-} + 0.5NO + 1.25O_2 + 0.75N_2 \quad (5)$$

Because of the high stability of nitrogen at the prevailing conditions (1 atm, 620 °C), the relevance of the nitrogen molecule in the stabilization of the molten salt is still questionable.

To prove the effect of nitrogen on the oxide ion equilibrium position, a set of storage experiments without nitrogen but with argon in the purge gas were performed in this study. It is emphasized that this investigation was conducted prior to the system optimization. Thus, this dataset is directly comparable to the published data of Sötz [26]. The result of the anion composition development over time in comparison to the equilibrium values (red lines) of the reference study is shown in Figure 2.

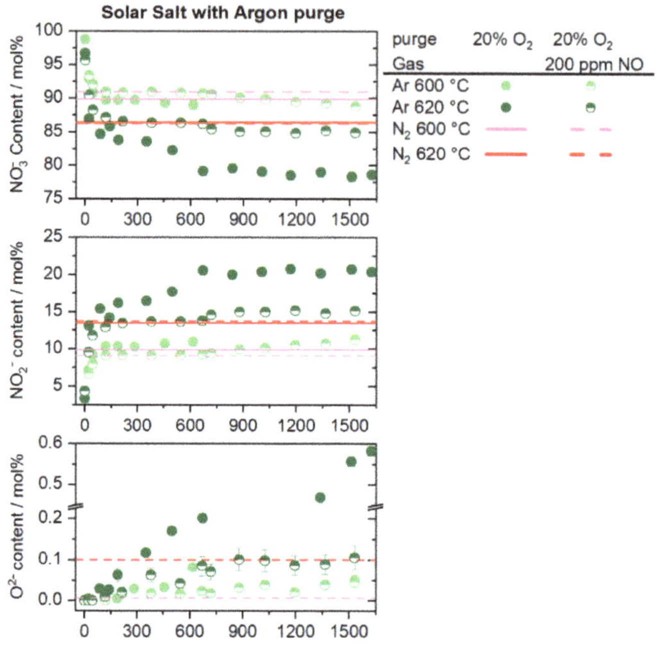

Figure 2. Anion content over time of solar salt stored at 600 (light green) and 620 °C (dark green) purged with synthetic air (full symbols) and synthetic air with 200 ppm NO (half symbols). Compared to the equilibrium values from Sötz et al., 2020 (red and light red lines), nitrogen was substituted with argon in this work (green and light green symbols).

For all experiments, the same partial pressure of oxygen (0.2 atm) was present in the purge gas. Consequently, the nitrate–nitrite equilibriums at 600 °C and 620 °C were expected to be identical to the published values from Sötz [26]. As can be seen in Figure 2 (top and middle graphs), three out of the four experiments align well with the published data performed in the presence of nitrogen gas (red line). Solely the experiment performed at 620 °C in 20% O_2 with argon showed a shift of the equilibrium composition towards the product side of Equation (1) with 78.9 ± 0.9 mol% nitrate and 20.4 ± 0.5 mol% nitrite compared to the reference at the same partial pressure of oxygen (86.4 mol% and 13.5 mol%). Additionally, the same salt samples contained significant amounts of chromate ions (up to 0.5 ± 0.1 mol%), which is a known indicator for the presence of corrosion reactions during the experiment [24]. This deviation can most likely be ascribed to more severe corrosion reactions of the salt with a metal stirrer compared to the other experiments. In the bottom graph of Figure 2, the oxide ion content is depicted. The experiment purged with 200 ppm NO (20% O_2 in argon) shows stable oxide ion concentrations after 700 h and is close to the published data at 620 °C (0.10 ± 0.08 mol%). At 600 °C, the data with argon purge are above the published results (0.06 ± 0.04 mol%) (dotted red line).

At 600 °C, another study [27] performed in the same test rig is available for comparison. Here, after the regeneration of solar salt with identical experimental conditions, an oxide ion concentration of 0.11 ± 0.02 mol% was obtained, which is above even the result at 620 °C

(0.10 ± 0.08 mol%). These findings indicate that impurities and corrosion side reactions inflict the oxide ion data with a significant error and at such a low concentration that data need careful analysis. However, the results allow for a couple of conclusions concerning the stabilization of solar salt. Firstly, the corrosion reaction did alter the nitrate–nitrite equilibrium and clearly demonstrates the necessity of the optimization of the experimental setup. Secondly, it was found that nearly equivalent oxide ion equilibrium values could be achieved with and without the presence of nitrogen in the purge gas.

The latter result is in contradiction to the oxide equilibrium in Equation (5) where NO, O_2, and N_2 are absolutely required for the stabilization of the nitrite decomposition. In consequence, the postulated equilibrium description (Equation (5)) does not reflect the present reaction mechanism of the experiments discussed in this paper. However, the detection of N_2 during the thermal decomposition of nitrate salt in several studies [9,18] remains; therefore, it is likely that the decomposition of nitrite proceeds under NO (Equation (2)) as well as N_2 (Equation (3)) emissions. From a thermodynamic perspective, Yang et al. [6] stated that N_2 evolution (Equation (3)) is more favorable than NO evolution. Thus, the reverse direction of Equation (3), a regeneration of the salt with N_2, is unlikely and can be understood as an irreversible reaction path in the prevailing conditions. Consequently, a stable oxide ion concentration over time, as shown in Figure 2, is only feasible if the irreversibly produced oxide ions (as in Equation (3)) are compensated by NO regeneration through Equation (2).

Apart from N_2 and NO, the gases NO_2 and N_2O were also detected to be reaction products of nitrate salt decomposition [9,18]. The next part of the study focuses on answering which nitrous gas species is the most relevant in controlling the stabilization of solar salt at elevated temperatures.

3.3. Impact of NO, NO_2, and N_2O

The following set of experiments was performed with 20% O_2 and 200 ppm NO and NO_2 or N_2O in N_2 at 600 °C or 620 °C. To accelerate the equilibration process, the initial conditions were modified by either adding nitrite salt (10 mol%), purging with pure nitrogen, or overheating (630 °C) the salt. However, the modifications did not affect the final equilibrium composition of solar salt and are comparable to each other, as has been stated by Nissen and Meeker [11]. To reveal the influence of the different nitrous gases on the equilibrium composition, especially the oxide ion concentration, the anion concentrations over time of the respective experiments are summarized in Figure 3.

The equilibrium concentrations for nitrate as well as for nitrite for experiments at 600 °C and at 620 °C are in close agreement. A thermodynamic evaluation and equilibrium data are given in the upcoming Section 3.5.

In the bottom graph of Figure 3, the oxide ion concentration evolution over time of the different experiments is shown. It is immediately visible that the oxide ion concentration increases constantly (1E-06 mol%/h) for molten salt purged with 200 ppm N_2O and that it did not reach an equilibrium state. In order to obtain oxide equilibrium from the opposing side in the 600 °C experiment (Figure 3, blue cross), the NO_2 purge gas was subsequently simply switched on after 300 h to initially increase oxide ion formation. In comparison to solar salt purged with N_2O, the oxide content without nitrous gases after 300 h was at 0.090 ± 0.002 mol%, whereas with N_2O purge gas after 300 h, it was significantly less at 0.045 ± 0.008 mol%. For experiments purged with 200 ppm NO and NO_2 at 600 °C (light blue), the oxide equilibrium state was approached from initially higher values of oxide ion, which subsequently became reduced via regeneration reactions (Equation (2)). In particular, with NO_2, the regeneration was quicker than with NO (Figure 3, dashed line: 0.032 ± 0.008 mol%) and led to the lowest oxide ion content (Figure 3, dash dot: 0.016 ± 0.002 mol%) under these conditions. The same trend is visible for the experiments at 620 °C, where the oxide equilibrium of salt purged with NO_2 (0.029 ± 0.002 mol%) is slightly below that purged with NO (0.045 ± 0.002 mol%).

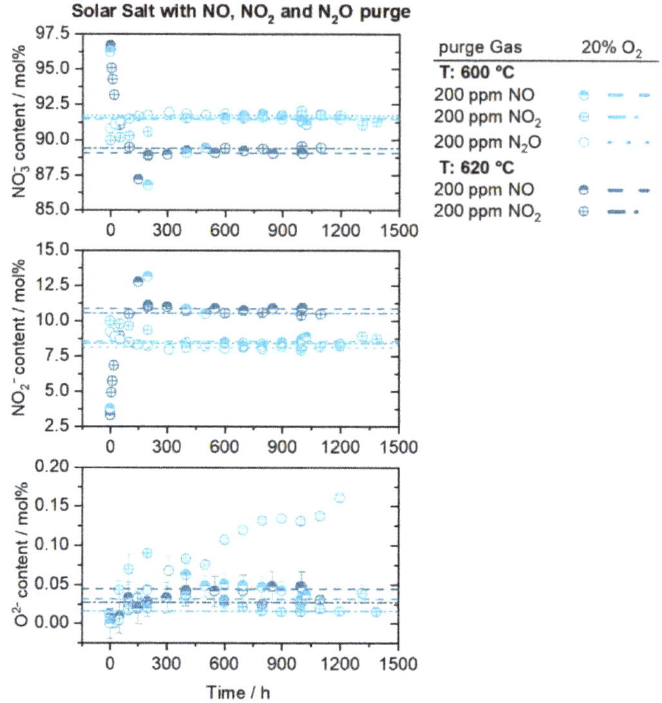

Figure 3. Anion content over time of solar salt stored at 600 °C (light blue) and 620 °C (dark blue) then purged with 20% O_2 and 200 ppm of either NO, NO_2, or N_2O. Horizontal lines mark the equilibrium content of the respective experiment.

Based on these experiments, it was found that the stabilizing effect of N_2O (20% O_2 with 200 ppm N_2O) on solar salt is significantly lower compared to that of NO and NO_2. However, with respect to the nitrate–nitrite equilibrium, similar results for all of the NO_x gases (at the same temperatures) were obtained. It is concluded that all NO_x gas species (200 ppm) reduce the extent of the nitrite decomposition (Equation (2)), which allows for equilibration of the nitrate–nitrite reactions without the impact of oxide ion formation. This is the first evidence for the suggestion in the literature [26,28] that above 600 °C, nitrite decomposition needs to be suppressed to achieve a proper thermodynamic nitrate–nitrite equilibrium (see Section 3.5, Equation (7)). When comparing 200 ppm of NO and NO_2, both lead to stable oxide ion concentrations at 600 °C and 620 °C, whereas NO_2 gives lower oxide ion values than NO.

Interestingly, a remarkable difference when analyzing the composition of the waste gas of molten salt purged with NO and NO_2 was detectable. The raw data are found in the supplementary section, exemplary for two of the previously discussed experiments at 620 °C with NO or NO_2 purge gas (Figure S2). To validate the extent of the temperature-dependent gas–gas reaction (Equation (4)) of NO, O_2, and NO_2, as mentioned in the introduction, a blank experiment without salt at 620 °C and different NO_x concentrations of 200–1200 ppm was performed (Figure S2). For an inlet gas containing 200 ppm NO, 20 ± 5 ppm was oxidized to NO_2, whereas from an initial value of 200 ppm NO_2, 50 ± 5 ppm was reduced to NO after passing the hot chamber. During every molten salt experiment, the measured concentrations of NO or NO_2 in the exhaust gas typically settle at a steady value after a few hundred hours, which is considered in the following discussion. The measured composition thereby correlates with the originally set purge gas composition (e.g., 200 ppm). With 200 ppm NO purged molten salt (620 °C, 20% O_2), a value of 174 ± 6 ppm NO and

21 ± 5 ppm NO_2 was measured (t = 700–1000 h). In contrast, the gas composition for molten salt (620 °C, 20% O_2) purged with 200 ppm NO_2 contained 108 ± 5 ppm NO and 81 ± 5 ppm NO_2 (t = 800–1100 h), which is not in agreement with the blank experiment (Figure S2). However, the total NO_x concentration (NO and NO_2 combined) measured matches that of the supply gas, which ensures no net consumption of nitrous gas species during the experiments.

In general, solar salt with a constant composition of nitrate, nitrite, and oxide ions was achieved with NO and NO_2. This state is only feasible when forward and reverse reactions (Equations (1) and (2)) occur at an equal rate. In other words, the decomposition of solar salt is stopped by the simultaneous regeneration of the decomposition products (NO_2^-, O^{2-}) with O_2 and NO or NO_2. This state is defined as an equilibrium where the reactant concentration is constant, including the consumption and emission of gases. With NO purge, this was the case, and the exhaust gas composition was almost the same as the incoming gas (200 ppm NO) with some partial oxidation to NO_2. In this regard, the temperature-dependent gas–gas reaction (Equation (4)) of NO, O_2, and NO_2 comes into play and was validated in a blank experiment without salt. In the hot reaction chamber (620 °C), NO gas is stable but cools down (25 °C) on its way to the gas analyzer, and consequently suffers from oxidation to NO_2, which corresponds to the measured concentration of 26 ± 5 ppm NO_2. Without the consumption of NO_x gas after passing the molten salt, it is reasonable to assume that an equilibrium state was attained.

In the case of NO_2, 107 ± 5 ppm NO is found after leaving the reaction vessel with molten salt, which is significantly more than expected from the blank experiment (50 ppm, Figures S1 and S2). In the following, a potential reaction mechanism is suggested to explain the outcome of the experiments purged with NO_2. NO_2 gas is known to react with the nitrite ion and form nitrate according to the following reaction in Equation (6) [30].

$$NO_2^- + NO_2 \rightarrow NO_3^- + NO \tag{6}$$

Because the concentration of nitrite is typically two orders of magnitude higher (see Figure 3) than oxide, it is statistically more likely for the NO_2 molecule to react with a nitrite ion than an oxide ion. In general, it is believed that reactions such as in Equation (6) but also the reverse direction of the equilibrium reactions in Equations (1) and (2) are dominantly occurring at a liquid–gas interface. This is because gas solubility in molten salts is typically low due to the lack of intermolecular interactions [31]. After the reaction of NO_2 at the liquid–gas interface, NO molecules are subsequently formed via Equation (6). It is reasonable to assume that a regeneration reaction of oxide ions (Equation (2)) with NO, which is produced directly at the reaction interface, is more effective compared to NO, which is obstructed by gas diffusion towards the reaction interface. This hypothesis could explain why NO_2 purge gas shows better performance compared to NO in the stabilization of solar salt at 600 and 620 °C.

3.4. Boundaries for Solar Salt Stabilization

The next set of experiments was performed to identify potential boundary conditions for the stabilization of solar salt with 200 ppm NO at 600 °C and 620 °C. Boundary conditions are defined as critical process parameters that, under equilibration such as those of solar salt in terms of stable nitrate, nitrite, and oxide ion content, are no longer feasible. Thereby, the effect of higher nitrite concentrations (achieved by a reduction in the oxygen partial pressure) on the position of the nitrite oxide equilibrium was investigated. At the same time, 200 ppm NO was not varied to isolate the effect of the nitrite concentration on the nitrite–oxide equilibrium. The results of the salt composition data are summarized in Figure 4.

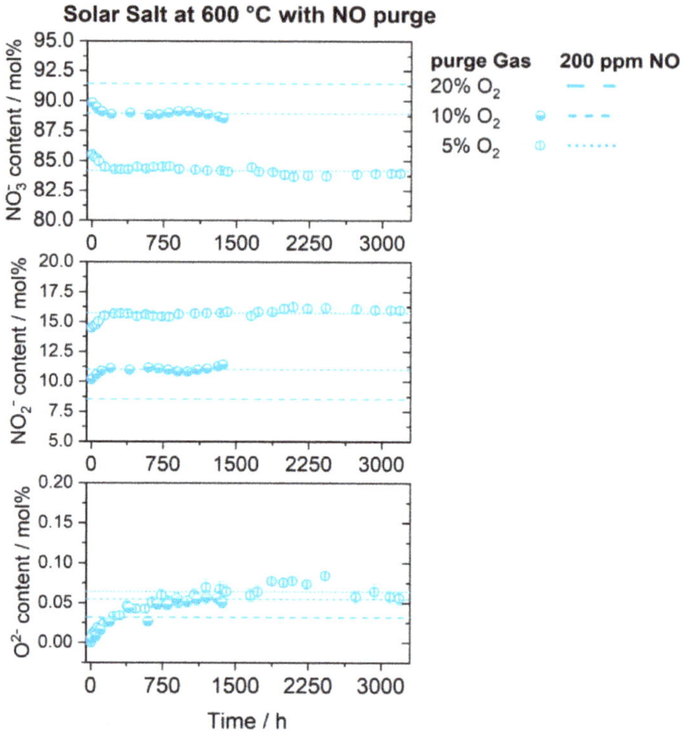

Figure 4. Anion content over time of solar salt stored at 600 °C (light blue) purged with 200 ppm NO in N_2 and 20% (data from Figure 3, not shown for clarity), 10%, or 5% O_2. Horizontal lines mark the equilibrium content of the respective experiment.

Both experiments attained nitrate–nitrite equilibrium after 300 h (Figure 4, top and middle), and this equilibrium composition was extracted for thermodynamic evaluation (Table 1). An oxide ion equilibrium was accomplished with 5% and 10% O_2 (200 ppm NO), at 0.06 ± 0.01 mol% and 0.055 ± 0.008 mol%.

With these results, it becomes clear that the content of NO_2^- changes the position of the oxide ion equilibrium, according to Equation (2). However, the relation is not exactly as written in the reaction mechanism (Equation (2)) because the oxide to nitrite ratio theoretically should be constant at a single temperature, which is not reflected by the experimental data (see Table 1). This behavior can potentially be explained by the strong regeneration of NO and the resultant low equilibrium concentration of oxide ions, in which case detection is hardly possible without a significant error. Still, the combined data from Figures 3 and 4 are the first to link the oxygen and NO_x partial pressure with a stable oxide ion concentration at 600 and 620 °C. The results are summarized in Section 3.5. and are recommended to be used for the modeling of salt composition evolution in any solar salt-based system operating under similar conditions.

To challenge the stabilization of solar salt with 200 ppm NO, a set of experiments at 620 °C was performed. As a reference, for solar salt without stabilization, the reaction chamber was purged with 100% N_2. Additionally, the mixing velocity was varied, and in another experiment, the effect of a pressurized system (1 bar overpressure) on the stabilization of solar salt at 620 °C was investigated. The data are summarized in Figure 5.

Table 1. Equilibrium nitrite to nitrate ratios and extracted $\ln(K_1)$ values as well as oxide to nitrite ratios at 600 °C and 620 °C. Oxid ion formation rate (mol%/h) is given for experiments not in equilibrium.

Temp. [°C]	P_{O_2}[atm]	c_{NOx}[ppm]	$\frac{[NO_2^-]}{[NO_3^-]}$ [mol·mol^{-1}]	$\ln(K_1)$	$\frac{[O^{2-}]}{[NO_2^-]}$ [mol·mol^{-1}]	$\frac{[O^{2-}]}{h}$ [mol·h^{-1}]	Source
600	0.2		1.12×10^{-1}	−2.9955	-	-	Bonk [25] & Sötz [26]
	0.2	200 (NO)	1.00×10^{-1}	−3.1016	6.56×10^{-3}	-	Sötz [26]
	0.2	200 (NO)	1.30×10^{-1}	−3.1756	3.72×10^{-3}	-	This work
	0.2	200 (NO_2)	9.18×10^{-2}	−3.1928	1.90×10^{-3}	-	This work
	0.2	200 (N_2O)	8.85×10^{-2}	−3.2300	-	1×10^{-6}	This work
	0.1	200 (NO)	1.24×10^{-1}	−3.2389	4.99×10^{-3}	-	This work
	0.05	200 (NO)	1.87×10^{-1}	−3.1743	3.84×10^{-3}	-	This work
620	0.2		1.56×10^{-1}	−2.6624	-	-	Sötz et al. [26]
	0.2		1.36×10^{-1}	−2.8030	-	2×10^{-6}	This work
	0.2	200 (NO)	1.56×10^{-1}	−2.6601	7.40×10^{-3}	-	Sötz et al. [26]
	0.2	200 (NO)	1.30×10^{-1}	−2.8448	3.87×10^{-3}	-	This work
	0.2	200 (NO)	1.22×10^{-1}	−2.9064	4.10×10^{-3}	-	This work
	0.2	200 (NO_2)	1.18×10^{-1}	−2.9390	2.60×10^{-3}	-	This work
	Additional experimental design modifications						
	0	0	-	-	-	8×10^{-6}	This work (pure N_2)
	0.2	200 (NO)	1.23×10^{-1}	−2.9037	-	6×10^{-7}	This work (1 bar rel.)
	0.2	200 (NO)	1.28×10^{-1}	−2.8639	-	7×10^{-7}	This work (120 rpm)
	0.05	200 (NO)	3.01×10^{-1}	−2.6988	-	6×10^{-7}	This work (0 rpm)

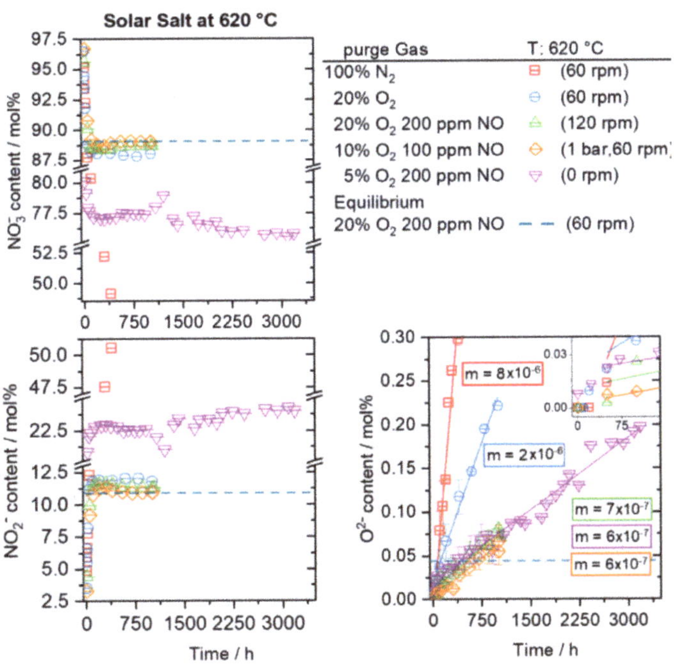

Figure 5. Anion content over time of solar salt stored at 620 °C. Purge gas was pure nitrogen (red color), 20% O_2 (blue color), 20% O_2 200 ppm NO and 120 rpm mixing velocity (green color), 10% 100 ppm NO 1 bar overpressure (orange color), and 5% O_2 200 ppm and 0 rpm (purple color). Linear regression fit of the oxide ion formation severs as a guide to the eye with the respective slope (m).

The system purged with pure N_2 shows the continuous decomposition of nitrate and the formation of nitrite and oxide ions. Experiments purged with oxygen attain a stable nitrate and nitrite composition after 500 h. With 5% O_2 at 1200 h, fluctuations of the nitrate and nitrite content are visible and originate from pressure built up in the chamber due to the clogging of a waste gas filter with salt powder. Despite the presence of nitrous gases in the gas stream, no stable oxide ion concentration was attained over the course of the different experiments. At 620 °C and 5% O_2 (200 ppm NO), oxide ion concentration increased over 3000 h. In the case of an accelerated mixing velocity (120 rpm), there was no significant difference in the nitrate–nitrite equilibrium compared to the reference (with identical purge gas) present (see Table 1), but the oxide ion composition did not reach a stable value within 1000 h.

Despite the fact that none of the experiments reached an oxide ion equilibrium, a close examination of the oxide ion formation in Figure 5 (bottom right) gives insights into the effect of the parameter variation. The equilibrium values for the nitrate, nitrite, and oxide ions of the experiment from Section 3.3. (Figure 3) at 620 °C (20% O_2, 200 ppm NO) are shown as dashed blue lines for comparison. At the start of the experiment, the oxide formation rates of NO purged experiments under atmospheric pressure (Figure 5, green and purple color; *t*: 0–75 h) are fast and close to those of the experiment purged with 20% oxygen (blue color). Hereafter, the rates decrease and are approximated with the linear regression fit. Interestingly, the pressurized experiment (orange color) did not show the same behavior but remained low and constant over 1000 h. In this case, the final oxide composition is close to the equilibrium composition of the reference (dotted blue line), which could indicate that the experiment was not long enough to reach a stable oxide ion composition.

The current data highlight the importance of proper gas phase management for the stabilization of solar salt at 620 °C. As shown in Figure 5, in addition to the temperature, the mixing velocity, the present pressure, and the specific overlaying gas, the atmosphere is crucial for the equilibration process. Thereby, the dominant factor for the oxide ion formation rate is the presence of nitrous gases rather than mixing velocity. Pressurizing the test rig seems to have had a positive effect on the stabilization of the molten salt because the oxide ion formation was immediately reduced to a low rate. These results could be explained by the equilibrium in Equation (2). Here, the oxide ion formation correlates with the emission of NO gas molecules, which in turn is suppressed if an external pressure, e.g., 1 bar overpressure, is applied, but further studies are necessary for final proof. Additionally, oxide formation rates with the same NO partial pressure are quite similar (Figure 5, bottom right: 6×10^{-6} mol%/h) and are not affected by the nitrite ion concentration. This finding agrees with the result of the previously mentioned study [28], where the oxide formation rate did not correlate with the nitrite concentration but with the temperature. In addition, in a similar way, the oxide ion equilibrium levels show a weak impact on the nitrite concentration (Figure 4). As already mentioned, it seems likely that reactions with oxide species and NO mainly take place at the liquid gas interface and should correlate with the surface to bulk ratio. Due to the 100 g scale of the experiments, the surface to bulk ratio is larger compared to a TES tank and even increases during the experiments from 12% to 25% due to sample extraction. Still, no significant effect of this change on the salt composition was found, probably because the mixing of the salt was sufficient to prevent any mass transport limitations. However, without mixing (Figure 5, purple color), oxide ion concentration increased for over 3000 h, which could indicate the presence of a mass transport limitation in the regeneration of oxide ions with nitrous gases. Especially at a larger scale, where the surface to bulk ratio becomes orders of magnitude smaller (e.g., in a TES tank), such mass transport effects could have an effect on the salt's stability.

3.5. Thermodynamic Evaluation

In the previous study [28], the equilibrium values of nitrate and nitrite at temperatures above 600 °C were postulated to be afflicted with a significant error because of oxide ion formation. It was mentioned that solar salt purged with NO gas counters this effect by the regeneration of decomposition products in the molten salt. To compare the thermodynamic data of different studies, a van't Hoff diagram is typically used, which consists of the natural logarithm of the equilibrium constant (K_1) plotted against the inverse of the temperature. According to Equation (7), the slope of a linear regression of the temperature-dependent equilibrium constants allows for the extraction of Gibb's free energy (ΔG_R).

$$\ln K_1 = -\frac{\Delta G_R}{RT}; \text{with } K_1 = \frac{[NO_2^-]}{[NO_3^-]} P_{O_2}^{0.5} \quad (7)$$

Figure 6 gives the nitrate–nitrite equilibrium data of several studies in comparison to the new set of data of this study (see Table 1) for 600 °C (0.00115 K^{-1}) to 620 °C (0.00112 K^{-1}). For completeness, all generated data are summarized in Table 1. For experiments where an oxide ion equilibrium was attained, the nitrite to oxide ion ratio is given. Further studies may relate this ratio to the oxide equilibrium constant (K_2) and NO_x partial pressure, according to Equation (8).

$$K_2 \sim \frac{[O^{2-}]}{[NO_2^-]^y} P_{NO_x}^y \quad (8)$$

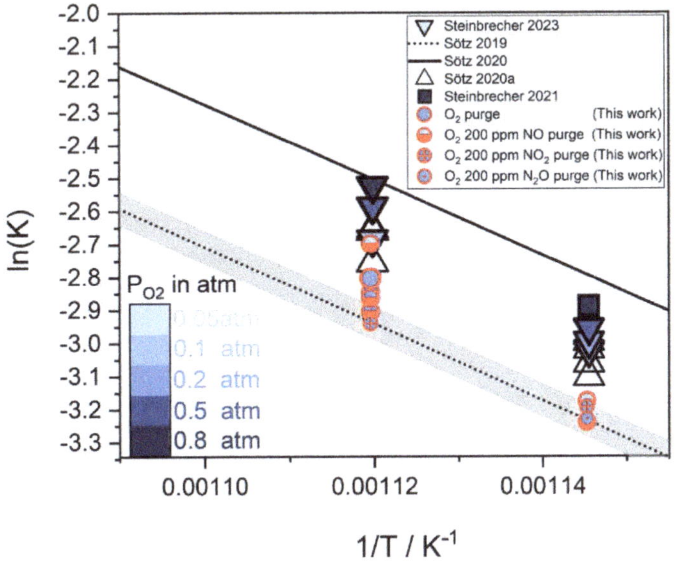

Figure 6. van't Hoff plots of nitrate and nitrite equilibrium individual data points (column 4, Table 1) obtained in this work and shown in Figures 1–4, as well as literature data from Steinbrecher et al. 2021, 2023and Sötz et al. 2010, 2020, 2020a for comparison. Color code indicates oxygen partial pressure of the individual experiments. Reprinted/adapted with permission from Ref. [27] 2021, MDPI, [12] 2019, Elsevier B.V., [32] 2020 Elsevier B.V., [26], 2020, International Solar Energy Society, [28] 2023, Elsevier B.V.)

The rate of oxide formation for experiments without stable anion content over time is given in mol% per h.

The solid line in Figure 6 is reprinted from the literature data of Sötz et al. 2020 [32] and presents an upper boundary for the nitrate–nitrite equilibrium composition. This study was performed with 1 g scale sample volumes and uncontrolled salt decomposition, or more precisely, strong oxide ion formation. In a recent study [28], it was shown that above 600 °C, also at a 100 g scale, oxide ion formation leads to a shift of the equilibrium position to the nitrite side and higher values of ln(K_1) (Equation (7)). The dotted line is extrapolated from low-temperature data from another study by Sötz et al. 2019 [12] without measurable amounts of oxide ions. This line is interpreted as the ideal nitrate–nitrite equilibrium composition, without the influence of oxide ion formation and without mass transport effects (at a mg sample scale). The entire dataset presented herewith (red symbols), which was performed in an experimental setup with a reduced corrosion impact, now, indeed, shows a significant shift towards lower values of $\ln(K_1)$ compared to the published data with steel components (black symbols). Consequently, the present paper confirms that solar salt is more stable than previous studies suggested, and the reported reaction enthalpy ($\Delta H = 95$ kJ/mol) and entropy ($\Delta S = 85$ J/molK) published by Sötz [12] are also accurate for 600 and 620 °C. Another verification of this concept is given with the experiment performed at 620 °C and 20% oxygen (without NO). Here, the $\ln(K_1)$ value is the second highest (−2.8030) of the study because oxide ion formation occurred unopposed. The experiment with the highest $\ln(K_1)$ value of −2.6988 was performed without mixing at 5% oxygen and 200 ppm NO. Even though oxide formation was reduced due to the low concentration of oxygen and the eliminated stirrer, nitrite regeneration, the reverse direction of Equation (1), was probably limited by oxygen mass transport, which is known to have an impact on the reaction rate [11]. This result supports the theory that a molten solar salt system that suffers either from mass transport issues, oxide formation, or both will give $\ln(K_1)$ values above the ideal equilibrium line (Figure 6, dotted black line).

The investigation carried out here leads to at least five major factors that determine the stability of solar salt at temperatures beyond 600 °C:

- Corrosion reactions lead to increased salt decomposition and reduce solar salt stability.
- Nitrogen gas has a subordinate role in the stabilization of solar salt.
- Nitrous gases' impact on salt stability increases from N_2O to NO to NO_2.
- At 600 °C, solar salt is stable over time when in contact with 200 ppm NO and an oxygen partial pressure of at least 0.05 atm.
- At 620 °C, the surface to volume ratio seems to become of increasing importance, and for stabilization, a higher concentration of NO_x gas is advisable.

The presented data are highly relevant for large-scale TES units as well as for any corrosion study performed at temperatures where oxide formation can be expected (e.g., 600 °C). Figure 6 can serve as a stability indicator for solar salt. More precisely, the nitrate–nitrite equilibrium data of this study are understood as the benchmark for a stable salt composition. It was shown that any deviation towards larger values of $\ln(K_1)$ indicates the presence of congruent reactions. With this information, simple monitoring of the anion composition will give deep insight into the long-term behavior (e.g., stability, corrosivity) of any solar salt-based system.

4. Conclusions

One goal of this study was to obtain molten solar salt with stable oxide ion concentrations at 620 °C. By the successful practical exclusion of corrosion side reactions with metallic components, a solar salt equilibrium, in terms of a stable nitrite and oxide ion composition, was attained compared to previously published work. In addition, the fraction of the decomposition product in the equilibrium state at 620 °C was reduced by 16%, proving solar salt to be more stable than previously anticipated. Nitrogen gas was identified as having a subordinate role in the stabilization of molten salt and seemingly not taking part in a regeneration reaction. Concerning the different nitrous gas species, their positive impact on salt stability was found to increase in the order of N_2O to NO to NO_2. Mass transport limitations were not proven with either an increased mixing velocity or excess pressure. However, for static molten salt, indications were found that, at a larger scale, where the surface to bulk ratio becomes smaller, this effect might come into play. This is because regeneration reactions are thought to primarily occur at the liquid–gas interface, whereas decomposition is considered a bulk reaction. In summary, the combined set of data is significant for use in the development of high-temperature TES systems and molten salt heat transfer applications, where parameters such as surface to bulk ratio and mass transport change substantially. With this study, we deliver the first extended set of solar salt equilibrium data that are inevitably required in the future to understand and quantify the corrosiveness of molten salt under certain atmospheric conditions. With this knowledge, the operation of a solar salt-based TES unit at temperatures above 600 °C is realizable if the salt and gas phase compositions are monitored and controlled carefully.

Supplementary Materials: The following supporting information can be downloaded at: https://www.mdpi.com/article/10.3390/en16145241/s1, Figure S1: Blank experiment in the autoclave test rig at 620 °C without molten salt, purged with 20% O_2 and different concentrations of NO_2 (left) and NO (right). NO_x (cNO + cNO_2) concentration measured from the exhaust gas versus the input gas concentration; Figure S2: Concentration over time of NO and NO_2 in the exhaust gas of the purged reaction chamber containing molten solar salt. Experiment purged with 200 ppm NO_2, 20% O_2 (left) and 100% N_2 purged for the first 100 h, subsequently switched to purge gas with 200 ppm NO, 20% O_2 (100–1000 h; right).

Author Contributions: Conceptualization, J.S. and A.B.; Funding acquisition, A.B. and T.B.; Investigation, J.S. and M.B.; Methodology, J.S., M.B., and S.K.; Project administration, A.B. and T.B.; Resources, J.S. and M.B.; Supervision, J.S., M.B., and A.B.; Validation, J.S., S.K., and A.B.; Visualization, J.S.; Writing—original draft, J.S.; Writing—review and editing, A.B., S.K., and T.B. All authors have read and agreed to the published version of the manuscript.

Funding: This work was funded by the German Ministry of Economic Affairs and Climate Action (Contract No. 0325497 A).

Data Availability Statement: The data presented in this study are available on request from the corresponding author.

Acknowledgments: We kindly thank Andrea Hanke for laboratory work and great expertise in experimental methods.

Conflicts of Interest: The authors declare no conflict of interest. The funders had no role in the design of the study; in the collection, analyses, or interpretation of data; in the writing of the manuscript; or in the decision to publish the results.

References

1. Bauer, T.; Odenthal, C.; Bonk, A. Molten Salt Storage for Power Generation. *Chem. Ing. Tech.* **2021**, *93*, 534–546. [CrossRef]
2. Tracey, T.R. *Conceptual Design of Advanced Central Receiver Power System*; Final Report; Martin Marietta Corp.: Denver, CO, USA, 1978.
3. Pacheco, J.E.; Bradshaw, R.W.; Dawson, D.B.; De la Rosa, W.; Gilbert, R.; Goods, S.H. *Final Test and Evaluation Results from the Solar Two Project*; SAND2002-0120; Sandia National Laboratories (SNL-NM): Albuquerque, NM, USA; Sandia National Laboratories (SNL-CA): Livermore, CA, USA, 2002.
4. Bauer, T.; Pfleger, N.; Laing, D.; Steinmann, W.-D.; Eck, M.; Kaesche, S. 20—High-Temperature Molten Salts for Solar Power Application. In *Molten Salts Chemistry*; Lantelme, F., Groult, H., Eds.; Elsevier: Oxford, UK, 2013; pp. 415–438.
5. Durth, M.; Prieto, C.; Rodriguez-Sanchez, A.; Patino-Rodriguez, D.; Cabeza, L.F. Effects of sodium nitrate concentration on thermophysical properties of solar salts and on the thermal energy storage cost. *Sol. Energy* **2019**, *182*, 57–63. [CrossRef]
6. Yang, C.; Wei, X.; Wang, W.; Lin, Z.; Ding, J.; Wang, Y.; Yang, J. NO_x emissions and the component changes of ternary molten nitrate salts in thermal energy storage process. *Appl. Energy* **2016**, *184*, 346–352. [CrossRef]
7. Wei, X.; Yang, C.; Lu, J.; Wang, W.; Ding, J. The mechanism of NO_x emissions from binary molten nitrate salts contacting nickel base alloy in thermal energy storage process. *Appl. Energy* **2017**, *207*, 265–273. [CrossRef]
8. Wei, X.L.; Wang, Y.; Peng, Q.; Yang, J.P.; Yang, X.X.; Ding, J. NO_x emissions and NO_2^- formation in thermal energy storage process of binary molten nitrate salts. *Energy* **2014**, *74*, 215–221. [CrossRef]
9. Wright, S.; Tran, T.; Chen, C.; Olivares, R.I.; Sun, S. Thermal stability of potassium and sodium nitrate molten salt mixtures above 500 °C. In Proceedings of the Ninth International Conference on Molten Slags, Fluxes and Salts, Beijing, China, 27–30 May 2012.
10. Bradshaw, R.W.; Siegel, N.P. Molten nitrate salt development for thermal energy storage in parabolic trough solar power systems. In Proceedings of the ASME 2008 2nd International Conference on Energy Sustainability Collocated with the Heat Transfer, Fluids Engineering, and 3rd Energy Nanotechnology Conferences, American Society of Mechanical Engineers, Jacksonville, FL, USA, 10–14 August 2008; pp. 631–637.
11. Nissen, D.A.; Meeker, D.E. Nitrate/nitrite chemistry in sodium nitrate-potassium nitrate melts. *Inorg. Chem.* **1983**, *22*, 716–721. [CrossRef]
12. Sötz, V.A.; Bonk, A.; Forstner, J.; Bauer, T. Microkinetics of the reaction $NO_3^- \rightleftharpoons NO_2^- + 0.5\ O_2$ in molten sodium nitrate and potassium nitrate salt. *Thermochim. Acta* **2019**, *678*, 178301. [CrossRef]

13. Abe, O.; Utsunomiya, T.; Hoshino, Y. Acid—Base and redox reactions in fused sodium nitrite. *Thermochim. Acta* **1984**, *74*, 131–141. [CrossRef]
14. Stern, K.H. High temperature properties and decomposition of inorganic salts part 3, nitrates and nitrites. *J. Phys. Chem. Ref. Data* **1972**, *1*, 747–772. [CrossRef]
15. Xianwei, H.; Yue, L.; Zongxin, Y.; Bingliang, G.; Zhongning, S.; Zhaowen, W. Thermal stability of sodium nitrate–sodium nitrite melts: A Raman spectra study. *Spectrosc. Lett.* **2018**, *51*, 350–355. [CrossRef]
16. Laue, W.; Thiemann, M.; Scheibler, E.; Wiegand, K.W. Nitrates and nitrites. In *Ullmann's Encyclopedia of Industrial Chemistry*; John Wiley & Sons, Inc.: Hoboken, NJ, USA, 2000.
17. Tsukahara, H.; Ishida, T.; Mayumi, M. Gas-phase oxidation of nitric oxide: Chemical kinetics and rate constant. *Nitric Oxide* **1999**, *3*, 191–198. [CrossRef]
18. Olivares, R.I. The thermal stability of molten nitrite/nitrates salt for solar thermal energy storage in different atmospheres. *Sol. Energy* **2012**, *86*, 2576–2583. [CrossRef]
19. Villada, C.; Bonk, A.; Bauer, T.; Bolivar, F. High-temperature stability of nitrate/nitrite molten salt mixtures under different atmospheres. *Appl. Energy* **2018**, *226*, 107–115. [CrossRef]
20. Bauer, T.; Pfleger, N.; Breidenbach, N.; Eck, M.; Laing, D.; Kaesche, S. Material aspects of Solar Salt for sensible heat storage. *Appl. Energy* **2013**, *111*, 1114–1119. [CrossRef]
21. Bonk, A.; Braun, M.; Hanke, A.; Forstner, J.; Rückle, D.; Kaesche, S.; Sötz, V.A.; Bauer, T. *Influence of Different Atmospheres on Molten Salt Chemistry and Its Effect on Steel Corrosion*; SolarPACES; AIP Publishing: Santigo, Chile, 2017; p. 090003.
22. Summers, K.; Chidambaram, D. Corrosion in Molten Salts for Solar Thermal Power. *Electrochem. Soc. Interface* **2021**, *30*, 63–66. [CrossRef]
23. Fernandez, A.G.; Cabeza, L.F. Molten salt corrosion mechanisms of nitrate based thermal energy storage materials for concentrated solar power plants: A review. *Sol. Energy Mater. Sol. Cells* **2019**, *194*, 160–165. [CrossRef]
24. Federsel, K.; Wortmann, J.; Ladenberger, M. High-temperature and corrosion behavior of nitrate nitrite molten salt mixtures regarding their application in concentrating solar power plants. *Energy Procedia* **2015**, *69*, 618–625. [CrossRef]
25. Bonk, A.; Braun, M.; Sotz, V.A.; Bauer, T. Solar Salt—Pushing an old material for energy storage to a new limit. *Appl. Energy* **2020**, *262*, 114535. [CrossRef]
26. Sötz, V.A.; Bonk, A.; Steinbrecher, J.; Bauer, T. Defined purge gas composition stabilizes molten nitrate salt—Experimental prove and thermodynamic calculations. *Sol. Energy* **2020**, *211*, 453–462. [CrossRef]
27. Steinbrecher, J.; Bonk, A.; Sötz, V.A.; Bauer, T. Investigation of Regeneration Mechanisms of Aged Solar Salt. *Materials* **2021**, *14*, 5664. [CrossRef] [PubMed]
28. Steinbrecher, J.; Hanke, A.; Braun, M.; Bauer, T.; Bonk, A. Stabilization of Solar Salt at 650 °C—Thermodynamics and practical implications for thermal energy storage systems. *Sol. Energy Mater. Sol. Cells* **2023**, *258*, 112411. [CrossRef]
29. Bonk, A.; Rückle, D.; Kaesche, S.; Braun, M.; Bauer, T. Impact of Solar Salt aging on corrosion of martensitic and austenitic steel for concentrating solar power plants. *Sol. Energy Mater. Sol. Cells* **2019**, *203*, 110162. [CrossRef]
30. White, S.H.; Twardoch, U.M. *Study of the Interactions of Molten Sodium Nitrate-Potassium Nitrate 50 mol% Mixture with Water Vapor and Carbon Dioxide in Air*; Final report; Sandia National Labs.: Livermore, CA, USA; EIC Labs., Inc.: Newton, MA, USA, 1981.
31. Tomkins, R.P.T.; Bansal, N.P. (Eds.) *Gases in Molten Salts*; Elsevier: Morristown, NJ, USA, 1991.
32. Sötz, V.A.; Bonk, A.; Bauer, T. With a view to elevated operating temperatures in thermal energy storage—Reaction chemistry of Solar Salt up to 630 °C. *Sol. Energy Mater. Sol. Cells* **2020**, *212*, 110577. [CrossRef]

energies

Article

Thermal Management of Short-Range Distribution of Perishable Food Products Using Phase Change Materials in Packaging: Real-Time Field Data Acquisition

Martim Aguiar [1,2], Pedro Dinis Gaspar [1,2] and Pedro Dinho da Silva [1,2,]*

1 Rua Marquês d'Ávila e Bolama, Department of Electroechanical Engineering, Faculty of Engineering, University of Beira Interior, 6201-001 Covilha, Portugal; martim.aguiar@ubi.pt (M.A.); dinis@ubi.pt (P.D.G.)
2 C-MAST—Centre for Mechanical and Aerospace Science and Technologies, 6201-001 Covilha, Portugal
* Correspondence: dinho@ubi.pt

Abstract: Maintaining a stable temperature is critical in ensuring the longevity of perishable foods, and frequent fluctuations due to short-range distribution conditions can negatively affect this stability. To mitigate these variations, an innovative modular packaging system utilizing phase change materials (PCMs) was employed in the transport and storage of horticultural products. This study's real-time thermal condition data, collected using a wireless data acquisition system inserted in the packaging, demonstrated the efficacy of PCM in increasing temperature stability within the crates of horticultural products. The field tests conducted over 8 h showed that PCM-equipped packaging boxes exhibited a temperature variation of less than 1 °C, compared to non-PCM boxes, which saw variations up to 3 °C. This marked reduction in temperature fluctuation signifies the potential of PCM in improving thermal and logistics management in food conservation, thus reducing food waste. However, it is essential to implement a system for PCM alveoli reuse to avoid adverse environmental impacts. Future research should focus on the PCM alveoli autonomy and quantity requirements for specific conditions, and integrate sensors to monitor transport dynamics to enhance the understanding of temperature stability in perishable food transportation.

Keywords: phase change materials; PCM; passive temperature control; temperature fluctuation; perishable food products; food preservation

Citation: Aguiar, M.; Gaspar, P.D.; da Silva, P.D. Thermal Management of Short-Range Distribution of Perishable Food Products Using Phase Change Materials in Packaging: Real-Time Field Data Acquisition. *Energies* **2023**, *16*, 5191. https://doi.org/10.3390/en16135191

Academic Editor: Ann Lee

Received: 9 June 2023
Revised: 26 June 2023
Accepted: 29 June 2023
Published: 5 July 2023

1. Introduction

Addressing the escalating food demands of an increasing global population is a challenge of unparalleled importance. As per the Food and Agriculture Organization [1], the world's population is projected to hit 9.7 billion by 2050. Consequently, the need for a sustainable and secure food supply is more critical than ever. Regrettably, nearly a third of the food produced worldwide for human consumption, approximately 1.3 billion tons annually, is wasted [2]. These wastage levels lead to serious financial and environmental repercussions. They contribute to climate change through notable greenhouse gas emissions during food production and distribution. Additionally, the decay of food loss and waste releases methane, further worsening the situation [3–9]. Beyond these environmental implications, food wastage undermines food security, as it signifies inefficient use of resources dedicated to food production. Therefore, a reduction in food waste is a vital goal in our pursuit of a sustainable and equitable food system [10].

Refrigeration is fundamental for food preservation. It reduces the growth of microorganisms and enzymes, lengthening the shelf life of perishable products, ultimately reducing food waste and sustaining food safety [11]. Thus, efficient storage and transportation of food at regulated temperatures are key in minimizing food losses and promoting food security [10].

Prior studies indicate that temperature fluctuations during food transportation [8] and storage [12–14] can be detrimental to food quality and shelf life. These fluctuations, caused by a host of factors, including ambient temperature changes, subpar refrigeration, and operational procedures, such as door openings, need to be controlled effectively to maintain food quality. Despite some of these factors being unavoidable, our research proposes a means to minimize their impact on temperature stability.

Thermal energy storage is primarily categorized into Sensible Heat Storage (SHS), Latent Heat Storage (LHS), and Thermochemical Storage. While SHS is the most basic form of energy storage, relying on the material's specific heat capacity and density, it offers limited energy storage. LHS, conversely, leverages phase transitions (solid to liquid, liquid to gas, and solid to solid) for energy storage, resulting in a higher storage density at specific temperatures. This is where phase change materials (PCMs) come into play. Thermochemical storage, although having the highest energy storage capacity, grapples with long-term stability issues in controlled environments and has a complex and costly development process [15].

PCMs have garnered considerable attention in the field of refrigeration systems, particularly in the realm of preserving temperature stability [16,17]. This interest is primarily owed to their remarkable latent heat capacity, which can be leveraged to maintain consistent temperatures and, hence, enhance the preservation of food during transportation [18]. The potential of PCMs in curbing food waste and securing food quality has been rigorously studied and is increasingly acknowledged as a promising solution [19].

As shown in Figure 1, PCMs can be broadly segregated into two main types: solid–solid and solid–liquid, each exhibiting distinct thermodynamic properties. The former includes materials such as polymers and polyalcohols, which undergo a phase transition between different solid states. While this type of PCM has its applications, the solid–liquid class is most frequently employed due to its significantly higher latent heat of fusion. This facilitates greater energy storage and release during the phase transition process [15].

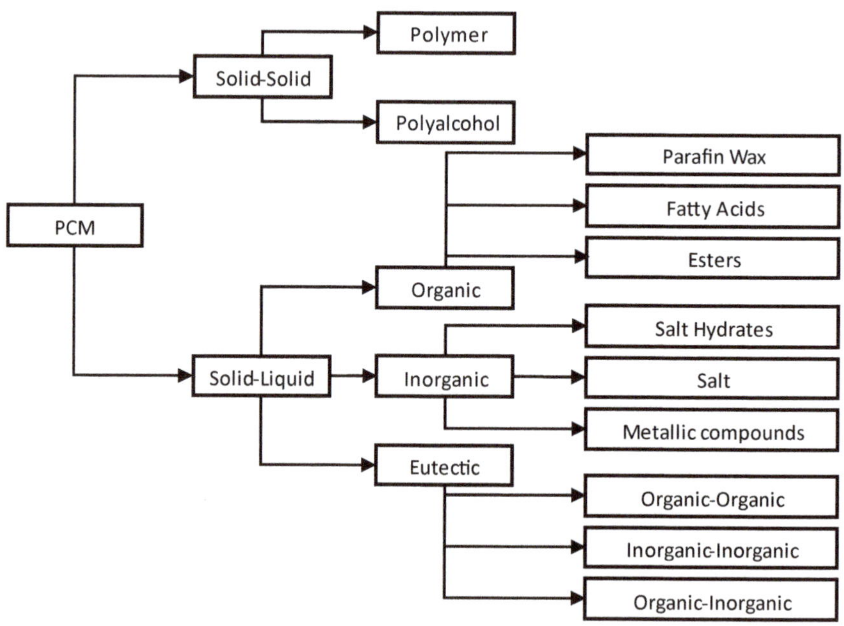

Figure 1. Schematic representation of PCM categorization, adapted from [15].

Solid–liquid PCMs can be further bifurcated into organic, inorganic, and eutectic categories, each with unique characteristics and applications [20,21]. Organic PCMs encompass substances such as paraffin waxes, fatty acids, and esters. Inorganic PCMs, on the other hand, primarily include salt hydrates, salts, and metallic compounds. Lastly, eutectic PCMs are mixtures that form a unique composition that melts and solidifies at a single temperature and can be comprised of organic–organic, inorganic–inorganic, or organic–inorganic combinations. This diversification of PCM categories allows for a wide range of potential applications, tailoring their usage according to the specific thermal management requirements. Several studies have explored the thermal performance of various fruit packaging crates [22–26]. Studies also analyzed the utility of PCM pouches for active and intelligent packaging [27,28], and the measurement of temperatures and relative humidity during the transportation of horticultural goods [8].

In [28], the authors deployed a transient three-dimensional computational fluid dynamics model to examine the application of PCM in packaging alveoli for fruit storage. This study showed that such alveoli can effectively slow the heating rate of fruits when exposed to non-ideal temperatures. The PCM, acting as a "thermal buffer", was found to increase the shelf life of the produce by maintaining optimal temperatures over extended periods.

Meanwhile, reference [24] provides experimental evidence of the thermal response of two types of food alveoli, comparing their performance with and without the integration of PCM. The authors demonstrated that aluminum foil alveoli, in conjunction with PCM, delayed the heating process effectively. These findings imply that PCM could be a promising solution for the preservation of cold storage food products in packaging.

In [29], researchers undertook an in-depth investigation of the use of PCM panels in a freezer subjected to repeated power loss over a 2-week period. The findings revealed that freezers equipped with PCM panels helped to maintain product temperatures closer to the optimal range, even during power loss incidents, leading to minimal impacts on food quality. This research not only underscores the potential of PCM to enhance the preservation of frozen food during power outages, but also evidences the potential for PCM to enhance overall food quality in regular storage conditions.

These studies underline the potential advantages of integrating PCM into food packaging for the preservation of perishable goods, making it a topic worth further exploration in real-world conditions.

This study aims to apply insights from previous works to real-world situations, evaluating the effectiveness of PCM in maintaining temperature stability for horticultural products during transportation. The transition from controlled environments to real-life scenarios is crucial, as it tests novel technologies under practical conditions, revealing challenges otherwise unseen. Specifically, for PCM, studying its performance in real-world transportation provides evidence of its efficiency, fostering its broader adoption in food preservation and helping mitigate food waste.

2. Materials and Methods

The study was conducted in the context of a distribution route originating from ALBIFRUTAS, a company located in Castelo Branco, Portugal. The focus was to evaluate the impact of PCM on maintaining temperature stability during the transportation of horticultural products. The experimental setup involved crates of oranges, wherein half of the crates were equipped with PCM-filled alveoli, and the rest were loaded with empty alveoli, serving as controls.

2.1. PCM Alveoli and Crates

The crates used for the study boasted an innovative design with enhanced thermal attributes, enabling improved temperature stability. Their modularity facilitated efficient transportation, storage, and public sale display [22–26]. Fabricated from polypropylene, each crate measured 600 mm in width, 400 mm in depth, and 90 mm in height. Figure 2 provides a visual representation of the crates: Figure 2a illustrates an assembled empty crate; Figure 2b shows the crate equipped with PCM alveoli; Figure 2c displays the crate with PCM alveoli and an attached lid.

(**a**) (**b**) (**c**)

Figure 2. Crate (**a**), with PCM alveoli (**b**), and lid (**c**).

The main novelty of these crates consists of the PCM alveoli, depicted in white in Figure 2b, comprising a polypropylene macro-encapsulation of the RT5HC PCM. The RT5HC PCM is an organic material, recognized for its high latent heat capacity and tight temperature range, making it ideal for enhancing temperature stability during horticultural goods' transportation. The key properties of the RT5HC PCM are presented in Table 1.

Table 1. Properties of the RT5HC PCM [30].

Property	Value
Melting temperature amplitude [°C]	5 to 6
Congealing temperature amplitude [°C]	6 to 5
Heat storage capacity ±7.5% [kJ/kg]	250
Latent and sensible heat in a temperature range of −2 °C to 13 °C [Wh/kg]	70
Density solid [kg/L]	0.88
Density liquid [kg/L]	0.76
Heat conductivity (both phases) [W/(m·K)]	0.2
Volume expansion [%]	13
Flash point [°C]	115
Max. operation temperature [°C]	30

As shown in Figure 3, the PCM alveoli were sized to fit the crate's bottom, enhancing temperature stability during the transportation of horticultural goods. In this experiment, half the crates included alveoli filled with the RT5HC PCM, while the remaining crates held empty alveoli for control comparison.

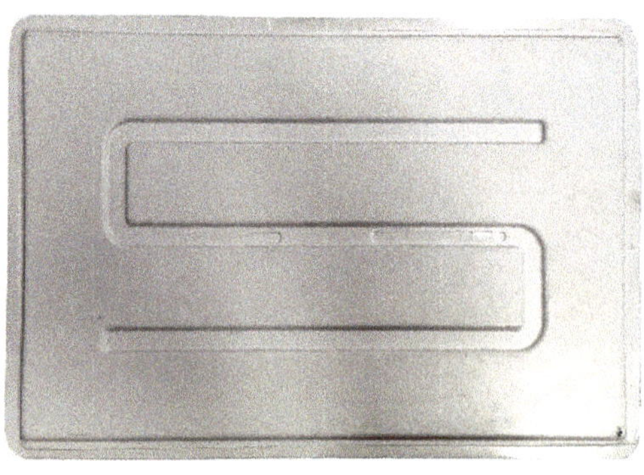

Figure 3. PCM alveoli.

2.2. Cargo Distribution Characterization

The investigation involved monitoring crates filled with oranges, as depicted in Figure 4. The crates, having been stored at approximately 5 °C overnight, were transported to the ALBIFRUTAS warehouse. Here, they were placed among the cargo prepared for real-time distribution.

Figure 4. Crate outfitted with PCM alveoli, loaded with oranges, and equipped with sensors for data acquisition.

As illustrated in Figure 5, the crates were strategically positioned on the van's floor, adjacent to the right wall and near the rear door. Each of the two sets of crates, one with PCM and the other serving as a control without PCM, were equally loaded with 48 oranges.

Figure 5. Placement of the experimental and control crates within the overall cargo configuration.

The cargo comprised a variety of horticultural products—fruits and vegetables—destined for a range of establishments, such as restaurants, hotels, and local grocery stores. Figure 6 provides a visual representation of the diverse cargo arrangement. Notably, the cargo was grouped based on client orders rather than produce type.

Figure 6. Overall arrangement of the diverse horticultural produce within the van's cargo area.

2.3. *Distribution Vehicle and Refrigeration System*

Examining the impact of common routines, such as the opening of doors for unloading and the effect of sun exposure during transportation, is crucial for understanding the temperature behavior within the crates. Thus, data can be gathered for the field performance of the PCM alveoli. This information is fundamental in assessing the packaging's effectiveness in maintaining temperature stability, thus preserving the quality of the produce during transportation and storage.

Data were gathered during a delivery run conducted by an IVECO Daily van, shown in Figure 7, which belongs to the ALBIFRUTAS distribution company. The van was tasked with the delivery of a wide variety of horticultural products to multiple establishments, including restaurants and local grocery stores.

Figure 7. An IVECO Daily van from the ALBIFRUTAS fleet.

The vehicles are equipped with a Hwasung HT-250RT-ESC refrigeration system. However, as the measurements were conducted during the winter, the refrigeration system was turned off. To minimize thermal losses and enhance temperature control under varying external conditions, the vans are also fitted with insulation.

The delivery of the produce took place between 8:00 h and 14:05 h. The route, illustrated in Figure 8, traversed the northwestern region of Castelo Branco, Portugal.

Along this route, there were six stops, during which the doors were opened for unloading cargo. This routine resulted in temperature fluctuations, as observed in the Section 3. The weather in Castelo Branco during this period was sunny with no cloud cover, with temperatures ranging from 1 °C (around 8:00 h) to 10 °C (around 14:00 h). The sun exposure, coupled with the refrigeration system being turned off, created ideal conditions for testing the potential of PCM as a solution to mitigate the impacts of unforeseen conditions that are typical in real-world scenarios.

Figure 8. Delivery run route.

2.4. *Data Acquisition System and Final Test Assembly*

The primary goal of this study is to emphasize the role of PCM in maintaining temperature stability during transportation. To track temperature changes and systematically interpret the collected data, a data measurement and transmission system was established. The study will compare temperature fluctuations in crates with and without PCM, thereby demonstrating the effectiveness of PCM in maintaining temperature stability.

2.4.1. Sensing Modules

The study measures the temperature and relative humidity around different horticultural products during their transportation journey. Other approaches for agricultural and agrifood parameters monitoring were followed in the past [31]. For this purpose, a SHT30 sensor, shown in Figure 9a, capable of reading the air temperature with a precision of ± 0.3 °C, is utilized. The sensor is connected to a LOLIN D1 mini pro V2.0 microcontroller, shown in Figure 9b, through a LOLIN SHT30 v2.1.0 shield. The system is powered by a 3.7 V 500 mAh lithium battery, as shown in Figure 9c. This setup is assembled as shown in Figure 9d and housed in a 3D-printed protective box showcased in Figure 9e. The sensor box is designed with numerous perforations, ensuring adequate ventilation. This feature facilitates a direct interaction between the sensors and the ambient air, thereby enabling precise and accurate measurement of the atmospheric conditions. More details can be found in [32].

The sensors are numbered from 1 to 20 to ensure easy identification, as shown in Figure 10. Alongside this, the sensors are also labeled with a smaller number. That is the sensor reference used in the data communication system.

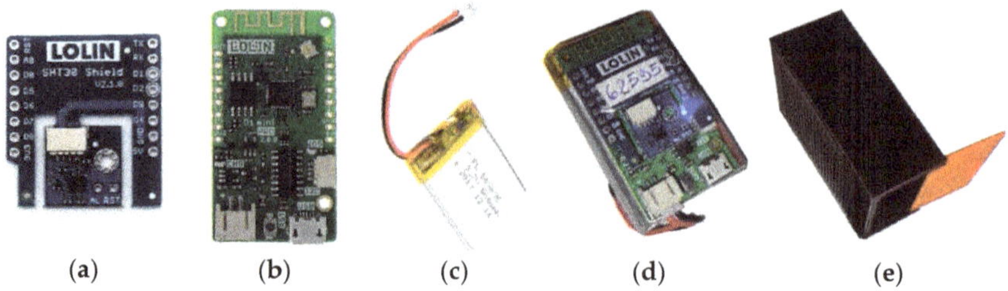

Figure 9. Sensing Module Components: Temperature and humidity sensor (**a**), microcontroller (**b**), battery (**c**), assembled sensing module (**d**), and sensing module box (**e**).

Figure 10. Sensor placement among produce.

2.4.2. Data Communication System

The sensing modules are connected to a gateway, shown in Figure 11a. This gateway includes an Industruino microcontroller and a GSM/GPRS module, enabling real-time data transmission with GPS location tracking. The readings are collected and transmitted every five minutes and can be accessed live through the iTrack platform, specifically designed for this purpose. The overall system schematics are illustrated in Figure 11b.

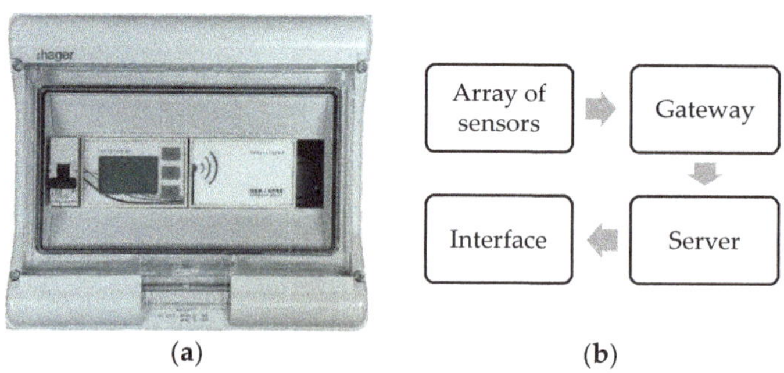

Figure 11. (**a**) Gateway module and (**b**) System diagram.

2.4.3. Temperature Monitoring and Data Communication System

The sensing modules' proximity to the horticultural products ensures accurate temperature readings, providing a more precise representation of the conditions during the distribution journey. For a clearer understanding of the results, Figure 12 illustrates the testing setup and the positioning of crates and sensing modules during the experiments.

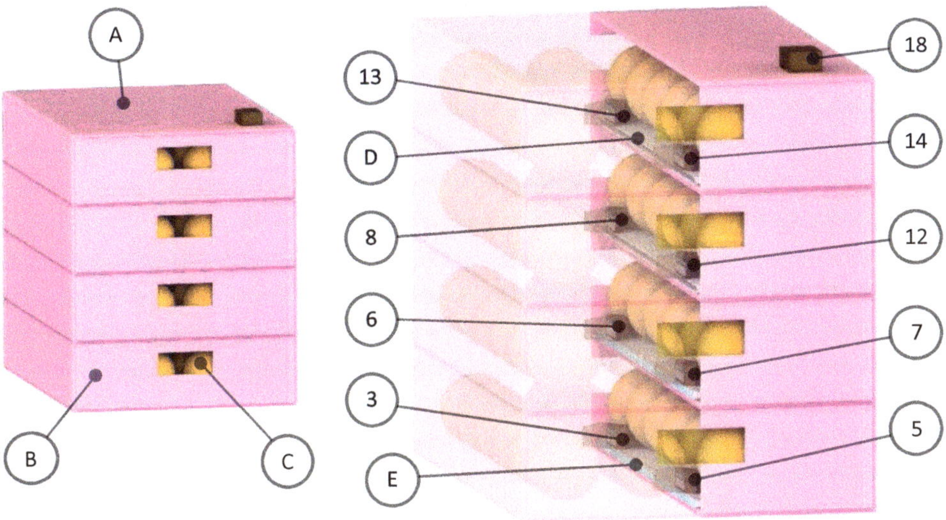

Figure 12. Arrangement of the crates, sensing modules, alveoli, and horticultural products as explained in Table 2.

Table 2. Elements of the testing arrangement.

Component	Description
A	Crate lid
B	Crate
C	Orange
D	Empty alveoli
E	PCM alveoli
3, 5, 6, 7, 8, 12, 13, 14, 18	Sensing module

The components displayed in Figure 12 are detailed in Table 2. The sensor reference number corresponds to the sensor ID in the iTrack platform.

Sensor positions within the crate are outlined in Table 3.

Table 3. Sensor position within the crates.

	Rear Side	Front Side
Crates with PCM alveoli	3 6	5 7
Crates with empty alveoli	8 13	12 14
Ambient temperature (cargo area)		18

The sensor placement within the crates is vital for results analysis since the rear sensors are more exposed when the rear doors are open. Furthermore, sensors near PCM alveoli or empty alveoli are expected to have different temperature readings. The comparison of ambient temperature with the measurements within the crates underscores the importance of monitoring temperature within produce crates for precise monitoring of food decay. The lid on the top crate aids in ensuring a more uniform temperature is measured across the column of crates.

3. Results

Crates equipped with PCM alveoli were observed to maintain more consistent temperatures during transportation than those without. Temperature spikes were noted during unloading events. Figure 13a,b compare temperature measurements between crates with PCM alveoli and those with empty alveoli. The graphs demonstrate temperature stability in crates equipped with PCM alveoli during transportation.

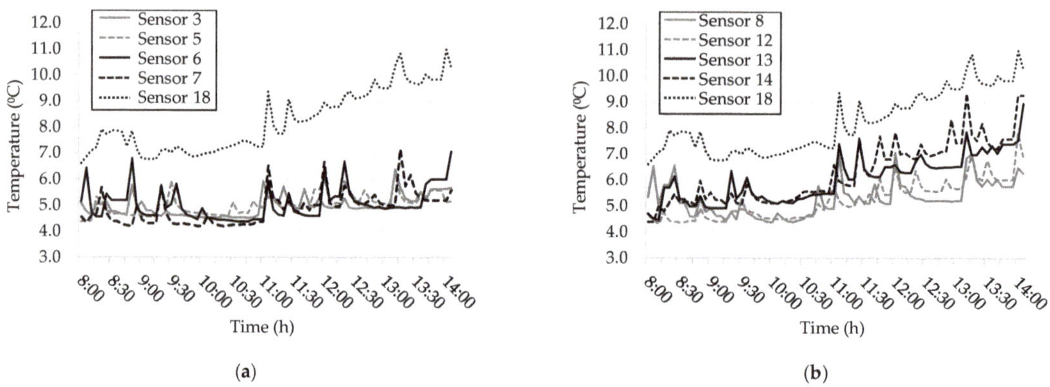

Figure 13. Temperature readings in crates containing alveoli with (**a**) and without (**b**) PCM.

To analyze the stability of temperatures between crates with PCM and empty alveoli, the averages of all temperature readings in the PCM alveoli and empty alveoli were plotted. A four-measurement simple moving average was applied to mitigate the visual impact of temperature spikes during unloading, allowing for an easier overall visual comparison. Figure 14a,b display the average temperature and four-measurement moving average comparisons for both types of alveoli.

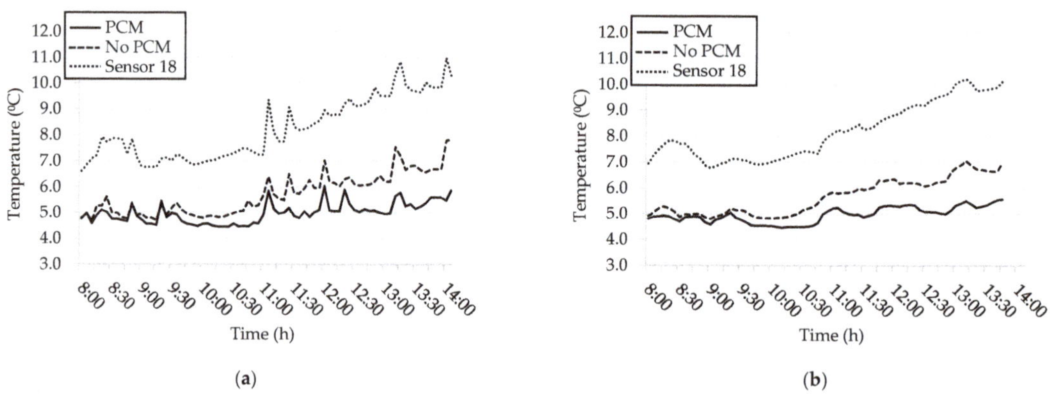

Figure 14. Comparison of average temperatures (**a**) and four-measurement moving average temperatures (**b**) for the van cargo area, in crates with alveoli containing PCM and empty alveoli.

4. Discussion

The results clearly demonstrate that the application of PCM alveoli in crates significantly enhances temperature stability during transportation. In Figure 14a, there is a noticeable difference of approximately 2 °C by the journey's end, marking a considerable deviation. Furthermore, the 4-measurement moving average of the temperatures shown in Figure 14b indicates that the temperature of the produce in crates with PCM alveoli remains mostly within the melting temperature range of the selected PCM (5 °C to 6 °C).

This implies that the phase change process within the alveoli was not finished by the end of the journey. It is important to consider that the temperature spikes caused by the stops move the four-measurement moving average temperatures.

In contrast, there is considerable variation among crates with empty alveoli, as depicted in Figure 13b. However, Figure 13a highlights a relative uniformity among the temperature measurements within the crates containing PCM alveoli, when the temperature fluctuations caused by the door openings during unloads are not considered. This uniformity suggests that PCM's thermal buffering capacity significantly contributes to temperature stability within the crates.

Of note, while the peak amplitudes during unloads are similar for crates with and without PCM, there is a distinct difference in temperature recovery post-unloading. Crates with PCM tend to return to temperatures near 5 °C after each stop, whereas those without PCM show a significant increase in temperature. This observation underscores the effectiveness of PCM in limiting temperature fluctuations, particularly in scenarios where external conditions may introduce thermal stress, such as during unloading.

This study, in line with previous research, supports the necessity of temperature monitoring directly among the food products, as opposed to single-point measurements, such as sensor 18. The results highlight that temperature readings of the cargo area alone are inadequate for accurately representing the conditions experienced by the perishable food products within the crates. Therefore, direct measurement within the crates is crucial for effective monitoring of food storage quality during transportation.

5. Conclusions

The results from this study clearly highlight the efficacy of utilizing PCM in ensuring temperature stability during the transportation of horticultural products. Comparisons between average temperature readings in crates equipped with PCM alveoli and those with empty alveoli underscore the significant benefits of using PCM in maintaining consistent adequate temperatures. As a result, the incorporation of PCM alveoli can significantly improve the quality of food storage, leading to a reduction in food loss and waste.

This research emphasizes the advantages of PCM integration in maintaining temperature stability within the transportation crates. The data suggest that the widespread implementation of PCM across all transported crates could enhance overall temperature control within the cargo area, mitigating temperature spikes caused by door openings and achieving faster temperature stabilization post-unloading. Nonetheless, additional research is necessary to substantiate these findings and to fully explore the potential of PCM in large-scale applications for improved temperature control in produce transportation and storage.

However, the benefits of using PCM come with certain caveats. The use of PCM alveoli in a reusable manner is a significant point to address. The environmental impact of discarding used PCM alveoli can potentially outweigh the benefits of increased preservation if a system for collecting and reusing them is not implemented. These alveoli could follow the same reuse cycle that reusable horticultural crates follow.

Future research should aim to determine the autonomy of the PCM alveoli and investigate the required quantity of PCM for specific produce types, distribution routes, and climatic conditions. Increased frequency of temperature readings, while reducing sensor battery life, could offer more detailed insights into temperature fluctuations, particularly during unloading events. This will necessitate a careful balance between sensor battery life and the accuracy of temperature readings, potentially utilizing a dynamic measuring rate in correlation with van movement and cargo area luminosity.

To this effect, integrating an accelerometer and a light sensor could provide valuable data on van movement and door opening events, respectively. Correlating these occurrences with temperature fluctuations could yield a more comprehensive understanding of how transportation events impact temperature stability.

Author Contributions: Conceptualization, M.A., P.D.G. and P.D.d.S.; methodology, M.A., P.D.G. and P.D.d.S.; software, M.A.; validation, P.D.G. and P.D.d.S.; formal analysis, M.A., P.D.G. and P.D.d.S.; investigation, M.A., P.D.G. and P.D.d.S.; resources, P.D.G.; data curation, M.A.; writing—original draft preparation, M.A.; writing—review and editing, M.A., P.D.G. and P.D.d.S.; supervision, P.D.G. and P.D.d.S.; funding acquisition, P.D.G. All authors have read and agreed to the published version of the manuscript.

Funding: The authors would like to express their gratitude to Fundação para a Ciência e Tecnologia (FCT) and C-MAST (Centre for Mechanical and Aerospace Science and Technologies) for their support in the form of funding, under the project UIDB/00151/2020. This study was conducted within the activities of project "Pack2Life—High performance packaging", project IDT in consortium No. 33792, call No. 03/SI/2017, Ref. POCI-01-0247-FEDER-033792, promoted by COMPETE 2020 and co-funded by FEDER within Portugal 2020.

Data Availability Statement: Research data can be found in https://docs.google.com/spreadsheets/d/1dElPPMlGjcHwgnMF0mWOd97PoZcnnl51/edit?usp=sharing&ouid=110715179057307420800&rtpof=true&sd=true.

Acknowledgments: The authors would also like to express their gratitude to Albifrutas—Produtos Horticolas, Lda. and José António for granting us access to perform real-life measurements and to António, the van driver, for his cooperation and patience throughout the experimentation process during the transportation route.

Conflicts of Interest: The authors declare no conflict of interest.

References

1. FAO. *The State of Food and Agriculture 2020, Overcoming Water Challenges in Agriculture*; Food and Agriculture Organization of the United Nations: Rome, Italy, 2020. [CrossRef]
2. Kaza, S.; Yao, L.C.; Bhada-Tata, P.; Van Woerden, F. *What a Waste 2.0—A Global Snapshot of Solid Waste Management to 2050*; World Bank: Washington, DC, USA, 2018.
3. Pires Gaspar, J.; Dinis Gaspar, P.; Dinho da Silva, P.; Simões, M.P.; Santo, C.E. Energy Life-Cycle Assessment of Fruit Products—Case Study of Beira Interior's Peach (Portugal). *Sustainability* **2018**, *10*, 3530. [CrossRef]
4. Fernandez, C.M.; Alves, J.; Gaspar, P.D.; Lima, T.M. Fostering Awareness on Environmentally Sustainable Technological Solutions for the Post-Harvest Food Supply Chain. *Processes* **2021**, *9*, 1611. [CrossRef]
5. Gaspar, P.; Silva, P.; Nunes, J.; Andrade, L. Characterization of the Specific Electrical Energy Consumption of Agrifood Industries in the Central Region of Portugal. *Appl. Mech. Mater.* **2014**, *590*, 878–882. [CrossRef]
6. Silva, P.D.; Gaspar, P.D.; Nunes, J.; Andrade, L. Specific Electrical Energy Consumption and CO_2 Emissions Assessment of Agrifood Industries in the Central Region of Portugal. *Appl. Mech. Mater.* **2014**, *675–677*, 1880–1886. [CrossRef]
7. Morais, D.; Gaspar, P.D.; Silva, P.D.; Andrade, L.P.; Nunes, J. Energy consumption and efficiency measures in the Portuguese food processing industry. *J. Food Process. Preserv.* **2020**, *46*, e14862. [CrossRef]
8. Aguiar, M.L.; Gaspar, P.D.; Silva, P.D.; Domingues, L.C.; Silva, D.M. Real-Time Temperature and Humidity Measurements during the Short-Range Distribution of Perishable Food Products as a Tool for Supply-Chain Energy Improvements. *Processes* **2022**, *10*, 2286. [CrossRef]
9. Proença, A.P.; Gaspar, P.D.; Lima, T.M. Lean Optimization Techniques for Improvement of Production Flows and Logistics Management: The Case Study of a Fruits Distribution Center. *Processes* **2022**, *10*, 1384. [CrossRef]
10. Gaspar, P.D.; da Silva, P.D.; Andrade, L.P.; Nunes, J.; Santo, C.E. Technologies for Monitoring the Safety of Perishable Food Products. In *Research Anthology on Food Waste Reduction and Alternative Diets for Food and Nutrition Security*; Hershey, E.P., Ed.; IGI Global: Hershey, PA, USA, 2021; p. 1318. [CrossRef]
11. Rodrigues, C.; Gaspar, P.D.; Simões, M.P.; Silva, P.D.; Andrade, L.P. Review on techniques and treatments toward the mitigation of the chilling injury of peaches. *J. Food Process. Preserv.* **2020**, *46*, e14358. [CrossRef]
12. Gaspar, P.D.; Gonçalves, L.C.C.; Pitarma, R.A. Three-Dimensional CFD modelling and analysis of the thermal entrainment in open refrigerated display cabinets. In Proceedings of the ASME 2008 Heat Transfer Summer Conference (HT 2008), Jacksonville, FL USA, 10–14 August 2008; Volume 2, pp. 63–73, ISBN 0-7918-3832-3. [CrossRef]
13. Gaspar, P.D.; Gonçalves, L.C.; Pitarma, R. Experimental analysis of the thermal entrainment factor of air curtains in vertical open display cabinets for different ambient air conditions. *Appl. Therm. Eng.* **2011**, *31*, 961–969. [CrossRef]
14. Gaspar, P.D.; Gonçalves, L.C.C.; Pitarma, R.A. Detailed CFD Modelling of Open Refrigerated Display Cabinets. *Model. Simul. Eng.* **2012**, *2012*, 973601. [CrossRef]
15. Ilangovan, A.; Hamdane, S.; Silva, P.D.; Gaspar, P.D.; Pires, L. Promising and Potential Applications of Phase Change Materials in the Cold Chain: A Systematic Review. *Energies* **2022**, *15*, 7683. [CrossRef]
16. Ahmed, M.; Meade, O.; Medina, M.A. Reducing heat transfer across the insulated walls of refrigerated truck trailers by the application of phase change materials. *Energy Convers. Manag.* **2010**, *51*, 383–392. [CrossRef]

17. Fioretti, R.; Principi, P.; Copertaro, B. A refrigerated container envelope with a PCM (Phase Change Material) layer: Experimental and theoretical investigation in a representative town in Central Italy. *Energy Convers. Manag.* **2016**, *122*, 131–141. [CrossRef]
18. Tassou, S.; De-Lille, G.; Ge, Y. Food transport refrigeration—Approaches to reduce energy consumption and environmental impacts of road transport. *Appl. Therm. Eng.* **2009**, *29*, 1467–1477. [CrossRef]
19. Mehling, H. Use of Phase Change Materials for Food Applications—State of the Art in 2022. *Appl. Sci* **2023**, *13*, 3354. [CrossRef]
20. Hekimoğlu, G.; Sarı, A. A review on phase change materials (PCMs) for thermal energy storage implementations. *Mater. Today Proc.* **2022**, *58*, 1360–1367. [CrossRef]
21. Kahwaji, S.; White, M.A. Organic Phase Change Materials for Thermal Energy Storage: Influence of Molecular Structure on Properties. *Molecules* **2021**, *26*, 6635. [CrossRef]
22. Curto, J.; Ilangovan, A.; Gaspar, P.D.; Silva, P.D.; Alves, N. Numerical study of the impact on cooling behavior of vent-holes design of fruit packaging boxes. In Proceedings of the 14th World Congress in Computational Mechanics (WCCM) ECCOMAS Congress 2020, Paris, France, 11–15 January 2021. [CrossRef]
23. Leitão, F.; Silva, P.D.; Gaspar, P.D.; Pires, L.C.; Duarte, D. Experimental Study of Thermal Performance of Different Fruit Packaging Box Designs. *Energies* **2021**, *14*, 3588. [CrossRef]
24. Leitão, F.; Madhan, S.K.; Silva, P.D.; Gaspar, P.D. Experimental testing of the thermal response of different food alveoli solutions for packaging boxes. In Proceedings of the World Congress on Engineering 2021 (WCE 2021)—Lecture Notes in Engineering and Computer Science, London, UK, 7–9 July 2021; pp. 268–272. Available online: http://www.iaeng.org/publication/WCE2021/WCE2021_pp268-272.pdf (accessed on 28 March 2022).
25. Ilangovan, A.; Curto, J.; Gaspar, P.D.; Silva, P.D.; Alves, N. CFD Modelling of the Thermal Performance of Fruit Packaging Boxes—Influence of Vent-Holes Design. *Energies* **2021**, *14*, 7990. [CrossRef]
26. Nanga, R.; Curto, J.; Gaspar, P.D.; Silva, P.D.; Cruz, J.M.S. Numerical Parametric Study of the Influence of Fruit Packaging Boxes Arrangement on Fluid Flow and Heat Transfer. *Int. J. Energy Clean Environ.* **2023**, *24*, 45–59. [CrossRef]
27. Madhan, S.; Santo, C.E.; Andrade, L.P.; Silva, P.D.; Gaspar, P.D. Active and intelligent packaging with phase change materials to promote the shelf life extension of food Products. In Proceedings of the ICEUBI2019—International Congress on Engineering—Engineering for Evolution, Covilhã, Portugal, 27–29 November 2019; pp. 232–241.
28. Curto, J.; Ilangovan, A.; Gaspar, P.D.; Silva, P.D. CFD modelling of PCM alveoli for fruit packaging boxes. *Int. J. Energy Clean Environ.* **2023**, *24*, 61–79. [CrossRef]
29. Gin, B.; Farid, M.M. The use of PCM panels to improve storage condition of frozen food. *J. Food Eng.* **2010**, *100*, 372–376. [CrossRef]
30. RubiTherm. *RUBITHERM®RT5HC Data Sheet*; Rubitherm Technologies GmbH: Berlin, Germany, 2020.
31. Varandas, L.; Faria, J.; Gaspar, P.D.; Aguiar, M.L. Low-Cost IoT Remote Sensor Mesh for Large-Scale Orchard Monitorization. *J. Sens. Actuator Netw.* **2020**, *9*, 44. [CrossRef]
32. Morais, D.; Aguiar, M.L.; Gaspar, P.D.; Silva, P.D.; Alves, N. Development of a monitoring device of fruit products along the cold chain. *Procedia Environ. Sci. Eng. Manag.* **2021**, *8*, 195–204. Available online: http://procedia-esem.eu/pdf/issues/2021/no1/22_02.22.Gaspar.01_21.pdf (accessed on 25 March 2022).

energies

Article

Impact of Stearic Acid as Heat Storage Material on Energy Efficiency and Economic Feasibility of a Vacuum Tube Solar Water Heater

K. Chopra [1,2], V. V. Tyagi [3,*], Sudhir Kumar Pathak [3], Apaar Khajuria [1], A. K. Pandey [4,5], Nazaruddin Abd Rahman [2,6], Muhamad Mansor [2,6] and Ahmet Sari [7,8]

1 School of Mechanical Engineering, Shri Mata Vaishno Devi University, Katra 182320, India; kapil.chopra@smvdu.ac.in (K.C.)
2 Institute of Power Engineering, Universiti Tenaga Nasional, Kajang 43000, Malaysia
3 School of Energy Management, Shri Mata Vaishno Devi University, Katra 182320, India
4 Research Centre for Nano-Materials and Energy Technology (RCNMET), School of Engineering and Technology, Sunway University, Kuala Lumpur 46150, Malaysia
5 Center for Transdisciplinary Research (CFTR), Saveetha Institute of Medical and Technical Sciences, Saveetha University, Chennai 600077, India
6 Department of Electrical and Electronics Engineering, College of Engineering, Universiti Tenaga Nasional, Kajang 43000, Malaysia
7 Department of Metallurgical and Material Engineering, Karadeniz Technical University, Trabzon 61080, Turkey
8 Interdisciplinary Research Center of Renewable Energy and Power Systems (IRC-REPS), King Fahd University of Petroleum and Minerals, Dhahran 31261, Saudi Arabia
* Correspondence: v.tyagi@smvdu.ac.in

check for updates

Citation: Chopra, K.; Tyagi, V.V.; Pathak, S.K.; Khajuria, A.; Pandey, A.K.; Rahman, N.A.; Mansor, M.; Sari, A. Impact of Stearic Acid as Heat Storage Material on Energy Efficiency and Economic Feasibility of a Vacuum Tube Solar Water Heater. *Energies* **2023**, *16*, 4291. https://doi.org/10.3390/en16114291

Academic Editor: Antonio Rosato

Received: 30 March 2023
Revised: 3 May 2023
Accepted: 17 May 2023
Published: 24 May 2023

Abstract: The overheating of heat pipes, poor transfer of heat across the absorber and finned heat pipes, and inability to provide hot water in the late evening hours are major problems associated with conventional heat pipe vacuum collector systems. The amalgamation of highly conductive storage material between the absorber tube (heat collecting surface) and the heat pipe is an effective way to overcome these problems. In this study, a stearic acid amalgamated vacuum tube solar collector system was designed and fabricated and its thermal output compared with a conventional vacuum tube system without storage material under the same environmental conditions. The experimental results showed that the amalgamation of stearic acid as storage material enhanced the thermal output of the solar system compared to the conventional one. The desired heat gain of the solar system with storage material increased by 31.30, 23.34, and 18.78% for Test 1_40 °C, Test 2_45 °C, and Test 3_50 °C, respectively. The technoeconomic analysis showed that almost 118.80 USD in revenue could be earned by the proposed solar system at the end of 15 years. The total running cost of ELG and the developed solar system was observed to be 202.62 and 86.70 USD, respectively. On average, the cost of hot water production using the solar system and ELG was found to be 0.0016 and 0.004 USD/L, respectively. The value of LEC was found to be 0.062 USD/electricity unit, which was much lower than the LEC value of ELG (0.116 USD/electricity unit). The value of NPW (73.73 USD) indicated high acceptability of the proposed system. The payback time is lower than the life of the system, indicating its suitability for use in the commercial sector. Therefore, the proposed solar system is highly recommended over conventional water heating systems in urban and rural areas.

Keywords: energy; vacuum tube solar water heating; technoeconomic investigation; thermal energy storage material

1. Introduction

Due to global warming and changes in the ecosystem due to the increasing use of fossil fuels, it is necessary to shift to more efficient renewable energy systems. Among the nonconventional energy sources, solar energy is the best candidate due to its free cost and abundant availability. The adoption of solar energy for the production of useful energy can save millions of tons of carbon dioxide emissions into the atmosphere every year [1]. In the USA and European Union, approximately 18 and 14% of total energy, respectively, is required to complete domestic hot water demand for bathing, cooking, cleaning, etc. [2]. A report by the Indian New and Renewable Ministry (IMNRE) showed that hot water demand in the domestic sector in India is exponentially increasing, from 129 million/day in 2017 to double that amount in 2022 [3]. This large demand for hot water is fulfilled using electricity or fossil fuels. Due to vacuum insulation, low heat losses, low initial/maintenance cost, high thermal efficiency, and highly selective coating, vacuum tube solar collectors have become popular compared to other options available for solar water heating. Vacuum tube collectors are extensively used in the industrial and residential sectors due to these peculiar features.

Vacuum tube collectors are categorized into three categories based on their design: U-tube, heat pipe, and thermosiphon. Among these designs, the heat pipe is one of the most advanced and efficient due to comparatively low heat losses, low thermal stratification, high heat dissipation, etc. The performance of this type of collector can be further enhanced by modifying its design, such as making changes to the heat pipe working fluid and integrating heat storage material. The integration of storage material in a heat pipe system is one of the convenient and economical ways to not only enhance its performance but also supply useful heat during cloudy or non-sunny times. In this regard, Pawar et al. [4] experimentally examined the performance of a heat pipe vacuum tube water heater integrated with heat storage material entrenched with copper. They also compared the performance of the proposed collector with traditional solar water under the same environmental conditions and found that the highest energy efficiency values of the developed system and the conventional system were 85.64 and 36.91%, respectively. The authors also reported that after sunshine hours, the difference in the fin temperature between the developed system and the conventional system was 36.1 °C. Algarni et al. [5] studied the impact of the integration of nano-boosted heat storage material on the thermal output of a vacuum-tube-type solar water heater. The authors reported that the integration of 0.33 wt.% copper/storage material with a solar water heater increased its thermal output by 32%. The authors also found that a solar water heater integrated with copper/storage material could deliver hot water of 50 °C for almost 2 h longer than a conventional solar water heater. Essa et al. [6] compared the thermal output of a heat pipe solar water heater with and without helical fins. The authors observed that for 0.665 and 0.5 kg/min flow rate of water, the improvement in the energy efficiency of the system with helical fin compared to the system without helical fin was 13.6 and 15%, respectively. Olfian et al. [7] analyzed a serpentine-type vacuum tube water heater employed with heat storage material to store extra heat during the daytime. They selected three different diameters (6, 8, and 10 mm) of U-tube to transfer absorbed heat from the sun to flowing water. The authors concluded that the U-tube measuring 6 mm diameter showed the highest improvement of 13.5 and 25% in outlet temperature and liquid fraction, respectively. Li et al. [8] carried out a thermal performance and optimization analysis of a serpentine-type vacuum tube solar water heater embedded with thermal accumulating material. They observed that a high flow rate of water through the solar water heater reduced the hot water supply period. They also found that the water heater with heat storage material was able to supply hot water at 35 °C for approximately 160 min more than the system without storage. Wu et al. [9] identified the impact of specific heat capacity, density, thermal conductivity, phase change heat, and melting temperature of thermal accumulating material on the thermal output of vacuum tube water heaters. The results of the study revealed that storage material with high melting temperature was advantageous to upsurge the performance of the developed system. They also observed that a vacuum tube embedded with storage material with a density of 425 kg/m^3 increased the heat collection time compared to storage material with a density of 1700 kg/m^3. O'Neil and Sobhansarbandi [10] compared the performance of a heat pipe solar water heater and a U-tube solar water heater with and without storage material. The experimental results revealed that the U-tube type collector had higher thermal efficiency and hot water temperature compared to the heat pipe system. Senobar et al. [11] studied the thermal output of a vacuum tube system embedded with a heat storage medium (paraffin wax). The results showed that the heat transfer rate in the selected PCM increased 2–3 times more than pure PCM by employing copper metal foam into its structure. Bouadila et al. [12] conducted a comparative thermal analysis of three solar collector systems, in which the first one was the control collector, the second one had a re-insulated manifold, and the third one had a manifold enclosed by paraffin wax as the heat storage medium. The authors found that utilization of insulation around the manifold decreased the thermal losses from 8.18 to 6.05 W/m^2 K, enhanced the energy from 41 to 66%, and enhanced the exergy from 15 to 20%. Furthermore, the energy and exergy efficiencies of the third system were observed to be higher than the second system due to the reduction of overall heat losses. Kumar et al. [13] studied the impact of the integration of nano-boosted thermal accumulating material on the thermal output of the collector. The authors found that daily exergy and energy efficiencies of the system with nano-boosted PCM enriched by 13.72 and 28.78%, respectively, in comparison to the system integrated with pure PCM. Uniyal et al. [14] reviewed modifications in the latest designs, amalgamation of phase change material, and financial assessment with various vacuum-tube-based water heaters. They also addressed the problem of pure phase change materials and suggested mixing nanoparticles with pure PCM to enhance the heat transfer rate. Raza et al. [15] determined the different solar water heater potential in seven different districts of India. They found that the area of the solar collector should be determined in such a way that it can absorb solar fraction of almost 50%. The payback time of the selected collectors was in the range of 5–15 years. The simulation results showed that the state of Gujrat in India is one of the most appropriate locations to install solar water heaters. Olczak et al. [16] compared the performance of vacuum tube and flat plate solar collectors in terms of solar energy production. They observed that the solar energy productivity of the vacuum tube collector was much higher than the flat plate collector for the same absorbing area. The authors utilized a vacuum tube collector based solar water heater in mainly colder regions as they produce stable production of thermal energy. The results of the experimental investigation revealed that the efficiency of the vacuum tube collector with and without heat storage material was 58 and 38%, respectively. Jachura and Sekret [17] performed environment effect evaluation on a vacuum tube collector system and found that a PCM amalgamated collector reduced the harmful impact on the surrounding area by 17–24% in comparison to the collector without PCM.

As can be seen from Table 1, researchers have extensively carried out research work on vacuum-tube-based water heaters and improved the thermal output of the conventional solar water heater by incorporating phase change material with vacuum tube systems of different designs. In most of the investigations, authors either used pure paraffin wax and its nano-boosted composites as heat storage material. In some studies, the authors determined the economic practicality of the system before and after the amalgamation of thermal accumulating material. However, the economic viability of stearic acid as heat storage material amalgamated novel water storage heater combined with a vacuum tube collector system has not been investigated.

An economic assessment of the proposed solar collector system recommended its commercial and domestic utility in society. It was noted that almost 118.8 USD in revenue could be earned by the proposed solar system at the end of 15 years. On average, the cost of hot water production and LEC with the proposed solar system was found to be much less than ELG. Furthermore, the payback time was lower than the life of the system, making it suitable for use in the commercial and domestic sectors.

In addition to this, the proposed system reduced the issue of overheating in thermosiphon heat pipes, and its performance was enhanced by the incorporation of conductive stearic acid as the thermal energy storage material. This modified collector was also compared to the conventional system. The energy efficiency of the proposed solar water heater improved by 18–31% compared to the heat pipe system without storage.

Table 1. Comparison of the proposed study with previous studies.

Author (s)	Year	Type of System	Parameters Calculated	Increment in Energy Gain
Essa et al. [18]	2018	Serpentine	Energy and thermal output investigation	6.8–21.9%
Olfian et al. [19]	2020	Serpentine	Thermal and system efficiency measurement and impact of operation temperature	13.6–15%
Kumar et al. [13]	2021	Heat Pipe	Effect of different percentages of nano-boosted PCM and first and second law analysis	3–5%
Wong et al. [20]	2022	Heat Pipe	Heat transfer and thermal efficiency investigation	3–5%
Wu et al. [9]	2022	Serpentine	Impact of PCM's properties and thermal efficiency analysis	4.31%
Kumar et al. [21]	2022	Water in Glass	Effect of different percentages of nano-boosted PCM and energy and exergy investigation	7%
Proposed Study		**Experimental**	**Energy, Exergy, and Economic Analysis**	**18–31%**

2. Description of Experimental Setup and Methodology

The developed setup units were tested in the outside environment of Jammu and Kashmir, India. To judge the impact of heat storage material on the thermal output of the vacuum tube solar collector system, a conventional collector was also tested in parallel with a developed solar collector under the same environmental conditions. In this study, a solar collector (VTSC-I) was embedded with stearic acid as storage material, while a tube was left without storage material (VTSC-II).

As shown in Figure 1, in both systems (VTSC-I and VTSC-II), the hot water loaded tank was integrated with a heat pipe vacuum tube system to diminish the overall heat loss from the system. The water storage tank for both VTSC-I and VTSC-II could store up to 3.5 L of water. The experiment was conducted for three consecutive days for different temperature limits of hot water (40, 45, and 50 °C). During the first day, the highest limit of water in the header was set as 40 °C. On this day, water was allowed to heat from its initial temperature until it reached 40 °C, then drained out and stored in a tank for further usage. At the end of the day, the calculation was carried out according to the total collected hot water of 40 °C. The same methodology was adopted for temperature limits of 45 and 50 °C.

As shown in Figure 2, the temperature sensors T_1, T_2, and T_3 for VTSC-I were located at the bottommost, middle, and topmost portions of the insulated water tank. On the other hand, T_4, T_5, and T_6 for VTSC-II were located at the bottommost, middle, and topmost portions of the insulated water tank. The temperature of the heat storage material filled inside the tube of VTSC-I was measured by T_7 and T_8, while the temperature of the internal air of vacuum tubes of VTSC-II was measured by T_9 and T_{10}. The temperature of the outside air was determined by T_{11}. The properties of stearic acid embedded with a vacuum tube collector are given in Table 2. Table 3 presents the description of the specification of the experimental setup.

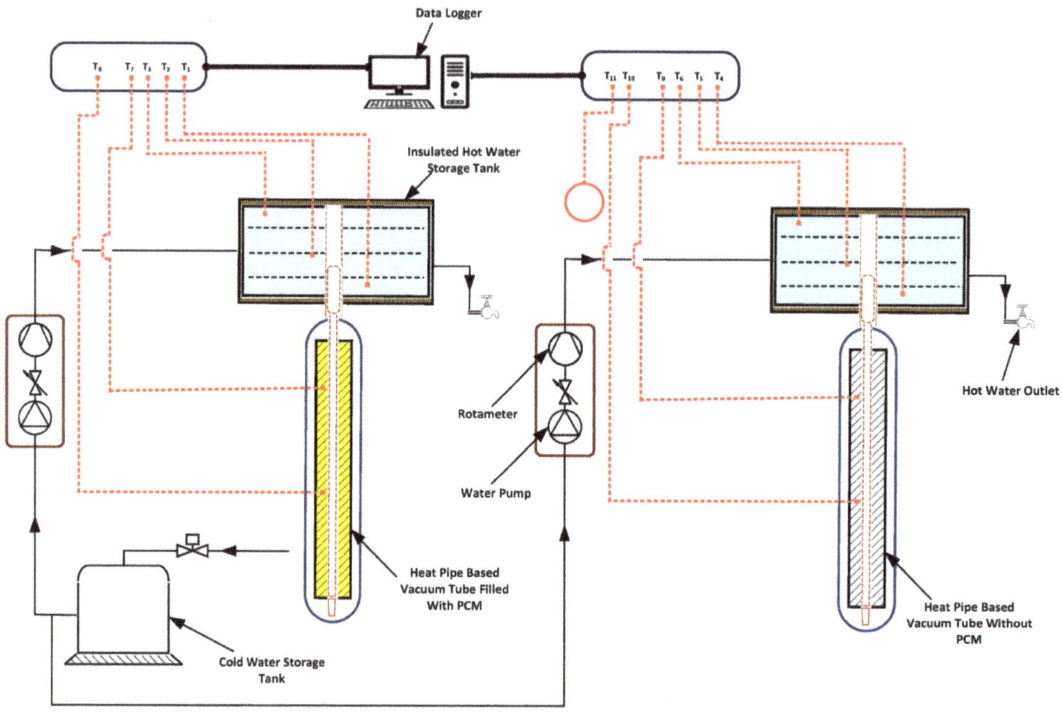

Figure 1. Schematic arrangement of the proposed vacuum tube collector systems.

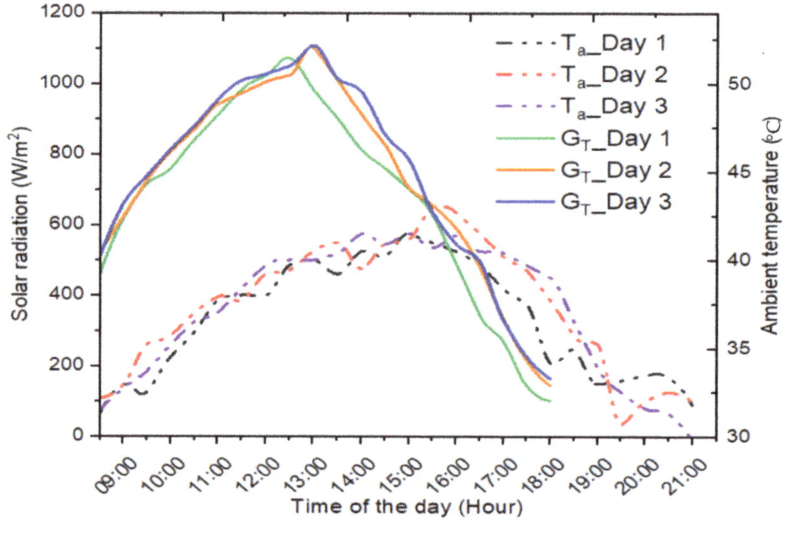

Figure 2. Solar radiation and ambient temperature variation with respect to time.

Table 2. Properties of stearic acid.

Property	Value/Specification
Color	Caramel white
Melting temperature range	60–62 °C
Latent heat range	185–190 J/g
Specific heat at 75 °C	2.45
Density at 75 °C	1.19
Purity	99.8%

Table 3. Specification of the proposed systems.

Item	Specification/Value
Number of vacuum tubes	1
Aperture area of collector	0.080
Diameter/length of an evaporator of heat pipe	9.5/1600 mm
Diameter/length of the condenser of heat pipe	14/63 mm
Material of heat pipe fin	Aluminium
Diameter of water storage header	130 mm
Length of water storage header	270 mm
Insulation on water storage header	Rockwool

3. Heat Transfer and Economic Analysis

This section provides the heat transfer and economic analysis of the design system and its comparison with an electrical heating system.

3.1. Heat Transfer Analysis

In the vacuum tube collector system, the incident solar insolation was gathered by the VTSC (vacuum tube solar collector) and transformed into heat energy by the absorber tube. After that, the heat energy collected by the absorber tube was transferred to phase change material or internal air for the systems with and without storage, respectively. Then, heat was conveyed to the finned heat pipe. Finally, heat was transferred to water flowing in the manifold through heat pipes. The incident energy gathered by the inner absorber of the tube was fragmented into two portions. The primary portion of the gathered heat was conveyed to the surrounding area as heat loss, while the other portion of collected thermal energy was conveyed to water in the puffed insulated manifold through a heat pipe.

To compare the thermal output of the developed and conventional solar collector systems under the same environment, the useful energy collected by each collector was calculated as follows [22]:

$$Q_{uf} = \rho_{wf} \times \dot{V}_{wf} \times C_{P,wf} \times (T_{wf,o} - T_{wf,in}) \tag{1}$$

where ρ_{wf} (kg/m^3) is the working fluid density; $\dot{V}_{wf}$ (m^3/s) is the working fluid flow rate; $C_{P,wf}$ (kJ/kgK) is the working fluid specific heat; $T_{wf,o}$ (°C) is the working fluid outlet temperature; and $T_{wf,in}$ (°C) is the working fluid inlet temperature.

For evaluation of the energy efficiency of the individual collector (developed or conventional solar collector system), incident solar radiation was determined by the following relation [22]:

$$Q_{in} = G_T \times A_{cr} \times \Delta\tau \tag{2}$$

where G_T (W/m^2) is the instantaneous solar insolation, and A_{cr} (m^2) is the area of the collector.

The daily energy performance (η_{EP}) of the solar water heating structure can be determined by the following equation [23]:

$$\eta_{EP} = \frac{\rho_{wf} C_{P,wf}}{A_{cr}} \sum \frac{\dot{V}_{wf} \times (T_{wf,o} - T_{wf,in})}{G_T} \tag{3}$$

3.2. Economic Viability Analysis

In the present study, along with heat transfer analysis, economic feasibility and financial benefits of the developed collector compared to a conventional solar water heater and an electric water heater were also studied. This analysis is advantageous in terms of the practical deployment of the project in the residential and commercial sectors.

As per the literature survey, three important factors help to assess the financial benefits and economic feasibility of any business model: levelized energy price (LEP), net present worth (NEPW), and payback time (PT). These three economic factors can be evaluated through the following relations.

The levelized energy price for solar water heater can be calculated by taking a proportion of the annual uniform charge from the produced hot water in the same year [24]:

$$LEP_N = \frac{YU_N}{AHP_N} \tag{4}$$

where YU_N ($) is the uniform cost of the system for the Nth year, and AHP_N (L) is the hot water production for the Nth year.

The uniform cost of the solar water heater includes yearly bank installments against debt for the installation of the system, maintenance and operation cost, and auxiliary energy cost during non-sunny days [24]

$$YU_N = YI_N + MO_N + AE_N \tag{5}$$

where YI_N ($) is the yearly installment for the Nth year; MO_N ($) is the maintenance and operation cost for the Nth year; and AE_N ($) is the auxiliary energy cost for the Nth year.

According to [25],

$$YI_N = C_I \times DE_r \times \left(\frac{1}{N_{Life}} + \left(\frac{N_{Life} - N}{N_{Life}} \right) \times it_r \right) \tag{6}$$

where C_I ($) is the capital investment in the system; DE_r (%) is the debt/equity ratio; N_{Life} (year) is the life of the system; and it_r (%) is the bank interest rate.

$$AE_N = \left(ER - \left[(N_s \times G_T \times A_{cr} \times \Delta\tau \times \eta_{th}) \times (1 - d_r)^N \right] \right) \times EP_N \tag{7}$$

where ER (kWh/year) is the energy required; N_s (days/year) is the number of clear sunny days per year; d_r (%) is the degradation rate of the performance of the collector; and EP_N ($/kWh) is the electricity price for the Nth year.

The net present worth (NEPW) provides the total income made by the installation of the business project. It is a very crucial factor whose value decides the viability of the installed project. The project is not acceptable if the NPEW is negative, otherwise, the project can be accepted [25].

$$NEPW = -C_I + \sum_{N=0}^{N} \frac{CFL_N}{(1 + DE_r)^N} \tag{8}$$

where CFL_N ($) is the cash flow for the Nth year.

Another important factor of the financial analysis is the payback time, which is beneficial for assessing the associated risk of initial capital investment of a project. This parameter indicates how much of the time invested in the system can be recovered [26].

$$PB_T = \frac{\sum_{N=0}^{N} CFL_N}{C_I} \tag{9}$$

4. Results and Discussion

This section presents and discusses a complete analysis of the different experimental outcomes. Measurable data are used to compare the energy efficiency and performance improvement ratio of a manifold cum water storage tank integrated single evacuated tube water heater with and without heat storage material. The subsections of this section are devoted to addressing the variation of input and output variables: solar radiation, water inlet, energy analysis, outlet temperature, and a useful amount of heat acquired by working fluid under different test situations and economic viability.

4.1. Disparity in Outside Air Temperature and Solar Energy for Different Test Days

Figure 2 depicts the disparity in outside air temperature and solar energy with respect to time for different test days. Each test was conducted in its unique setting. It can be observed from the figure that changes in solar radiation were relatively consistent between different test days. However, the intensity of solar radiation varied marginally on each of the three days, with the most variance occurring in the evening hours. According to the data, which were collected with the help of a solarimeter, solar radiation was at its highest in the morning and at its lowest in the afternoon. The investigations were conducted during periods of clear, bright weather conditions in April 2022.

The average daily maximum solar radiation was 1072 W/m^2 on day 1 (Test 1_50 °C), 1101 W/m^2 on day 2 (Test 2_45 °C), and 1108 W/m^2 on day 3 (Test 3_40 °C). As an additional note, the average daily solar radiation incident on the collector area was 24.34 MJ/m^2 on day 1, 26.03 MJ/m^2 on day 2, and 26.61 MJ/m^2 on day 3. Figure 2 also shows the variation in ambient temperature with respect to time for all three test days. The average ambient temperature was 36.71, 37.41, and 37.16 °C for days 1, 2, and 3 respectively. It can be seen that the highest value of ambient temperature for all tests surpassed 40–43 °C in the afternoon. The temperature variation shows the composite dry environmental conditions in which the experiment was performed.

4.2. Disparity in Hot Water Temperature of Control and Testing Systems

In the present investigation, the integrated header for both the control (reference) system (without thermal accumulating material) and the testing system (embedded with storage material) was built to be able to store 3.5 L of water. For Test 1_50 °C, the freshwater was assimilated inside the integrated storage tank of the vacuum tube system and heated until it reached 50 °C. Then, it was emptied and gathered in an insulated tank so that it could be used for domestic applications. A similar procedure was carried out for water temperatures of 45 °C (Test 2_45 °C) and 40 °C (Test 3_40 °C).

Figures 3–5 show the tank water temperature variations for the control and testing systems for Test 1_50 °C, Test 2_45 °C, and Test 3_40 °C, respectively. During each experiment, the freshwater in the 31–32 °C temperature range was filled in the tank (header). From Figure 3, it can be observed that there were four refills (14 L) for the reference system and five refills (17.5 L) for Test 1_50 °C. Due to the utilization of highly effective heat storage material (stearic acid), the testing system yielded a higher amount of heat compared to the reference system (without PCM).

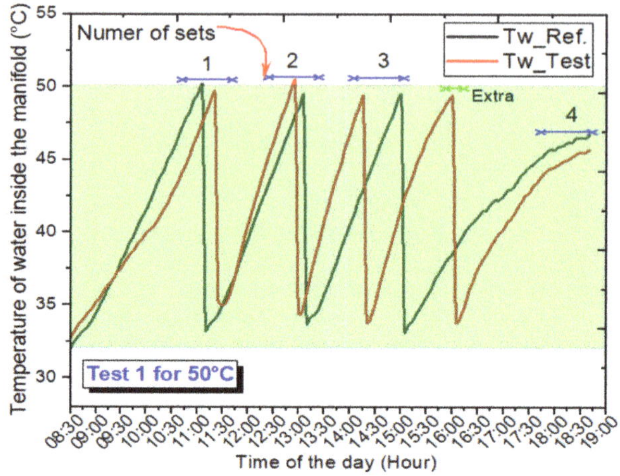

Figure 3. Hot water temperature variation inside the manifold of both systems for Test 1_50 °C.

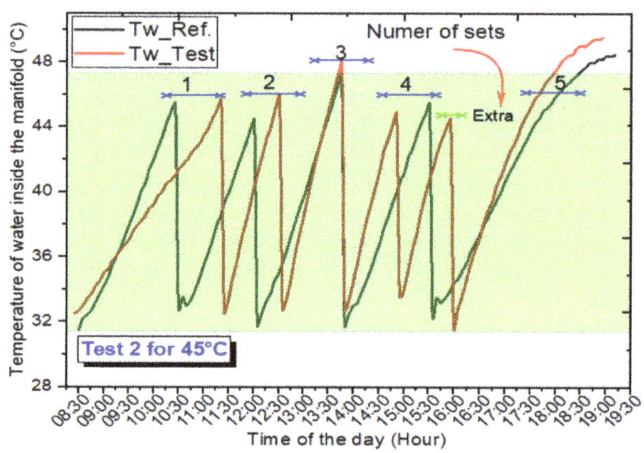

Figure 4. Hot water temperature variation inside the manifold of both systems for Test 2_45 °C.

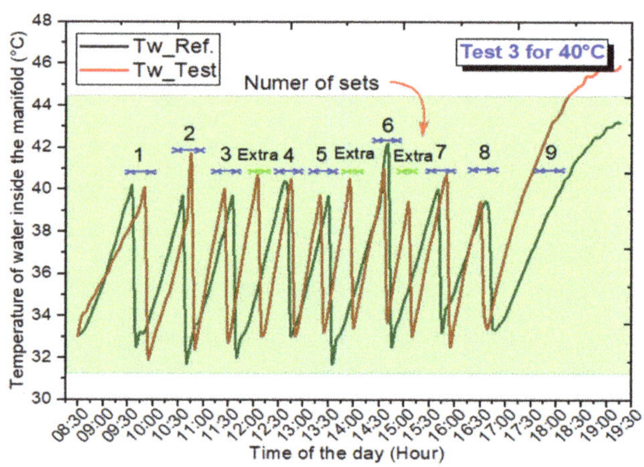

Figure 5. Hot water temperature variation inside the manifold of both systems for Test 3_40 °C.

The integration of PCM has several advantages, one of which is that it accumulates an excessive input thermal energy in the form of latent thermal energy in the course of its phase change and then discharges that heat in the late evening hours. Because the testing system employed phase change material, it could operate during the late evening hours when solar radiation was no longer available.

As shown in Figures 3 and 4 for both the reference and test systems, the variation in hot water temperature during Test 2_45 °C was similar to Test 1_50 °C. The freshwater (used for refill) temperature range was determined to be between 29 and 32.5 °C. Both the reference and test systems were found to be operational from 8:30 to 19:30. For Test 2_45 °C, the water was changed five times in the control system and six times in the developed testing system. The total amount of hot water (at 45 °C) provided by the control system was 17.5 L, whereas the amount of hot water provided by the developed testing system was 21 L.

The stored water temperature variations in the manifolds of the reference and test systems for Test 3_40 °C are shown in Figure 5. In both setups, the water was heated to 40 °C in a single-tube header before being drained and stored as hot water. As shown in Figure 5, the 40 °C hot water was changed nine times in the reference system and 12 times in the testing system. As per the calculation, 31.5 L of hot water was provided by the control system, whereas 42 L of water was collected from the testing system. The operation hours for the control and developed testing systems were determined to be 08:30 to 19:30. Hence, the testing system with thermal accumulating material could provide more hot water for an extended period after sunset than the control system without thermal accumulating material.

4.3. Temperature Variation in Internal Air and Heat Storage Material for Different Tests

This subsection discusses the variability in temperature inside the vacuum tube of both testing and reference systems for different tests under consideration (as shown in Figures 6–8). To conduct the experiment, stearic acid (the phase change substance) was poured into the testing system's tube, while the reference system tube was left empty. It is important to keep in mind that the designated PCM started melting at about 10:00 a.m. The PCM in the lower section had completed the phase transition process and was experiencing significant temperature increases, while the PCM in the upper section was still undergoing it. It was determined that the temperature of the PCM was consistently less than the air within the vacuum tube of the control (reference) system because PCM has a higher heat requirement for phase shift.

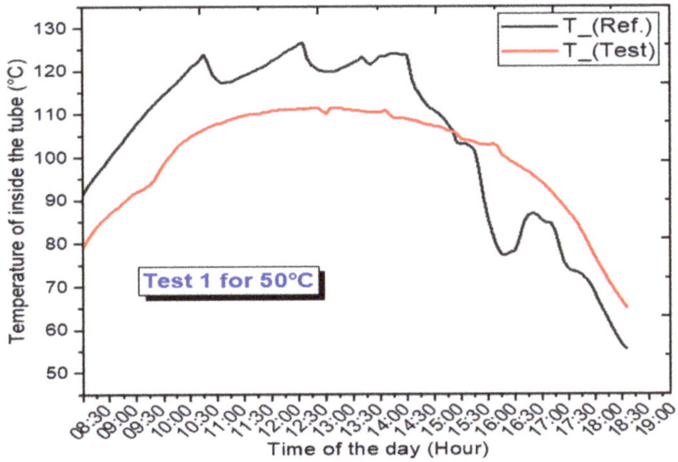

Figure 6. Temperature variation inside the vacuum tube of both systems for Test 1_50 °C.

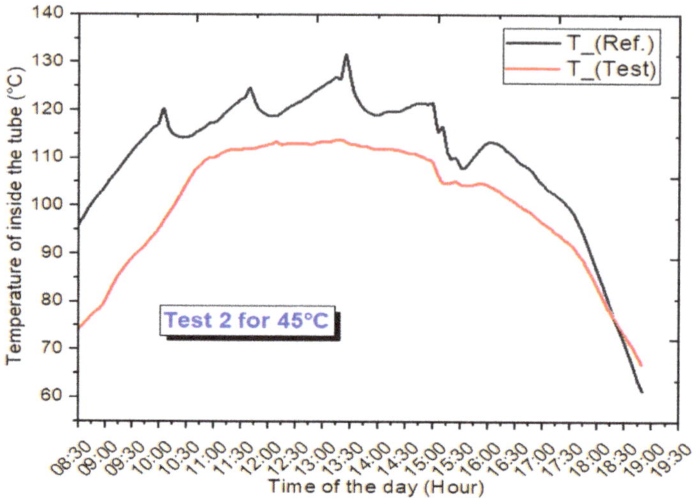

Figure 7. Temperature variation inside the vacuum tube of both systems for Test 2_45 °C.

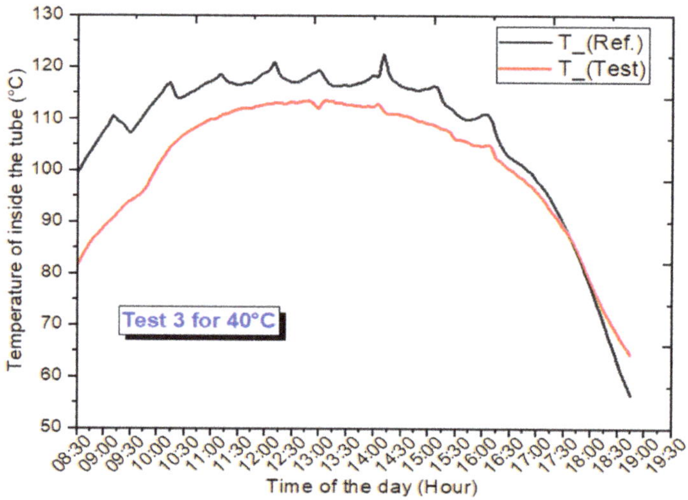

Figure 8. Temperature variation inside the vacuum tube of both systems for Test 3_40 °C.

Solar energy is accumulated within the heat storage material, and it is later transferred to the working fluid (water) when solar irradiation is weak or not available. Hence, in contrast to the situation in which PCM is absent, hysteresis takes place. The amalgamation of the thermal accumulating material with the developed testing system enables a reduction in the tube's highest temperature during high solar radiation. Moreover, this additional heat can be used to provide hot water during non-sunny hours.

4.4. Variation of Set-Wise Useful Heat for Different Tests at 50, 45, and 40 °C

The useful amount of heat collected by the reference and testing systems for Test 1_50 °C, Test 2_45 °C, and Test 3_40 °C conditions is shown in Figures 9–11.

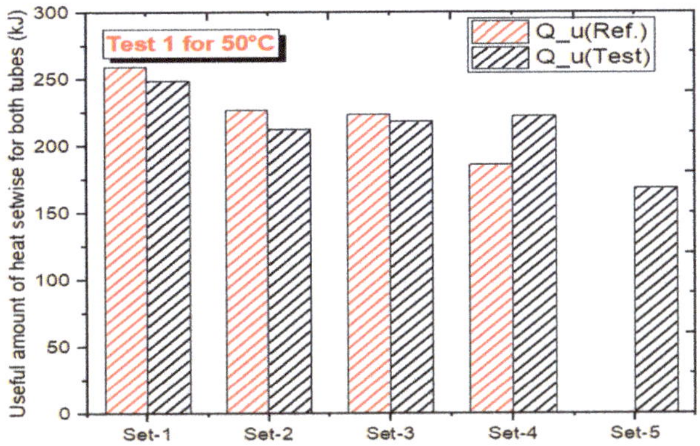

Figure 9. Set-wise useful amount of heat collected from both systems for Test 1_50 °C.

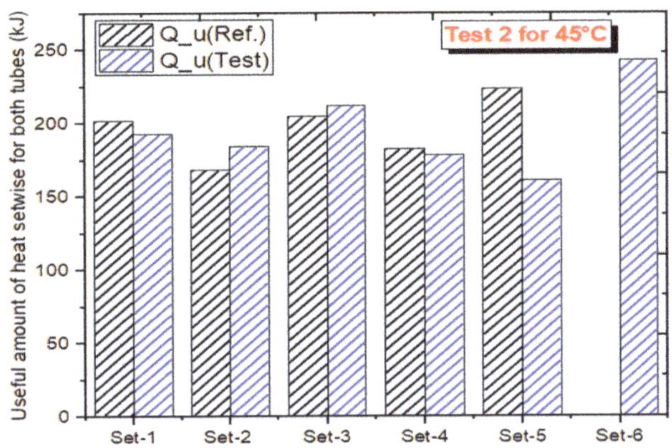

Figure 10. Set-wise useful amount of heat collected from both systems for Test 2_45 °C.

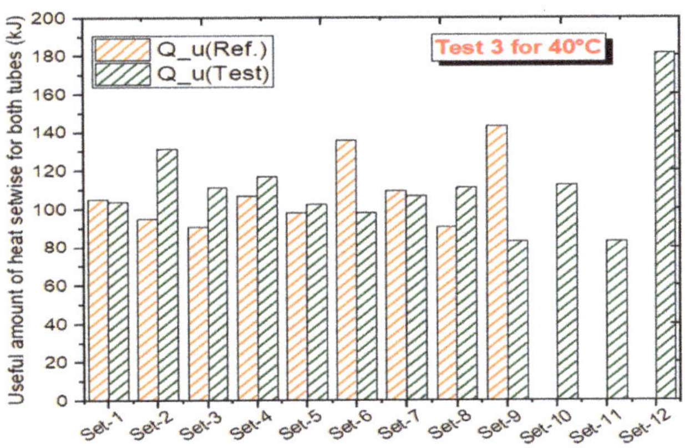

Figure 11. Set-wise useful amount of heat collected from both systems for Test 3_40 °C.

Figure 9 shows the amount of heat collected by water in different sets for day 1 to reach the set limit of temperature (50 °C) in both systems. It was observed that four sets of 50 °C were completed by the reference system with heat gain values of 258.95, 226.76,

223.83, and 185.801 kJ. In comparison, five sets of 50 °C (water temperature limit to refill) were completed by the testing system with heat gain values of 248.71, 212.13, 217.98, 222.37, and 168.24 kJ. The quantities of heat collected by water in both systems in different sets for day 2 to achieve the desired limit of temperature (45 °C) are depicted below in Figure 10.

The findings showed that the reference system completed five sets of 45 °C with heat gain values of 201.89, 168.24, 204.82, 182.87, and 223.83 kJ, respectively, for each set. In comparison, the testing system completed six sets of 45 °C with heat gain values of 193.11, 184.33, 212.13, 178.48, 160.93, and 242.85 kJ for each set.

Similarly, Figure 11 shows the amounts of heat collected by water in both systems in different settings on day 3 to reach the target temperature limit (40 °C). The result showed that the reference system completed nine cycles (45 °C water temperature) with heat gain values of 105.33, 95.09, 90.70, 106.79, 98.02, 136.05, 109.72, 90.70, and 143.37 kJ for each cycle. In comparison, the testing system performed 12 cycles of 45 °C water temperature with heat gain values of 103.87, 131.67, 111.18, 117.04, 102.41, 98.02, 106.79, 111.18, 83.39, 112.65, 83.39, and 181.41 kJ for each cycle.

4.5. Disparity in Daily Thermal Energy Input/Output and Energy Efficiency of the Proposed Systems for Different Tests

Every thermal system's energy efficiency depends on the quantity of absorbed/lost heat. In the current study, the testing system was embedded with heat storage material, while the reference system was without heat storage material or internal air between the absorber and finned heat pipe. The PCM possesses improved thermal properties than internal air. As a result of this, it was observed that the heat transfer rate across the blackened inner tube and the aluminum finned heat pipe was significantly higher, which resulted in a reduction in the overall heat loss and an increase in the generation of heat that can be used.

Figure 12 depicts the fluctuation in usable energy obtained by water and incident energy with error bars on the aperture area of a single evacuated tube. For Test 1_50 °C, the testing system obtained 1069.45 kJ of usable energy from water, whereas the reference system gained 895.35 kJ. Similarly, the desired thermal outputs acquired by working fluid (water) for the testing and control systems were 1171.86 and 975.82 kJ, respectively, for Test 2_45 °C and 1343.03 and 981.67 kJ, respectively, for Test 3_40 °C. The total energy incident on the aperture area of the single tube was 1840.34, 1968.53, and 2012.35 for Test 1_50 °C, Test 2_45 °C, and Test 3_40 °C, respectively.

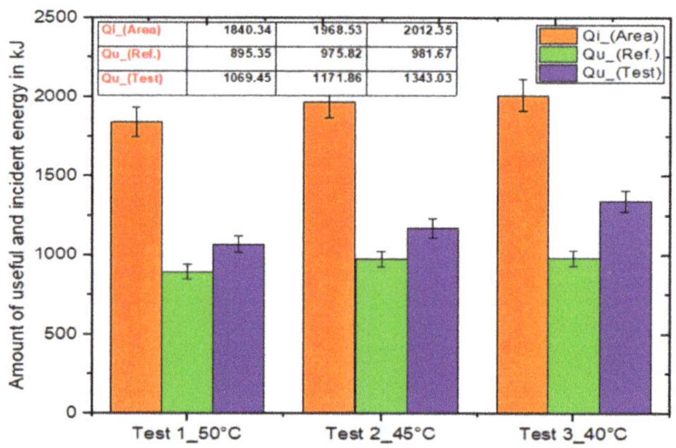

Figure 12. Daily heat input and outputs of both systems at 50, 45, and 40 °C.

Figure 13 shows the daily thermal (energy) performance of both systems (testing and reference) for different tests. As a result of the experimental research, it was determined that the daily thermal (energy) efficiency of the developed testing system was 66.78, 59.82, and 44.58% for Test 1_40 °C, Test 2_45 °C, and Test 3_50 °C, respectively. In contrast to this, the daily energy efficiency of the control (reference) system was 50.86, 48.50, and 37.53% for Test 1_40 °C, Test 2_45 °C, and Test 3_50 °C, respectively. The ratio of augmentation in the testing system's daily thermal efficiency was determined to be 31.30, 23.34, and 18.78% for Test 1_40 °C, Test 2_45 °C, and Test 3_50 °C, respectively. Thus, it could be concluded that the utilization of thermal accumulating material in the developed testing system resulted in an improvement in the daily energy efficiency for all tests.

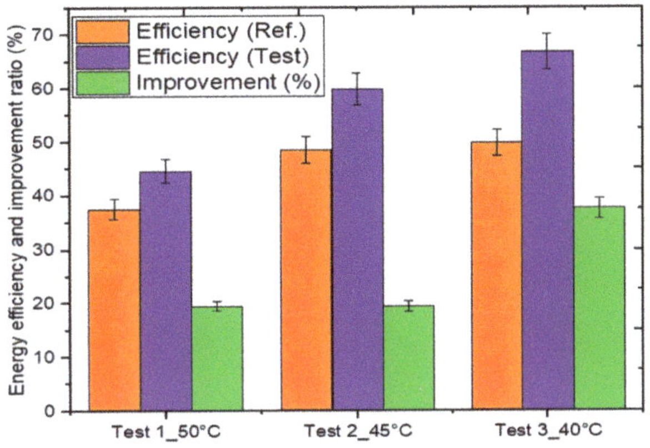

Figure 13. Daily thermal (energy) efficiency and enhancement ratio for different tests at 50, 45, and 40 °C.

The main goal of any water heating system based on solar energy is to offer water of the desired temperature even during cloudy, rainy, or non-sunny hours. Often, this is possible with the installation of a collector that is bigger than necessary, thus allowing the storage of additional hot water in an insulated tank. This increases both the overall cost and the heat loss from the water container.

Hence, the utilization of heat storage material in the water heating system proposed in this study along with the provision of hot water during periods of non-sunny hours enhances the system's output at a lower price. In contrast, commercial solar water heaters are incapable of producing hot water at these periods. Hence, the use of potential PCMs is a realistic option in solar water heaters.

4.6. Outcomes of Technoeconomic Analysis

In the present section, the economic practicality of the designed and developed vacuum collector systems is discussed and its results compared with the traditional thermal energy supplying system i.e., electric geyser (ELG). It may be noted that the economic practicality of the developed system was assessed by taking into consideration the economic factors of the local solar market of India. These economic factors are given in Table 4.

The loan cost, O&M cost, auxiliary energy cost, and gross solar system cost variation of the proposed solar collector system are depicted in Figure 14. It was found that the loan cost decreased over the years due to decreasing interest, whereas the maintenance and operational cost was found to increase due to increase in inflation. Similarly, the auxiliary energy cost was found to increase with time. This is because the auxiliary energy cost is dependent on the price of electricity, which is increasing with time. In a nutshell, gross solar system costs increased over the years due to the dominance of increasing O&M costs and auxiliary energy costs over decreasing loan costs. On average, the annual cost of a solar system was found to be 5.43 USD after 15 years of operation.

Table 4. Economic factors of the local solar market in India.

Factors	Value	Unit
Initial investment on proposed collector [27]	375	USD/m^2 area
Maintenance and operational cost of system [27]	1	%
Loan interest rate [28]	9	%
Loan term	15	years
Debt ratio	90	%
Price of electricity [27]	0.081	USD/unit
Life of system [28]	15	years
Rise in electricity price [27]	10	%/year
Discount/reinvestment rate [29]	5	%
Inflation rate [30]	4.5	%/year [27]
Rate of degradation of thermal energy	0.50	%/year
Average daily solar insolation based on the selected days during experiments	6.76–7.39	kWh/m^2
Average daily energy efficiency based on experiment	44–67	%
Number of sunny days [31]	300	days/year

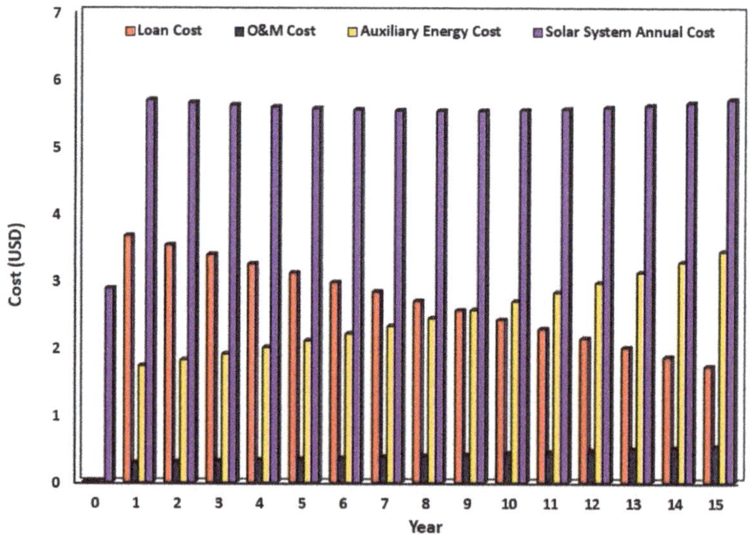

Figure 14. Variation of loan cost, O&M cost, auxiliary energy cost, and gross solar system cost.

The variation of hot water generation cost by solar and ELG, the fuel cost of the ELG system, and revenue earned by a solar system with reference to ELG are depicted in Figure 15. This figure also indicates the calculated value of the levelized cost of energy, the net present worth, and the payback time of the solar system after its 15 years of operation. The revenue earned by the solar system was observed to have increased due to the continuously increasing electricity prices. It was noted that at the end of 15 years, almost 9504 INR revenue could be earned by the proposed solar system. The annual fuel cost of the traditional thermal system (ELG) exponentially increased due to the tremendous increase in electricity prices. The total running cost of ELG and the developed solar system was observed to be 202.62 and 86.70 USD, respectively.

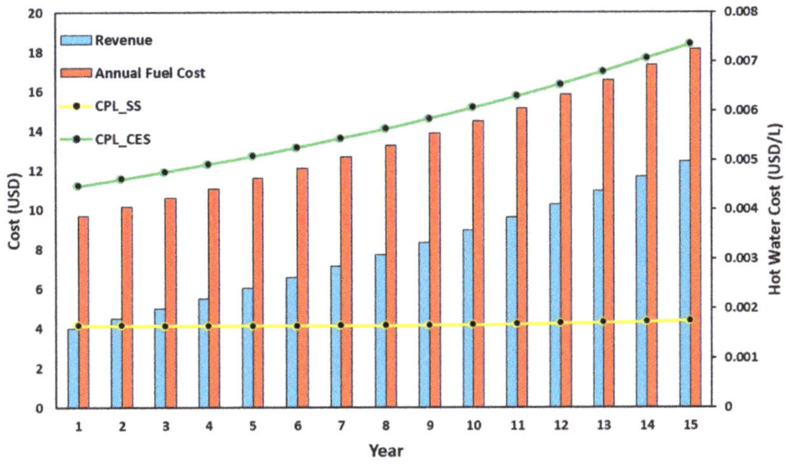

Figure 15. Variation of revenue, annual fuel cost, and cost per liter for the proposed systems.

On the other hand, the cost of hot water production using both the solar system and ELG increased with time due to the increase in electricity prices. On average, the cost of hot water production with the solar system and ELG was found to be 0.0016 and 0.004 USD/L, respectively. Therefore, the proposed solar system is highly recommended over ELG-based water heating systems in urban and rural areas.

After simulation through EES software, the three important economic factors were also evaluated to measure the economic practicality of the proposed system, i.e., levelized cost of energy (LEC), net present worth (NPW), and the payback time (PBT). The value of LEC was found to be 0.062 USD/electricity unit, which was much lower than the LEC value of ELG (0.116 USD/electricity unit). As shown in Figure 15, the value of NPW (73.73 USD) indicates the high acceptability of the proposed system. Furthermore, the payback time was less than the life of the system, making it suitable for use in the commercial sector.

5. Conclusions

The overheating of heat pipes, poor transfer of heat across the absorber and finned heat pipe, and inability to provide hot water in the late evening hours are major problems associated with conventional heat pipe vacuum collector systems. The amalgamation of highly conductive storage material between the absorber tube (heat collecting surface) and the heat pipe is an effective way to overcome these problems. In this study, a stearic acid amalgamated vacuum tube solar collector system was designed and fabricated and its thermal output compared with a conventional vacuum tube system without storage material under the same environmental conditions. The experimental results showed that the amalgamation of stearic acid as storage material enhanced the thermal output of the solar system compared to the conventional one. This study's experimental results led to the following conclusions.

- The testing system with PCM could produce hot water during late evening hours (non-sunny/night hours) and improve the collector's thermal output.
- The amount of total usable heat output produced by the testing system with PCM was 179.10, 196.04, and 361.36 kJ more than the system without storage for Test 1_50 °C, Test 2_45 °C, and 19.44% Test 3_40 °C, respectively.
- The maximum energy efficiency of the two systems was 66.78 and 50.86% for Test 3_40 °C.
- The improvement in the energy efficiency of the testing system was 37.67% for Test 1_50 °C, 19.37% for Test 2_45 °C, and 19.44% for Test 3_40 °C compared to the reference system.

- ❖ The amount of additional hot water received from the testing system was 3.5 L for Test 1_50 °C and Test 2_45 °C and 10.5 L for Test 3_40 °C.
- ❖ It was noted that at the end of 15 years, almost 118.8 USD in revenue could be earned by the proposed solar system. The total running cost of ELG and the proposed solar system was observed to be 202.62 and 86.70 USD, respectively.
- ❖ On average, the cost of hot water production with the solar system and ELG was found to be 0.0016 and 0.004 USD/L, respectively. Therefore, the proposed solar system is highly recommended over conventional water heating systems in urban and rural areas.
- ❖ The value of LEC was found to be 0.062 USD/electricity unit, which was much lower than the LEC value of ELG (0.116 USD/electricity unit). The value of NPW (73.73 USD) indicated high acceptability of the proposed system. Furthermore, the payback time was lower than the life of the system, making it suitable for use in the commercial sector.

Author Contributions: K.C.: writing—original draft, investigation, methodology, acquisition of data, review and editing, V.V.T.: writing, reviewing and editing, investigation, supervision, S.K.P.: methodology, writing, review and editing, A.K.: formal analysis, reviewing and editing, A.K.P.: writing—reviewing and editing, N.A.R.: formal analysis, writing—reviewing, M.M.: editing, supervision and reviewing, A.S.: methodology and formal analysis. All authors have read and agreed to the published version of the manuscript.

Funding: This work was supported by Tenaga Nasional Berhad (TNB) and UNITEN through the BOLD Refresh Postdoctoral Fellowships under the project code J510050002-IC-6 BOLDREFRESH2025-Centre of Excellence. It was also supported by Shri Mata Vaishno Devi University, Katra (J&K) through the research project with unique ID RP-141 (Japan International Cooperation Agency Project for AUN/SEED-NET and the University of Malaya Malaysia).

Data Availability Statement: Data should be made available on request.

Conflicts of Interest: The authors declare no conflict of interest.

References

1. Zhu, T.-T.; Zhao, Y.-H.; Diao, Y.-H.; Li, F.-F.; Quan, Z.-H. Experimental investigation and performance evaluation of a vacuum tube solar air collector based on micro heat pipe arrays. *J. Clean. Prod.* **2017**, *142*, 3517–3526. [CrossRef]
2. Fuentes, E.; Arce, L.; Salom, J. A review of domestic hot water consumption profiles for application in systems and buildings energy performance analysis. *Renew. Sustain. Energy Rev.* **2018**, *81*, 1530–1547. [CrossRef]
3. Chopra, K.; Tyagi, V.; Pandey, A.; Sari, A. Global advancement on experimental and thermal analysis of evacuated tube collector with and without heat pipe systems and possible applications. *Appl. Energy* **2018**, *228*, 351–389. [CrossRef]
4. Pawar, V.R.; Sobhansarbandi, S. Heat transfer enhancement of a PCM-porous metal based heat pipe evacuated tube solar collector: An experimental study. *Sol. Energy* **2023**, *251*, 106–118. [CrossRef]
5. Algarni, S.; Mellouli, S.; Alqahtani, T.; Almutairi, K.; Khan, A.; Anqi, A. Experimental investigation of an evacuated tube solar collector incorporating nano-enhanced PCM as a thermal booster. *Appl. Therm. Eng.* **2020**, *180*, 115831. [CrossRef]
6. Essa, M.A.; Rofaiel, I.Y.; Ahmed, M.A. Experimental and Theoretical Analysis for the Performance of Evacuated Tube Collector Integrated with Helical Finned Heat Pipes using PCM Energy Storage. *Energy* **2020**, *206*, 118166. [CrossRef]
7. Olfian, H.; Ajarostaghi, S.S.M.; Ebrahimnataj, M.; Farhadi, M.; Arıcı, M. On the thermal performance of evacuated tube solar collector integrated with phase change material. *Sustain. Energy Technol. Assess.* **2022**, *53*, 102437. [CrossRef]
8. Li, Y.; Liang, X.; Song, W.; Li, T.; Wang, D.; Liu, Y. Optimization and thermal performance of U-type evacuated tube solar collector filled with phase change material. *Energy Rep.* **2022**, *8*, 6126–6138. [CrossRef]
9. Wu, Y.; Tong, X.; Li, D.; Arıcı, M.; Liu, C.; Liu, Y.; Yang, R.; Yu, Y. Energy analysis of evacuated tube solar collector integrating phase change material in northeast China. *J. Energy Storage* **2022**, *55*, 105772. [CrossRef]
10. O'neil, T.J.; Sobhansarbandi, S. Thermal performance investigation of energy storage based U-pipe evacuated tube solar collector: An experimental study. *Sustain. Energy Technol. Assess.* **2022**, *52*, 102146. [CrossRef]
11. Senobar, H.; Aramesh, M.; Shabani, B. Evacuated tube solar thermal collector with enhanced phase change material thermal storage: An experimental study. *J. Energy Storage* **2021**, *46*, 103838. [CrossRef]
12. Bouadila, S.; Rehman, T.-U.; Baig, M.A.A.; Skouri, S.; Baddadi, S. Energy, Exergy and Economic (3E) analysis of evacuated tube heat pipe solar collector to promote storage energy under North African climate. *Sustain. Energy Technol. Assess.* **2023**, *55*, 102959. [CrossRef]

13. Kumar, P.; Mylsamy, K. A comprehensive study on thermal storage characteristics of nano-CeO_2 embedded phase change material and its influence on the performance of evacuated tube solar water heater. *Renew. Energy* **2020**, *162*, 662–676. [CrossRef]
14. Uniyal, A.; Prajapati, Y.K.; Ranakoti, L.; Bhandari, P.; Singh, T.; Gangil, B.; Sharma, S.; Upadhyay, V.V.; Eldin, S.M. Recent Advancements in Evacuated Tube Solar Water Heaters: A Critical Review of the Integration of Phase Change Materials and Nanofluids with ETCs. *Energies* **2022**, *15*, 8999. [CrossRef]
15. Raza, S.A.; Ahmad, S.S.; Ratlamwala, T.A.H.; Hussain, G.; Alkahtani, M. Techno-Economic Analysis of Glazed, Unglazed and Evacuated Tube Solar Water Heaters. *Energies* **2020**, *13*, 6261. [CrossRef]
16. Olczak, P.; Matuszewska, D.; Zabagło, J. The Comparison of Solar Energy Gaining Effectiveness between Flat Plate Collectors and Evacuated Tube Collectors with Heat Pipe: Case Study. *Energies* **2020**, *13*, 1829. [CrossRef]
17. Jachura, A.; Sekret, R. Life Cycle Assessment of the Use of Phase Change Material in an Evacuated Solar Tube Collector. *Energies* **2021**, *14*, 4146. [CrossRef]
18. Essa, M.A.; Mostafa, N.H.; Ibrahim, M.M. An experimental investigation of the phase change process effects on the system performance for the evacuated tube solar collectors integrated with PCMs. *Energy Convers. Manag.* **2018**, *177*, 1–10. [CrossRef]
19. Olfian, H.; Ajarostaghi, S.S.M.; Farhadi, M.; Ramiar, A. Melting and solidification processes of phase change material in evacuated tube solar collector with U-shaped spirally corrugated tube. *Appl. Therm. Eng.* **2021**, *182*, 116149. [CrossRef]
20. Ali, S.H.; Eidan, A.A.; Al Sahlani, A. Study the Effect of Phase Change Material on the Performance of Evacuated Tube Heat Pipe Solar Collector under Iraqi Climatic Conditions. *IOP Conf. Ser. Mater. Sci. Eng.* **2021**, *1094*, 012009. [CrossRef]
21. Kumar, P.M.; Mylsamy, K.; Alagar, K.; Sudhakar, K. Investigations on an evacuated tube solar water heater using hybrid-nano based organic phase change material. *Int. J. Green Energy* **2020**, *17*, 872–883. [CrossRef]
22. Bouadila, S.; Baddadi, S.; Rehman, T.-U.; Ayed, R. Experimental investigation on the thermal appraisal of heat pipe-evacuated tube collector-based water heating system integrated with PCM. *Renew. Energy* **2022**, *199*, 382–394. [CrossRef]
23. Chopra, K.; Tyagi, V.; Pandey, A.; Sharma, R.; Sari, A. PCM integrated glass in glass tube solar collector for low and medium temperature applications: Thermodynamic & techno-economic approach. *Energy* **2020**, *198*, 117238. [CrossRef]
24. Chopra, K.; Tyagi, V.; Pathak, S.K.; Tripathi, P.A.; Sharma, R.; Singh, G.; Pandey, A. Thermal and chemical reliability of paraffin wax and its impact on thermal performance and economic analysis of solar water heater. *Energy Sustain. Dev.* **2023**, *73*, 39–53. [CrossRef]
25. Chopra, K.; Tyagi, V.; Pandey, A.; Popli, S.; Singh, G.; Sharma, R.; Sari, A. Effect of simultaneous & consecutive melting/solidification of phase change material on domestic solar water heating system. *Renew. Energy* **2022**, *188*, 329–348. [CrossRef]
26. Zheng, Y.; Hatzell, K.B. Technoeconomic analysis of solar thermal desalination. *Desalination* **2020**, *474*, 114168. [CrossRef]
27. Chopra, K.; Tyagi, V.; Popli, S.; Pandey, A. Technical & financial feasibility assessment of heat pipe evacuated tube collector for water heating using Monte Carlo technique for buildings. *Energy* **2023**, *267*, 126338. [CrossRef]
28. Pathak, S.K.; Tyagi, V.; Chopra, K.; Sari, A. Thermal performance and design analysis of U-tube based vacuum tube solar collectors with and without phase change material for constant hot water generation. *J. Energy Storage* **2023**, *66*, 107352. [CrossRef]
29. Retail Domestic Term Deposits-Interest Rates, (n.d.). Available online: https://sbi.co.in/web/interest-rates/deposit-rates/retail-domestic-term-deposits (accessed on 28 April 2023).
30. Inflation in India 2028 | Statista, (n.d.). Available online: https://www.statista.com/statistics/271322/inflation-rate-in-india/ (accessed on 28 April 2023).
31. Velayudhan, S. Dissemination of solar photovoltaics: A study on the government programme to promote solar lantern in India. *Energy Policy* **2003**, *31*, 1509–1518. [CrossRef]

energies

Article

Experimental Investigation of Graphene Nanoplatelets Enhanced Low Temperature Ternary Eutectic Salt Hydrate Phase Change Material

B. Kalidasan [1], A. K. Pandey [1,2,*], Saidur Rahman [1], Kamal Sharma [3] and V. V. Tyagi [4]

1 Research Centre for Nano-Materials and Energy Technology (RCNMET), School of Engineering and Technology, Sunway University, Petaling Jaya 47500, Malaysia
2 Center for Transdisciplinary Research (CFTR), Saveetha Institute of Medical and Technical Services, Saveetha University, Chennai 602105, India
3 Department of Mechanical Engineering, Institute of Engineering and Technology, GLA University, Mathura 281406, India
4 School of Energy Management, Shri Mata Vaishno Devi University, Katra 182320, India
* Correspondence: adarshp@sunway.edu.my

Abstract: A sustainable approach to ensuring the thermal regulation of space is reliable with phase change materials (PCMs) operating at 15–25 °C. Henceforth, there is a need of a search of binary and ternary eutectic PCMs operating at desirable phase transition temperatures of 15–25 °C, high energy storage enthalpy (180–220 J/g), improved thermal conductivity and better absorptivity of solar energy. In this current research, we developed a ternary eutectic inorganic salt hydrate PCM intended for a low-temperature thermal regulation system. Based on the eutectic melting point theory, the phase transition temperature and proportion of sodium carbonate decahydrate (SCD), sodium phosphate dibasic dodecahydrate (SPDD) and sodium sulphate decahydrate (SSD) were determined. As per the calculated proportion, ternary eutectic PCM was experimentally prepared. Furthermore, to enhance the thermal property, graphene nanoplatelets (GNP) were dispersed at weight concentrations of 0.4%, 0.7% and 1.0%. The prepared nanoparticle-dispersed PCMs were characterized using an optical microscope, Fourier transform infrared (FT-IR) spectroscopy and a thermal conductivity meter, and a differential scanning calorimeter (DSC) was used to evaluate the morphology, chemical stability and thermal properties. The results showed increases in thermal conductivity and optical absorbance by 71.5% and 106.5%, respectively, with GNP at 1.0% weight concentration. Similarly, the degree of supercooling and transmissibility was reduced by 43.5% and 76.2% correspondingly. The prepared composite PCM is expected to contribute towards cooling, with an intention to contribute towards sustainable development.

Keywords: salt hydrate; phase change materials; graphene nanoplatelets; thermal energy storage

check for updates

Citation: Kalidasan, B.; Pandey, A.K.; Rahman, S.; Sharma, K.; Tyagi, V.V. Experimental Investigation of Graphene Nanoplatelets Enhanced Low Temperature Ternary Eutectic Salt Hydrate Phase Change Material. *Energies* **2023**, *16*, 1574. https://doi.org/10.3390/en16041574

Academic Editor: Ann Lee

Received: 4 January 2023
Revised: 30 January 2023
Accepted: 31 January 2023
Published: 4 February 2023

1. Introduction

The desire to resolve the problem of excess carbon emissions that cause global warming and serious environmental threats has led to developing and utilizing clean and renewable energy sources. Due to the intermittent nature of renewable energy sources like solar, wind and hydropower, there exists an imbalance between the energy demand and supply. Based on the active research over the past decade, it is observed that thermal energy storage (TES) systems are more likely to resolve the mismatch of energy supply and demand by effectively utilizing solar energy. Among the three (latent heat storage, sensible heat storage and chemical heat storage) [1] general categories of TES techniques we would focus more on latent heat storage. The benefits of latent heat storage using phase change materials (PCMs) are more reliable as they need neither to be controlled as in the case of chemical heat storage nor undergo the hindrance of less energy density as in sensible heat storage. Based on the composition, PCMs are mainly categorized as organic (paraffin, fatty acids and sugar alcohols), inorganic (salt hydrates) and eutectic (mixture of organic–organic, organic–inorganic or inorganic–inorganic) [2]. A few notable hindrances with organic PCMs are low thermal conductivity, flammable nature and high cost. In the case of inorganic PCMs, their thermal conductivity is high compared to that of organic PCMs, they are non-flammable, have larger energy storage density and are cheaper [3]. Eutectic PCMs are a tailored material, where the melting point and the thermal property of the PCM are designed as per the designated application. Due to the non-flammable and economic benefit of inorganic PCMs, they are more preferred for space (building) heating and cooling, solar water heating systems, waste heat recovery units and in energy regulation units.

Considering the advantage of eutectic PCMs in formulating the eutectic composition as per the requirement of energy storage application, numerous research works have been carried out. A few outstanding contributions are discussed briefly to identify and elaborate the uniqueness of the current research work. Cheng et al. [4] conducted an experimental investigation with a ternary eutectic nitrate composite PCM $LiNO_3.3H_2O - KNO_3 - NaNO_3$ by dispersing expanded graphite nanoparticles. The results showed a maximal melting enthalpy of 159 J/g with thermal conductivity of 4.565 W/m·K with 20% of modified expanded graphite nanoparticles. Although the phase transition temperature of the developed ternary composite PCM was 21.66 °C, the heat storage enthalpy was much lower. Like in the previous work, Tian et al. [5] developed a ternary eutectic carbonate salt operating at a very high temperature of 725 °C. The composite mixture of the ternary salt was $(Li_2CO_3 - Na_2CO_3 - K_2CO_3)$ with heat of fusion equal to 160 J/g. Based on Gibbs energy of fusion, Wang et al. [6] developed a high-temperature (420 °C) ternary eutectic PCM with salts of $LiF - Na_2CO_3 - K_2CO_3$. The designed high-temperature ternary PCM was applied for heat transfer fluid for effective TES. Liang and Chen et al. [7] developed a ternary eutectic PCM $(K_2HPO_4.3H_2O - NaH_2PO_4.2H_2O - Na_2S_2O_3.5H_2O)$ for cold storage application operating at a temperature of -14 °C, with latent heat capacity of 127 J/g. The research investigation was focused on the role of thickening and nucleating agents in resolving the phase instability and the supercooling issue. Sun et al. [8] conducted a TES experiment for buildings by developing a ternary eutectic salt hydrate mixture using 95% of $CaCl_2.6H_2O$, 4% of NH_4Cl and 1% of $SrCl_2.6H_2O$ with a phase transition temperature of 23.5 °C. In another research work, a ternary eutectic PCM was developed for air conditioning and building cooling [9]. The ternary eutectic PCM $(Zn(NO_3)_2.6H_2O - Mn(NO_3)_2.4H_2O - KNO_3)$ was designed to operate at 20 °C with a melting enthalpy of 110 J/g. This investigation also focused on determining the right nucleating agent for the developed PCM.

In other work, Zhang et al. [10] developed a lauric–myristic–palmitic acid ternary eutectic PCM at a eutectic mixture proportion of 55.24/29.74/15.02 operating at a temperature of 31.5 °C. To further enhance the thermophysical property, they expanded the graphite nanoparticles' weight concentration to 5.5% and achieved thermal conductivity of 1.67 W/m·K with compensation in latent heat. An organic-PCM-based ternary eutectic composite operating at a phase transition temperature of 17 °C was developed with 64.8% capric acid, 22.6% myristic acid and 12.6% palmitic acid [11]. Melting latent heat and thermal conductivity of the developed CA–MA–PA ternary eutectic PCM were 131.7 J/g and 0.149 W/m·K, respectively. To further enhance the thermal performance, 10% exfoliated graphite was dispersed and resulted in an increase of thermal conductivity to 20% with compensating the heat storage ability. Similarly, in another research work [12], a ternary eutectic composite of capric acid, palmitic acid and stearic acid was investigated by dispersing expanded vermiculite at a 30% weight concentration. The developed ternary eutectic PCM CA–PA–SA was designed to operate at a temperature of 19.3 °C with melting enthalpy of 117 J/g. For its thermal stability and noncorrosive properties, Ke et al. [13] developed a fatty-acids-based ternary eutectic PCM. Ternary eutectic PCMs were opted based on his earlier research work and were (a) CA–LA–PA (63.37/31.56/5.07), (b) CA–LA–SA (65.32/32.54/2.14), (c) CA–MA–PA (71.82/18.86/9.32), (d) CA–MA–SA (72.65/21.17/6.18) and (e) CA–PA–SA (83.82/10.19/5.99). The melting point and latent heat values of all the prepared form-stable PCMs were determined to be 15–35 °C and 120–139 J/g correspondingly. Luo et al. [14] developed a form-stable ternary eutectic PCM by impregnation (CA–PA–SA) into nano SiO_2. The developed form-stable ternary PCM was designed to operate at the temperature range of 17–26 °C with latent heat of 100 J/g. The intended application of the PCM was to store solar radiation and provide better thermal regulation of a building during the winter season. He et al. [15] developed two ternary eutectic PCMs with (a) lauric–myristic–palmitic acid (59/26/15) and (b) lauric–myristic–paraffin (51/28/21) operating at melting points of 32.3 °C and 31.5 °C. The enthalpies of the developed ternary PCMs were 160 J/g and 172 J/g, respectively. Based on the current literature analysis, it can be inferred that most of the developed low eutectic PCMs are used directly in various TES applications without inclusion of any further additives or nanomaterials. Likewise, a recent advancement in material science has led to the usage of carbon-based nanomaterials with higher thermal conductivity to be dispersed with the base PCM for enhancing the thermal property. Depending on the unique shape and structure, carbon nanomaterials are of 1D, 2D and 3D. Among them, 2D graphene nanoplatelets and nanomaterials offer better thermal enhancement due to higher specific surface area, ability to better accommodate foreign materials at the gaps of layers, and higher thermal conductivity. The major focus of the research work was to experimentally compare their performance with respect to the number of thermal cycling over a longer duration of operation. In the literature it can be found that the predominantly used PCM for developing a new ternary eutectic PCM was organic fatty acids, and there is a void of research works performed with inorganic PCMs, especially with low-temperature salt hydrates.

The current research investigation focuses on the design and development of a low-temperature ternary eutectic PCM with a mixture of inorganic salt hydrate sodium carbonate decahydrate (SCD), sodium phosphate dibasic dodecahydrate (SPDD) and sodium sulphate decahydrate (SSD). The ternary eutectic mixture proportion and melting point of SCD/SPDD/SSD PCM was designed using Schrader's equation, followed by experimental synthesis and analysis of their thermal properties. Melting (21.3 °C) and latent heat values (202.2 J/g) of the ternary eutectic PCM determined using differential scanning calorimetry (DSC) were in accordance with the numerically calculated values (T_m = 21.5 °C and ΔH_m = 207 J/g) at the eutectic mixture. It was detected that graphene nanoplatelets (GNPs) exhibited a two-dimensional flat and thin plate-like structure with high surface area. GNP nanomaterials are unsystematically organized with spacing between alternate layers that enhance the intermolecular force of attraction when dispersed with a eutectic composite. GNPs are more likely to contribute to the base ternary eutectic salt hydrate PCM by improving the thermophysical characteristics, particularly (a) melting enthalpy, (b) conductive nature by developing thermal networks and (c) enhanced optical absorptivity of solar radiation owing to the dark appearance. For further enhancement of thermal properties, the developed ternary eutectic PCM SCD/SPDD/SSD were dispersed with GNP to prepare three composite mixtures at weight concentrations of 0.4%, 0.7% and 1.0%. Chemical stability using Fourier transform infrared spectroscopy (FT-IR), optical absorptivity using ultraviolet visible spectroscopy (UV–vis), thermal conductivity using thermal hot bridge instrument and thermophysical properties like melting point, latent heat storage and degree of supercooling using a DSC instrument were characterized. Finally, to analyze the heat transfer rate between the developed ternary eutectic PCM and the GNP-dispersed ternary eutectic PCM, a numerical simulation using ANSYS software was conducted based on their thermal conductivity with variation in heat input. The developed ternary eutectic PCM was expected to improve the application of the inorganic salt hydrate PCM for thermal regulation of buildings and is more likely to contribute to the Sustainable Development Goals (SDGs)

2. Materials and Methods

2.1. Materials

The salt hydrates PCMs used in the current research investigations were sodium carbonate decahydrate (SCD), sodium sulphate decahydrate (SSD) and sodium phosphate dibasic dodecahydrate (SPDD). All the three inorganic salt hydrate PCMs, SCD with T_m 31 °C, SSD with T_m 33 °C and SPDD with T_m 37 °C were procured from Sigma Aldrich, Germany. Nanoparticle graphene nanoplatelets (GNPs) with a surface area of 300 m^2/g, bulk density of 200–400 kg/m^3 and particle size of <2 μm were acquired from US research Nanomaterials Inc. to further improve the thermal properties of the developed ternary eutectic PCMs.

2.2. Determination of Thermal Properties of Ternary Eutectic PCMs

The mixture of three different salt hydrate PCMs at different proportions formed ternary eutectic salt hydrate PCMs. For any PCM, T_m is isothermal in nature, and while developing a new eutectic PCM (mixing two or more pure PCMs with different T_m), a constant solitary melting point is very important. Henceforth, we adopted a numerical method to determine the proportion at which the salt hydrate PCMs (SCD, SSD and SPDD) had to be added for a common phase transition temperature of the developed ternary eutectic PCM. Based on the literature and reliability of Schrader' equation, we used Equation (1) derived as per phase equilibrium theory to determine the eutectic T_m and the eutectic mixture ratio. This numerical method reduced the number of tests required to determine the common melting point and is also an economical technique. As Schrader's equation [16] is well applied for a eutectic mixture of two components, we considered the property of two pure salt hydrate PCMs individually (component A and component B). Using Equation (1), we determined the eutectic melting point and their composition to develop a binary eutectic salt hydrate [17]. Later, considering the properties of the designated binary salt hydrate PCM as a single salt (component A), we included the third pure salt hydrate PCM (component B) to prepare the ternary eutectic salt hydrate PCM. Likewise, we used Equation (2) to determine the phase transition enthalpy of the designed ternary eutectic salt hydrate PCM [18].

$$T_m = \left[\frac{1}{T_i} - R\cdot \frac{\ln X_i}{\Delta H_{m,i}}\right]^{-1} \tag{1}$$

$$H_{eu} = T_{eu}\cdot \sum_{i=1}^{n}\left[\frac{X_i\cdot H_i}{T_i} + \left\{X_i\cdot(Cp_{li} - Cp_{si})\cdot \ln\frac{T_{eu}}{T_i}\right\}\right] \tag{2}$$

In the equations above, T_m is the phase transition temperature of the eutectic mixture in K, T_i represents the phase transition temperature of individual salt hydrates (A, B, C) in K, the proportion of the individual salt hydrate mixture is represented as X in mole fraction. ΔH_m and ΔH_{eu} specify the melting enthalpy of pure and eutectic salt hydrate PCM, respectively, in $kJ.kmol^{-1}$, and R is the ùniversal gas constant 8.314 $kJ\cdot mol^{-1}\cdot K^{-1}$. C_p represents the specific heat capacity in solid and liquid phases, respectively, in KJ/kgK.

2.3. Preparation of Ternary Eutectic PCMs

This section elaborates the step-by-step process of preparing the ternary eutectic PCM, and the nanoparticle-dispersed ternary eutectic salt hydrate PCM. The designed ternary salt hydrate PCMs of composition SCD/SPDD/SSD at weight fractions of x/y/z were prepared through melting followed by sonication. Figure 1 illustrates the step-by-step synthesis process of the ternary eutectic PCM, and the process of dispersing GNPs with a base ternary eutectic PCM. Based on the numerical design, we obtained the ternary eutectic composition of SCD/SPDD/SSD with the weight ratio of 44.8/21/34.2 and a melting point of 21.5 °C. At first, we melted 4.48 g of SCD salt hydrate PCM in a beaker using a hot plate maintained at 45 °C. Furthermore, 2.1 g of SPDD was added to the liquid phase SCD, and we obtained a binary eutectic PCM (SCD/SPDD) with a melting point of 26.5 °C. As a final point, we mixed 3.42 g of SSD salt hydrate to the binary eutectic PCM, and hereby we prepared the designed ternary eutectic PCM (SCD/SPDD/SSD). To further enhance the thermal property of the prepared ternary eutectic salt hydrate PCM, GNPs were dispersed at different weight concentrations (0.4 wt%, 0.7 wt% and 1.0 wt%) and coded as SCD/SPDD/SSD-0.4GNP, SCD/SPDD/SSD-0.7GNP and SCD/SPDD/SSD-1.0GNP correspondingly. At the end, the composite samples were sonicated using an ultrasonicator bath (EASY 60H, ELMASONIC) for 30 min to ensure uniform dispersion of GNPs with the ternary eutectic PCM.

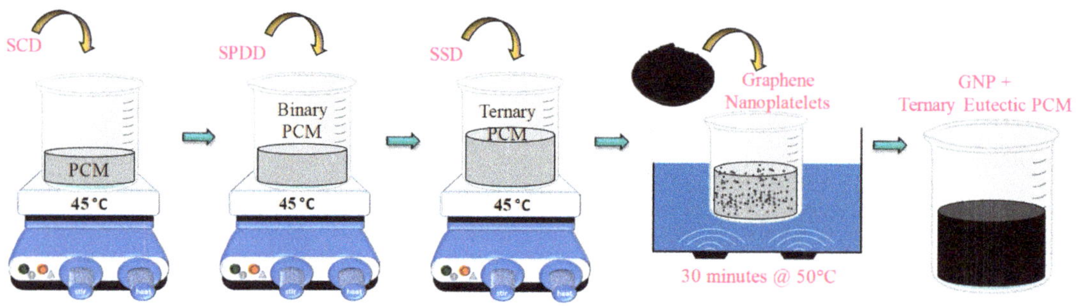

Figure 1. Synthesis of ternary eutectic salt hydrate PCM, with crystal structure visuals of pure salt hydrate PCM.

2.4. Instruments and Characterization Techniques

To characterize and experimentally evaluate the developed low-temperature ternary eutectic PCMs, we used a few sensitive instruments. The morphology of the individual pure salt hydrate PCMs was observed using an optical Stemi Microscope (Model STEMI2000, ZEISS). The microstructure of GNPs was examined with a scanning electron microscope (SEM) TESCAN VEGA 3. FT-IR spectroscopy (Perkin Elmer, Waltham, MA, USA) was used to detect spectral peaks of the ternary eutectic PCM and its composites with GNP. The wavenumber range of 400 to 4000 cm^{-1} was chosen to study the chemical composition, as the mid IR spectrum wavelength ranges between 2.5 and 25 μm. Optical absorbance and transmittance of the prepared sample in the solar spectrum region were examined using UV–vis spectroscopy (Model: LAMBDA 750, Perkin Elmer, USA). The readings were taken from a wavelength of 200 to 1400 nm at solid state. DSC instrument (DSC 3500 Sirius, NETZSCH) was used to analyze the melting temperature and latent heat properties of the TPCM composite. DSC melting and cooling curves were inspected between −15 and 45 °C under a N_2 atmosphere with a heating rate of 5 °C/min to determine the heating and cooling enthalpy, phase transition temperature and degree of supercooling. The base PCM and the developed nanocomposite PCM were numerically analyzed by ANSYS R16.0. All curves were plotted in Origin 2020 software.

3. Results and Discussion

A combination of numerical and experimental characterization was carried out in the current work. This research investigation was focused on developing a ternary eutectic salt hydrate PCM, designed to operate in low-temperature thermal energy regulation units. Here we designed, developed and characterize the synthesized ternary eutectic PCM with GNPs to ensure their chemical stability, optical absorptivity and thermal properties.

3.1. Design and Development of Ternary Eutectic Salt Hydrate PCM

Eutectic PCMs hold the advantage of tailoring the melting point of PCM as per the desired application. Here, a novel low-temperature ternary eutectic mixture PCM consisting of SCD, SPDD and SSD was designed numerically, followed by experimental synthesis and characterization for space cooling. Schrader's Equation (1) was chosen to numerically evaluate the eutectic weight ratio of the individual salt hydrate for obtaining a common phase transition point. At first, we considered the thermal properties of salt hydrate SCD (T_m = 31 °C; ΔH_m = 210 J/g) and SPDD (T_m = 37 °C; ΔH_m = 227 J/g) to evaluate the binary eutectic phase transition temperature (T_{eu}). The fraction of salt hydrate (X_i) in Equation (1) was varied to determine T_{eu}. On evaluation, it was observed that a binary eutectic mixture of SCD and SPDD exhibited a common phase transition temperature at 26.1 °C. Figure 2a represents T_{eu} of SCD/SPDD at different weight proportions, and it can be inferred that T_{eu} (26.1 °C) was obtained with 32% of SPDD and 68% of SCD. The obtained eutectic point is a temperature at which the individual component both starts to melt and undergoes phase transition. Eutectic phase transition temperature is always lower than the phase transition temperature of the individual component, owing to the increase in impurity (inclusion of another component), where entropy generation increases and tends to reduce the melting point of the mixture [19,20]. Both individual components undergo a decrease in temperature and converge at a common point, which is the eutectic point. Likewise, using Equation (2), we determined the heat storage enthalpy of the developed binary eutectic PCM. Figure 2b depicts the variation in the latent heat value of pure salt hydrate PCM and binary eutectic PCM at different weight compositions. On further calculation, the latent heat of the binary eutectic PCM was determined to be 210 J/g.

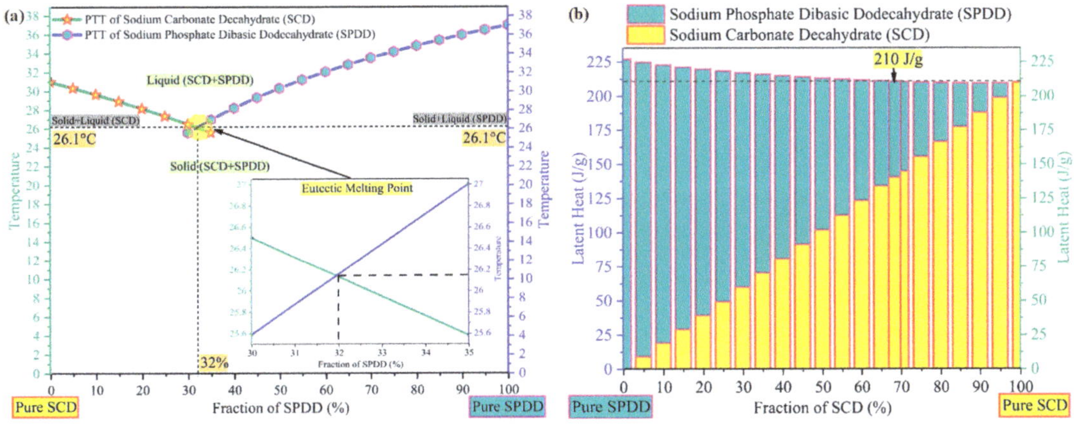

Figure 2. (**a**) Phase transition point of SCD/SPDD binary salt hydrate PCM; (**b**) heat storage enthalpy of SCD/SPDD binary salt hydrate PCM.

Considering the binary eutectic PCM SSD/SPDD (T_m = 26.1 °C; ΔH_m = 210 J/g) as a unique salt hydrate PCM (individual component) and pure SSD (T_m = 33 °C; ΔH_m = 217 J/g) salt hydrate PCM (second component), we adopted Equations (1) and (2) to calculate the eutectic melting point, composition ratio and latent heat for the ternary eutectic PCM. Figure 3a,b describes the variation in melting point and latent heat of the eutectic PCM, with variation in proportion of the component. On evaluation, it was observed that the ternary eutectic mixture of SCD, SPDD and SSD exhibited a common phase transition temperature at 21.5 °C. From Figure 3a it can be inferred that the eutectic phase transition temperature 21.5 °C was obtained at 34.2% of SSD and 65.8% of SCD/SPDD. SCD/SPDD exhibited a eutectic melting point at 68% of SCD and 32% of SPDD. Henceforth it was determined that the composition of the developed ternary eutectic PCM was 34.2% SSD, 44.75% SCD (65.8% of 68%SCD) and 21.05% SPDD (65.8% of 32%SPDD). Likewise, for SCD/SPDD/SSD (34.2/21.05/44.75), the latent heat value was determined to be 207 J/g.

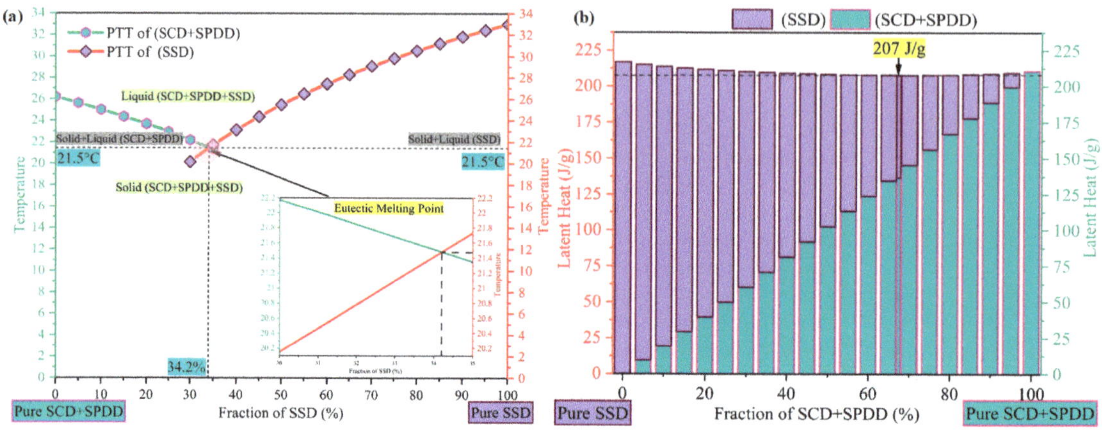

Figure 3. (**a**) Phase transition point of SCD/SPDD/SSD binary salt hydrate PCM; (**b**) heat storage enthalpy of SCD/SPDD/SSD binary salt hydrate PCM.

3.2. Morphological Behavior

The microstructure and morphology of the developed inorganic-salt-hydrate-based ternary eutectic PCM was examined using an optical microscope. Figure 4 shows the optical image visuals of pure salt hydrate SCD, SPDD and SSD for a better understanding on the crystal structure (bulk, coarse and fine). SCD salt hydrate displayed very fine salt crystals of uniform size, whereas SPDD salt hydrate PCM consisted of bulk and coarse salt crystals with irregular shape. Although SPDD is a dodecahydrate molecule, due to a high melting temperature (37 °C), the salt hydrate is complete in the solid state, without exhibiting any water molecule presence in the surface like SCD. On keen observation it could be inferred that the developed binary eutectic PCM (SCD/SPDD) and the ternary eutectic PCM (SCD/SPDD/SSD) replicated a uniform morphology in terms of the crystal structure and grain size and ensured a proper dispersion. High hydrated molecules of SCD/SPDD/SSD were also understandable from the optical visual image. The morphology and microstructure of nanoparticle GNPs were examined using a scanning electron microscope (SEM), and the images are provided in Figure 4. It can be inferred that GNP nanomaterials had two-dimensional irregular multiple layers of a sheet-like structure with some voids and gaps between subsequent layers. Nevertheless, we depicted the digital image of the base ternary eutectic PCM and GNP-dispersed ternary eutectic PCM for better consideration of the dispersion of GNP with the developed ternary eutectic PCM.

3.3. Chemical Stability

On investigating the developed ternary eutectic PCM under a FTIR spectrometer, the wavelength position and intensity of the absorption band imitated the characteristics of molecular structure, which is significant in understanding the presence of a chemical group. FTIR spectral curves of the developed ternary eutectic PCM (SCD/SPDD/SSD) and its composites with different fractions of GNP can be observed from Figure 5. In all the tested samples, we could observe wide and sharp spectral peaks with varying intensity. The wide peak around wavenumber 2900–3600 cm^{-1} represented the O-H stretching vibration due to the crystallization of water molecules [21]. Sharp peaks at wavenumber 990 cm^{-1} denoted the PO-H [22] bending vibration, ensuring the presence of the SPDD salt hydrate. The sharp peak at 1077 cm^{-1} represented $SO_4{}^{2-}$ [23], with asymmetric stretching vibration ensuring the presence of the SSD salt hydrate. The other higher intense sharp peak at 1383 cm^{-1} represented $CO_3{}^{2-}$ [24], indicating the asymmetric stretching vibration confirming the presence of the SCD component. Additionally, GNPs under exposure to the FTIR spectrum do not produce any peaks within the IR range, which ensures the non-IR active nature of GNPs [25]. In all the spectral curves of GNP-dispersed ternary eutectic PCM, no new absorption peaks were generated, and there were no missing peaks that were observed in the ternary eutectic PCM. This spectral curve as obtained from FTIR ensured that no chemical reaction occurred between the developed ternary eutectic PCM GNP at different concentrations due to any physical interaction.

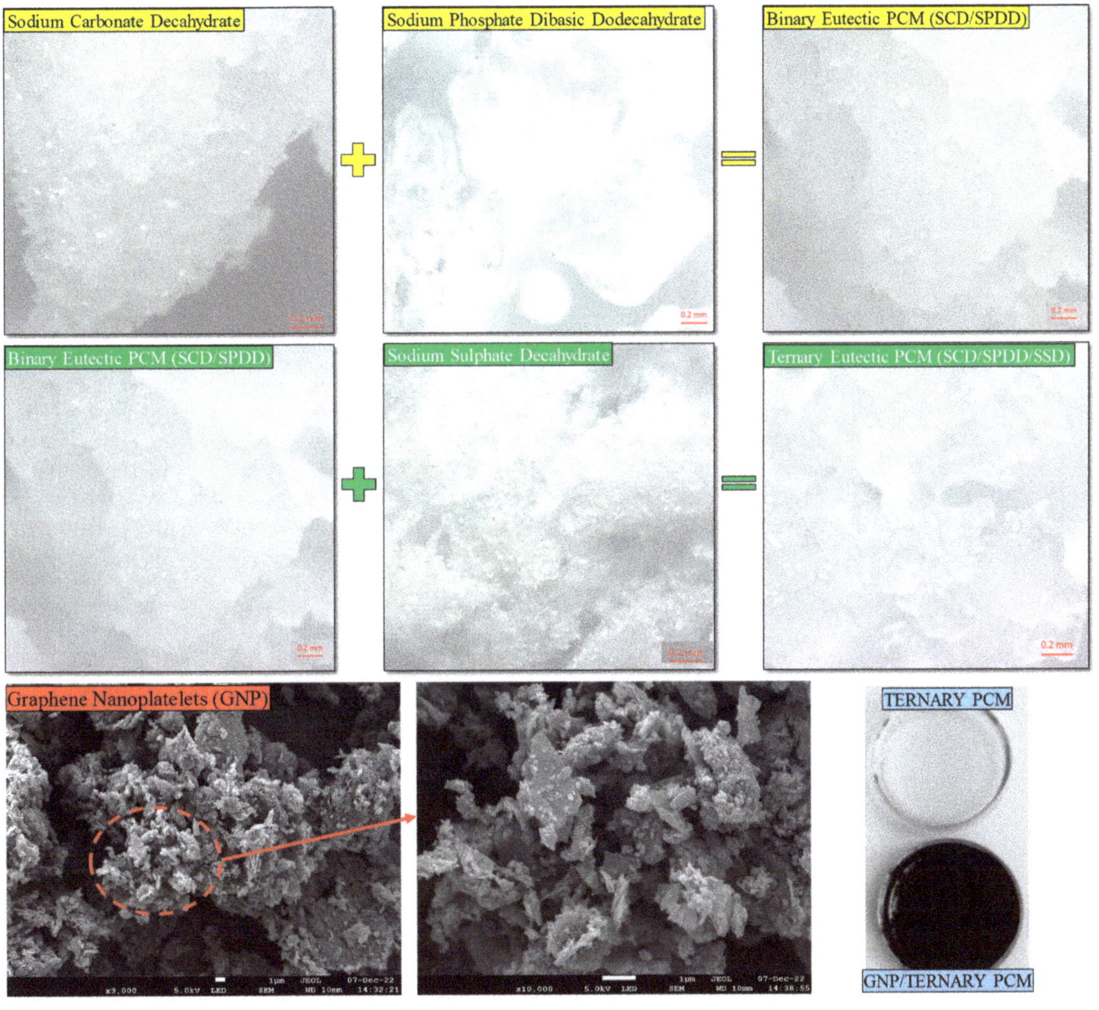

Figure 4. Crystal structure visuals of pure and eutectic salt hydrate PCM using an optical microscope; microstructure and morphological image of graphene nanoplatelets; and digital image of base ternary eutectic composite and GNP-dispersed ternary eutectic composite.

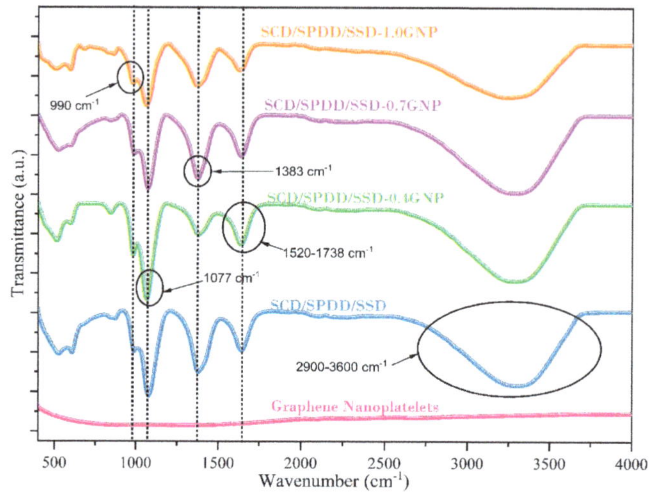

Figure 5. FTIR spectral curves of ternary eutectic PCM and GNP-dispersed ternary eutectic PCM.

3.4. Optical Property Analysis

The developed ternary eutectic PCM was designed to be operated for thermal regulation of buildings. PCM melting at 21.5 °C can be preferably used for building heating as well as building cooling. In general, for building heating, heat energy from solar power is absorbed and stored in the PCM, and later this energy is released into the heating space for thermal regulation [26]. On analysis it is noted that most of the inorganic materials are transparent and transmit more solar radiation than absorption. Hence in this section, the optical property of the developed ternary eutectic salt hydrate PCM was experimentally investigated under UV–vis spectroscopy. Figure 6a,b shows the absorbance and transmittance curves of the developed ternary eutectic salt hydrate PCM along with GNP at different weight concentrations with variation in ultraviolet–visible light waves.

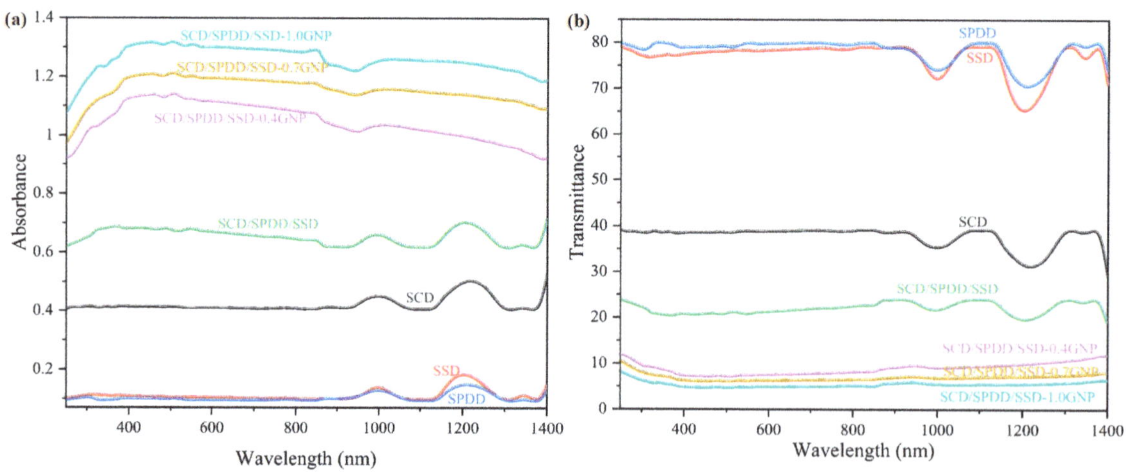

Figure 6. Solar spectral (**a**) absorbance and (**b**) transmittance of GNP-dispersed ternary eutectic PCM.

As can be seen in Figure 6a, the absorbance intensity of the developed ternary eutectic PCM SCD/SPDD/SSD (0.62 absorbance) was stronger in comparison with that of the individual salt hydrate PCM SCD (0.43 absorbance), SPDD (0.14 absorbance) and SSD (0.15 absorbance) in all regions (UV and visible). The higher optical absorptivity of SCD/SPDD/SSD could be ascribed to the stickier nature of the developed ternary PCM with higher content of water molecules. The crystal structures also contributed towards repeated scattering within the PCM and enhanced the absorptivity. In addition, the improvement in optical properties with dispersion of GNPs was further inferred. Spectral curves of GNP-dispersed SCD/SPDD/SSD at nanoparticle concentrations of 0.4%, 0.7% and 1.0% illustrated absorbances of 1.02, 1.18 and 1.28, respectively. Increases in optical absorptivity of GNP-dispersed ternary PCM were ascribed to the dark nature of the GNP-dispersed composite. The dark nature decreased the transmissibility of the PCM and improved the absorptivity. Besides absorbance, the transmissibility behavior of the developed ternary PCM is depicted in Figure 6b. The lower the transmittance, the better the absorbance. On further assessment, the optical absorptivity of the ternary salt hydrate PCM reduced from 21.8% to 8.4%, 6.7% and 5.2% for GNP weight concentrations of 0.4, 0.7 and 1.0, respectively. Greater absorptivity improves the solar energy input and contributes to higher TES at a comparatively faster rate.

3.5. Thermal Conductivity

PCM for any application is selected based on the heat energy storage potential and the rate at which the thermal energy charging takes place. Thermal conductivity indicates the rate at which the heat transfer occurs in any materials. In general, inorganic PCM offers improved thermal conductivity compared to that of organic PCM. Although this research work focuses on inorganic PCMs, to further increase the thermal properties of the developed ternary eutectic salt hydrate PCM, we dispersed 2D GNPs at concentrations of 0.4%, 0.7% and 1.0% to evaluate the thermal property enhancement.

As the developed ternary eutectic PCM was designed to operate at a low temperature, we used a transient hot-bridge thermal conductivity meter to find the thermal conductivity of the GNP-dispersed ternary eutectic PCM and the pure salt hydrate PCMs. The room temperature was maintained at 15 °C to ensure a solid state of the sample while conducting the characterization experiment. Figure 7 shows the bar plots of pure salt hydrate SCD, SPDD and SCD; developed ternary eutectic PCM SCD/SPDD/SSD and GNP-dispersed ternary eutectic PCM at different compositions. In solid materials, the significant cause for thermal conductivity is the movement of free electrons; however, inorganic salt hydrates have fewer free electrons as most of the electrons remain confined to the ionic crystal lattice [3]. Hence in inorganic salt hydrates, thermal conduction occurs due to crystal lattice vibration. Figure 7 depicts the thermal conductivity of base salt hydrate PCM SCD, SSD and SPDD as 0.441 W/m·K, 0.458 W/m·K and 0.513 W/m·K, respectively. As a new ternary eutectic PCM was developed, it was of utmost importance to evaluate the thermal conductivity of the synthetic PCM with their individual mixtures. The thermal conductivity of the synthesis PCM was 0.473 W/m·K, which was a nominal value between the thermal conductivity of SCD and SPDD, which could be ascribed to the distribution of the base salt hydrate. Furthermore, GNPs were dispersed at weight concentrations of 0.4%, 0.7% and 1.0% to advance the thermal conductivity of the synthesized PCM. It can be inferred that 0.4%, 0.7% and 1.0% of GNPs enhanced the thermal conductivity of the ternary PCM by 31.7%, 48.6% and 71.5%, respectively. The increment in thermal conductivity was ascribed to higher thermal conductivity of the GNPs and their higher surface area, which develops a well-connected thermal network for enhanced heat transfer rate [26]. On the whole, comparing the thermal conductivity (0.811 W/m·K) of SCD/SPDD/SSD-0.7GNP PCM with that of organic PCM (0.2 W/m·K), there was an increment of nearly 300%.

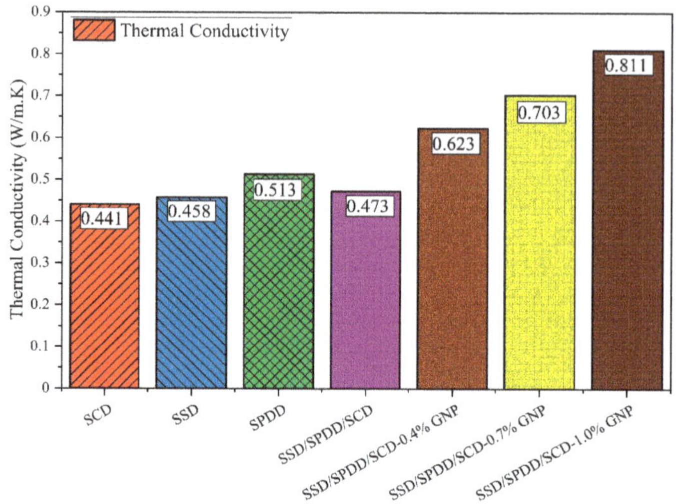

Figure 7. Thermal conductivity of GNP-dispersed ternary eutectic salt hydrate PCM.

3.6. *Melting Temperature, Latent Heat Storage and Degree of Supercooling*

PCM are preferred for TES owing to their ability to store higher energy within a small mass of PCM. Inorganic PCMs also offer the benefits of higher latent heat potential compared to that of organic PCMs. This section consolidates and compares the experimentally determined melting point, latent heat and degree of supercooling of the GNP-dispersed ternary eutectic PCM. To begin with, we compared the latent heat (207 J/g) value and melting temperature (21.5 °C) of the numerically designed ternary eutectic PCM in Section 3.1 with those of the experimentally determined PCM using a DSC instrument. Figure 8 shows the experimentally determined endothermic peaks of pure salt hydrate PCM (SCD, SPDD and SSD) and the binary eutectic PCM (SCD/SPDD). With further calculation it could be inferred that the latent heat values of SCD, SSD and SPDD salt hydrate were 208.2 J/g, 216.2 J/g and 226.1 J/g, respectively. All the experimental latent heat values were like the numerical values used for designing the binary and ternary eutectic PCMs. Similarly, the latent heat value of the binary eutectic PCM was 210.3 J/g, which was like the numerically calculated latent heat (210 J/g). The melting temperature of the binary eutectic PCM was inferred to be 25.6 °C and closely resembled the T_m of SCD/SPDD binary eutectic PCM designed based on Schrader's equation. Figure 9a shows the experimentally tested endothermic peak of the SCD/SPDD/SSD ternary eutectic PCM with heat storage potential of 203.2 J/g and a melting point of 21.3 °C. It was also observed that the eutectic points of the developed binary and ternary eutectic PCMs were lower compared to that of the individual salt hydrate, which was ascribed to the increase in impurity, where entropy generation increased and reduced the melting point of the mixture. It was also noticed that the latent heat value of the ternary eutectic PCM was slight lower than the numerical calculated value; the explanation to be noted is the increase in free energy ($\Delta G°$) due to the impure solid (mixture of two or more components), causing entropy generation resulting in a slight reduction of latent heat of the prepared eutectic PCM. This experimental investigation ensured the reliability of Schrader's equation in developing a new eutectic PCM for TES.

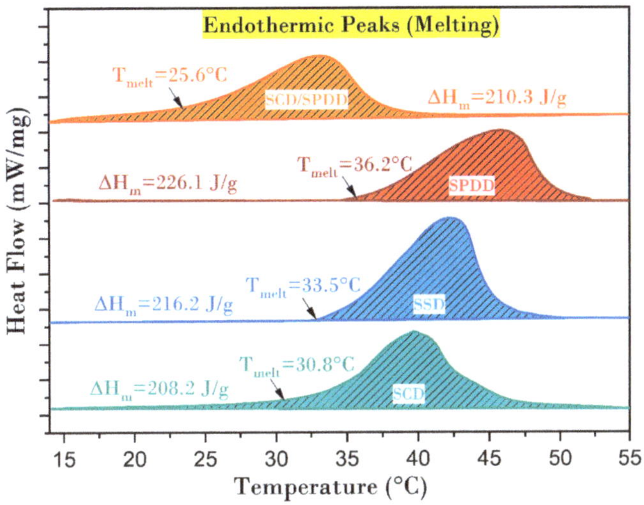

Figure 8. Heat storage enthalpy of pure salt hydrate PCM and binary eutectic salt hydrate PCM.

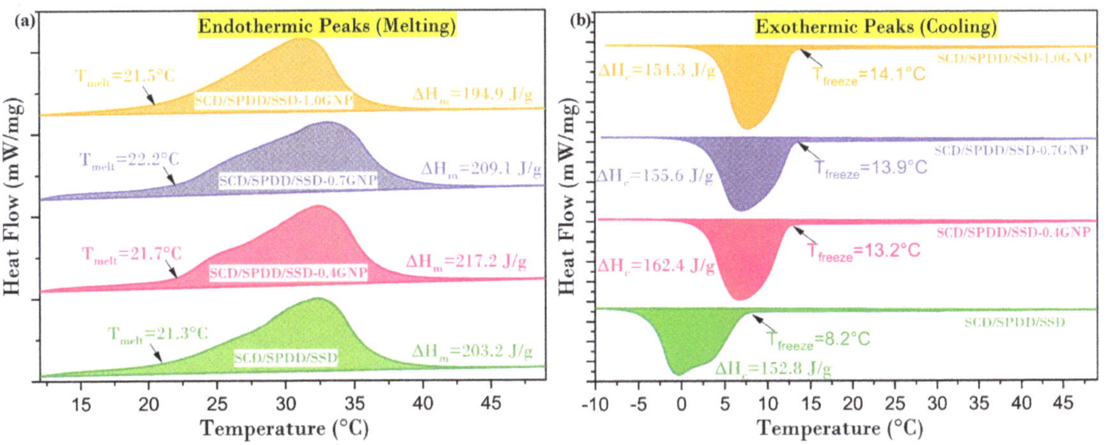

Figure 9. Heat storage enthalpy of GNP-dispersed eutectic salt hydrate PCM; (**a**) melting curve and (**b**) cooling curve.

DSC curves in Figure 9 depict the heating and cooling pattern of the ternary eutectic PCM and the GNP-dispersed ternary eutectic PCM. Three parameters to be discussed and inferred are (a) melting point, (b) heat storage enthalpy and (c) degree of supercooling. The developed ternary eutectic PCM was a combination of three different salt hydrates with melting points ranging between 31 and37 °C. Nevertheless, the synthesized eutectic PCM exhibited a single melting peak, which ensured the solid-to-liquid phase transition with uniform dispersion of the salt hydrate and the reliability of Schrader's equation. The ternary eutectic PCM's (SCD/SPDD/SSD) melting began at 21.3 °C, which was more like the numerically calculated eutectic point. The lower eutectic point of the developed PCM was due to the increase in impurity of the composition. On dispersion of GNPs, the melting point of the PCM increased slightly as the nanoparticles contributed towards higher thermal conductivity. Nevertheless, the phase transition temperature of the GNP-dispersed ternary eutectic PCM as in the range of 21.7–22.2 °C and is preferred for the thermal regulation of buildings.

Subsequently, the melting enthalpy and cooling enthalpy of the GNP-dispersed ternary eutectic PCMs were portrayed well in Figure 9a,b. It could be inferred that the melting enthalpies of the developed PCM with 0.4%, 0.7% and 1.0% of GNPs were 217.2 J/g, 209.1 J/g and 194.9 J/g, respectively. Latent heat describes the stronger intermolecular force of attraction between the PCM molecules. On the whole, the inclusion of nanoparticles with PCM increases the intermolecular attraction between the molecules of PCM and nanoparticles and accounts for storing higher heat energy. The increase in latent heat of the GNP-dispersed PCM was ascribed to the higher surface area of GNPs, which improved the intermolecular force of attraction between the nanoparticle and the ternary eutectic PCM. Nevertheless, the nanoparticle also replaced the mass of the PCM, and the material's ability to store heat energy drops at a higher concentration. Hence, at a low concentration of nanoparticles, the intermolecular interactions are dominant, and at a higher concentration, the drop in latent heat is accounted for by the mass of nanoparticles replacing the mass of PCM [27]. Likewise, the cooling plot in Figure 9b depicts that about 162.4 J/g, 155.6 J/g and 154.3 J/g of energy could be extracted from the energy that was stored during phase transition of the GNP-dispersed PCM from the liquid to solid state. On evaluation, the energy storage efficiency was 75.56% for the base ternary eutectic PCM, 75.46% for SCD/SPDD/SSD-0.4GNP, 74.0% for SCD/SPDD/SSD-0.7GNP and 78.4% for SCD/SPDD/SSD-1.0GNP.

One more important parameter to discuss with salt hydrate PCMs is their degree of supercooling (difference in temperature between the actual freezing point of PCM and the point at which PCM actually freezes). The major hindrance with usage of salt hydrate PCMs is their concern towards the issue of supercooling. During the phase transition of PCMs from liquid to solid state, they are expected to freeze at the temperature at which the PCM starts melting. With most salt hydrates failing to crystallize at their melting point, it reduced the heat storage ability. In this experimental investigation, the degree of supercooling was calculated as the difference in temperature between T_{melt} and T_{freeze}. From Figure 9a,b it was found that the degree of supercooling of the ternary eutectic PCM and its GNP-dispersed composite at concentrations of 0.4%, 0.7% and 1.0% were 13.1 °C, 8.5 °C, 8.3 °C and 7.4 °C, respectively. Although nucleating agents were not included in this investigation, GNPs performed the role of the nucleating agent and reduced the degree of supercooling by 35.2%, 36.655% and 43.52% with 0.4%, 0.7% and 1.0% weight concentrations, respectively.

3.7. Numerical Analysis of Ternary Eutectic PCM

A separate heat transfer analysis on the ternary eutectic PCM and GNP-dispersed PCM was conducted based on different heat energy input supply as in Figure 10. In this evaluation, we created a circular 2D enclosure to better understand the energy storage unit. Initially, we considered a circular PCM layer with their circumference open to a heat sink for constant heat input. The 2D PCM enclosure was maintained at a temperature of 20 °C, and the heat sink temperature was varied as 25, 30 and 35 °C, which was above the melting point (>21.6 °C) of the GNP-dispersed ternary eutectic PCM. We analyzed the developed ternary eutectic PCM with thermal conductivity of 0.473 W/m.K and 1.0 weight concentrated GNP-dispersed ternary eutectic PCM with thermal conductivity of 0.811 W/m.K.

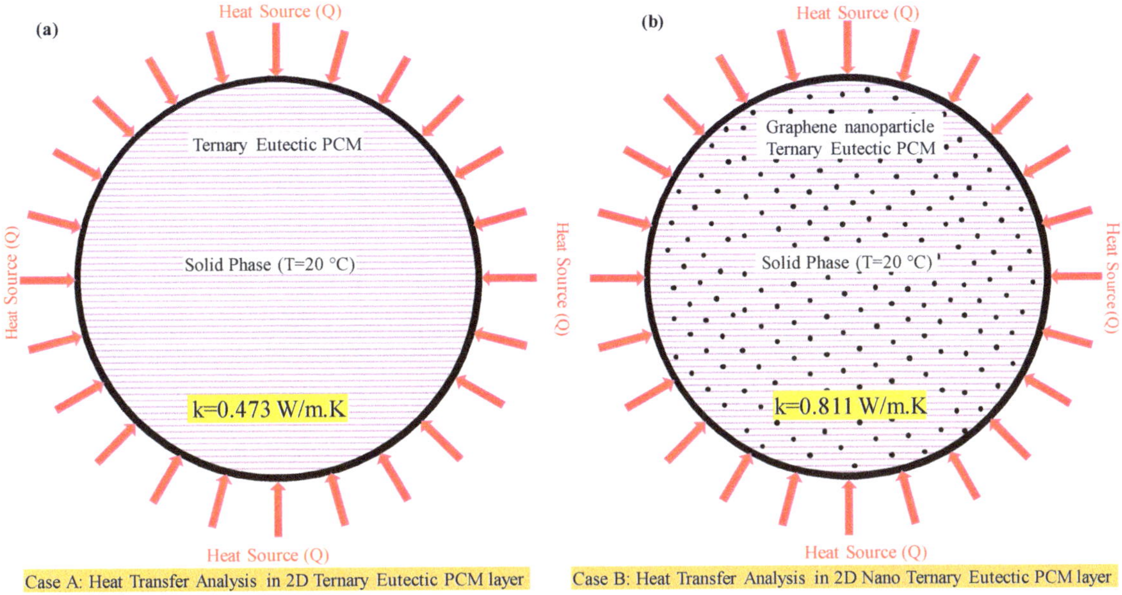

Figure 10. Boundary condition depiction of heat transfer investigation; (**a**) ternary eutectic PCM layer and (**b**) GNP-dispersed ternary eutectic PCM layer.

This numerical analysis as per the boundary condition in Figure 10 emphasizes the thermal energy propagation over the ternary eutectic PCM and GNP eutectic PCM. The heat transfer with temperature contour plots of the base and GNP-composite ternary eutectic PCMs with input heat energies of 40, 45 and 50 °C is shown in Figure 11. We conducted a steady-state heat transfer analysis for a circular PCM layer, with heat supply throughout the circumference layer. As the developed ternary eutectic PCM was specifically opted for low-temperature analysis, we supplied a low heat input with a temperature range of 40, 45 and 50 °C. Two major parameters to discuss are the thermal property of the PCM layer and the variation in supplied heat. It can be inferred from Figure 11a–c that with an increase in input energy source, the heat transfer and depth of penetration varied; however, the temperature change between the PCM and heat input was very minimal. The color scale depicts the heat penetration of the PCM. Likewise, enhancement of thermal conductivity with GNP dispersion further improved the heat penetration rate, which is clearly understandable from Figure 11 d–f. On analysis it was observed that for effective heat penetration the input source needed to be enhanced predominantly.

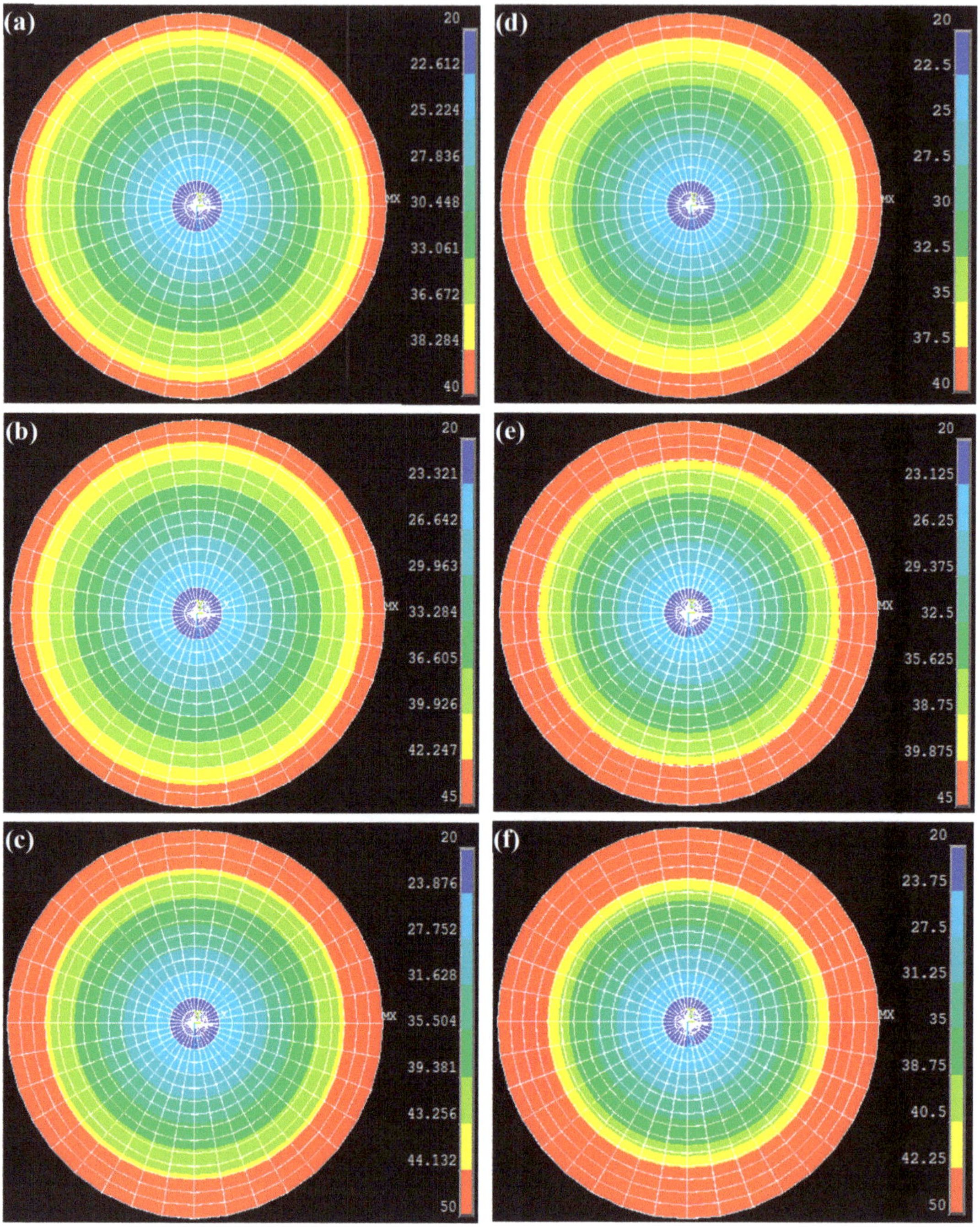

Figure 11. Temperature contour plots of ternary eutectic PCM with heat source of (**a**) 40 °C, (**b**) 45 °C (**c**) and 50 °C; and of 1.0 GNP-dispersed ternary eutectic PCM at heat input source of (**d**) 40 °C, (**e**) 45 °C and (**f**) 50 °C.

4. Conclusions

Energy storage using a eutectic salt hydrate is reliable with tailored thermal properties as per the desired application. In this current research work, a ternary eutectic salt hydrate PCM was designed numerically and developed experimentally to characterize its thermal properties. Ternary eutectic PCM SCD/SPDD/SCD showed a phase transition temperature of 21.5 °C, latent heat storage of 207 J/g, thermal conductivity of 0.473 W/m·K and optical absorptivity of 0.62 with a lower degree of supercooling compared to that of the individual salt hydrate. The developed ternary eutectic PCM was further investigated with dispersion of GNP and different weight concentrations. The results showed an increase in thermal conductivity and optical absorbance by 71.5% and 106.5%, respectively. Likewise, the degree of supercooling and transmissibility were reduced by 43.5% and 76.2% correspondingly. The GNP-dispersed composite ternary eutectic PCM was chemically stable and also exhibited a better enhancement in thermal and optical properties. The integration of PCMs with buildings to regulate effective utilization of thermal energy will contribute towards sustainable development.

Author Contributions: Conceptualization, B.K. and A.K.P.; Data curation, B.K.; Formal analysis, B.K.; Funding acquisition, A.K.P. and S.R.; Investigation, B.K., A.K.P. and V.V.T.; Methodology, B.K.; Project administration, A.K.P., S.R. and K.S.; Resources, A.K.P.; Supervision, S.R., K.S. and V.V.T.; Validation, B.K.; Writing—original draft, B.K.; Writing—review & editing, A.K.P., S.R. and V.V.T. All authors have read and agreed to the published version of the manuscript.

Funding: The authors acknowledge the financial assistance of Sunway University through Sunway University's International Research Network Grant Scheme 2.0 (IRNGS 2.0) 2021 (STR-IRNGS-SET-RCNMET-01-2021) for carrying out this research.

Institutional Review Board Statement: Not Applicable.

Informed Consent Statement: Not Applicable.

Data Availability Statement: The data presented in this study is available within the article.

Conflicts of Interest: The authors declare no conflict of interest.

References

1. Yu, K.; Liu, Y.; Yang, Y. Review on form-stable inorganic hydrated salt phase change materials: Preparation, characterization and effect on the thermophysical properties. *Appl. Energy* **2021**, *292*, 116845. [CrossRef]
2. Kalidasan, B.; Pandey, A.; Shahabuddin, S.; Samykano, M.; Thirugnanasambandam, M.; Saidur, R. Phase change materials integrated solar thermal energy systems: Global trends and current practices in experimental approaches. *J. Energy Storage* **2020**, *27*, 101118.
3. Kalidasan, B.; Pandey, A.; Saidur, R.; Samykano, M.; Tyagi, V. Nano additive enhanced salt hydrate phase change materials for thermal energy storage. *Int. Mater. Rev.* **2022**, 1–44. [CrossRef]
4. Chen, W.; Liang, X.; Wang, S.; Ding, Y.; Gao, X.; Zhang, Z.; Fang, Y. SiO_2 hydrophilic modification of expanded graphite to fabricate form-stable ternary nitrate composite room temperature phase change material for thermal energy storage. *Chem. Eng. J.* **2021**, *413*, 127549. [CrossRef]
5. Tian, H.; Du, L.; Wei, X.; Deng, S.; Wang, W.; Ding, J. Enhanced thermal conductivity of ternary carbonate salt phase change material with Mg particles for solar thermal energy storage. *Appl. Energy* **2017**, *204*, 525–530. [CrossRef]
6. Wang, T.; Mantha, D.; Reddy, R.G. Novel high thermal stability LiF–Na_2CO_3–K_2CO_3 eutectic ternary system for thermal energy storage applications. *Sol. Energy Mater. Sol. Cells* **2015**, *140*, 366–375. [CrossRef]
7. Liang, L.; Chen, X. Preparation and thermal properties of eutectic hydrate salt phase change thermal energy storage material. *Int. J. Photoenergy* **2018**, *2018*, 6432047. [CrossRef]
8. Sun, W.; Huang, R.; Ling, Z.; Fang, X.; Zhang, Z. Two types of composite phase change panels containing a ternary hydrated salt mixture for use in building envelope and ventilation system. *Energy Convers. Manag.* **2018**, *177*, 306–314. [CrossRef]
9. Efimova, A.; Pinnau, S.; Mischke, M.; Breitkopf, C.; Ruck, M.; Schmidt, P. Development of salt hydrate eutectics as latent heat storage for air conditioning and cooling. *Thermochim. Acta* **2014**, *575*, 276–278. [CrossRef]
10. Zhang, N.; Yuan, Y.; Wang, X.; Cao, X.; Yang, X.; Hu, S. Preparation and characterization of lauric–myristic–palmitic acid ternary eutectic mixtures/expanded graphite composite phase change material for thermal energy storage. *Chem. Eng. J.* **2013**, *231*, 214–219. [CrossRef]

11. Jebasingh, B.E. Preparation of organic based ternary eutectic fatty acid mixture as phase change material (PCM), optimizing their thermal properties by enriched solar treated exfoliated graphite for energy storage. *Mater. Today Proc.* **2016**, *3*, 1592–1598. [CrossRef]
12. Zhang, W.; Zhang, X.; Huang, Z.; Yin, Z.; Wen, R.; Huang, Y.; Wu, X.; Min, X. Preparation and characterization of capric-palmitic-stearic acid ternary eutectic mixture/expanded vermiculite composites as form-stabilized thermal energy storage materials. *J. Mater. Sci. Technol.* **2018**, *34*, 379–386. [CrossRef]
13. Ke, H. Investigation of the effects of nano-graphite on morphological structure and thermal performances of fatty acid ternary eutectics/polyacrylonitrile/nano-graphite form-stable phase change composite fibrous membranes for thermal energy storage. *Sol. Energy* **2018**, *173*, 1197–1206. [CrossRef]
14. Luo, Z.; Zhang, H.; Gao, X.; Xu, T.; Fang, Y.; Zhang, Z. Fabrication and characterization of form-stable capric-palmitic-stearic acid ternary eutectic mixture/nano-SiO_2 composite phase change material. *Energy Build.* **2017**, *147*, 41–46. [CrossRef]
15. He, Q.; Fei, H.; Zhou, J.; Du, W.; Pan, Y.; Liang, X. Preparation and characteristics of lauric acid-myristic acid-based ternary phase change materials for thermal storage. *Mater. Today Commun.* **2022**, *32*, 104058. [CrossRef]
16. Nazir, H.; Batool, M.; Ali, M.; Kannan, A.M. Fatty acids based eutectic phase change system for thermal energy storage applications. *Appl. Therm. Eng.* **2018**, *142*, 466–475. [CrossRef]
17. Ke, H. Phase diagrams, eutectic mass ratios and thermal energy storage properties of multiple fatty acid eutectics as novel solid-liquid phase change materials for storage and retrieval of thermal energy. *Appl. Therm. Eng.* **2017**, *113*, 1319–1331. [CrossRef]
18. Du, W.; Fei, H.; Pan, Y.; He, Q.; Zhou, J.; Liang, X. Development of capric acid-stearic acid-palmitic acid low-eutectic phase change material with expanded graphite for thermal energy storage. *Constr. Build. Mater.* **2022**, *320*, 126309. [CrossRef]
19. Nordlie, B. *Eutectic Melting. Petrology*; Kluwer Academic Publishers: Dordrecht, The Netherlands, 1989; pp. 155–157.
20. Nichols, L. *Organic Chemistry Laboratory Techniques*; Independent: Chicago, IL, USA, 2016.
21. Lin, N.; Li, C.; Zhang, D.; Li, Y.; Chen, J. Emerging phase change cold storage materials derived from sodium sulfate decahydrate. *Energy* **2022**, *245*, 123294. [CrossRef]
22. Shen, Z.; Oh, K.; Kwon, S.; Toivakka, M.; Lee, H.L. Use of cellulose nanofibril (CNF)/silver nanoparticles (AgNPs) composite in salt hydrate phase change material for efficient thermal energy storage. *Int. J. Biol. Macromol.* **2021**, *174*, 402–412. [CrossRef]
23. Zhang, Z.; Lian, Y.; Xu, X.; Xu, X.; Fang, G.; Gu, M. Synthesis and characterization of microencapsulated sodium sulfate decahydrate as phase change energy storage materials. *Appl. Energy* **2019**, *255*, 113830. [CrossRef]
24. Nickolov, Z.S.; Ozcan, O.; Miller, J. FTIR analysis of water structure and its significance in the flotation of sodium carbonate and sodium bicarbonate salts. *Colloids Surf. A Physicochem. Eng. Asp.* **2003**, *224*, 231–239. [CrossRef]
25. Kumar, R.; Samykano, M.; Ngui, W.; Pandey, A.; Kalidasan, B.; Kadirgama, K.; Tyagi, V. Investigation of thermal performance and chemical stability of graphene enhanced phase change material for thermal energy storage. *Phys. Chem. Earth Parts A/B/C* **2022**, *128*, 103250. [CrossRef]
26. Kalidasan, B.; Pandey, A.; Rahman, S.; Yadav, A.; Samykano, M.; Tyagi, V. Graphene–Silver Hybrid Nanoparticle based Organic Phase Change Materials for Enhanced Thermal Energy Storage. *Sustainability* **2022**, *14*, 13240. [CrossRef]
27. Wang, F.; Zheng, W.; Gou, Y.; Jia, Y.; Li, H. Thermal behaviors of energy storage process of eutectic hydrated salt phase change materials modified by Nano-TiO_2. *J. Energy Storage* **2022**, *53*, 105077. [CrossRef]

energies

MDPI

Article

Development of Self-Passivating, High-Strength Ferritic Alloys for Concentrating Solar Power (CSP) and Thermal Energy Storage (TES) Applications

Fadoua Aarab * and Bernd Kuhn †

Institute of Energy and Climate Research (IEK), Structure and Function of Materials (IEK-2), Forschungszentrum Jülich GmbH, 52425 Jülich, Germany
* Correspondence: f.aarab@fz-juelich.de; Tel.: +49-2461-61-4132
† Current address: Hochschule Rhein Main, Faculty of Engineering, 65428 Rüsselsheim, Germany.

Abstract: Concentrating solar power (CSP) and thermal energy storage (TES) based on molten salts still lacks economic feasibility, with the material investment costs being a major drawback. Ferritic stainless steels are a comparatively cheap class of materials that could significantly contribute to cost reductions. The addition of aluminum to ferritic steel can result in self-passivation by forming a compact Al_2O_3 top layer, which exhibits significantly higher corrosion resistance to solar salt compared to the Cr_2O_3 surface layers typically formed on expensive structural alloys for CSP and TES, such as austenitic stainless steels and Ni-base super alloys. However, to date, no ferritic stainless steel combining Al_2O_3 formation and sufficient structural strength is available. For this reason, cyclic salt corrosion tests under flowing synthetic air were carried out on seven Laves phase-forming, ferritic model alloys (17Cr2-14Al0.6-1Nb2.6-4W0.25Si), using "solar salt" (60 wt. % $NaNO_3$ and 40 wt. % KNO_3). The Al content was varied to investigate the influence on the precipitation of the mechanically strengthening Laves phase, as well as the impact on the formation of the Al-oxide top layer. The W and Nb contents of the alloys were increased to examine their influence on the precipitation of the Laves phase. The salt corrosion experiments demonstrated that simultaneous self-passivation against a molten salt attack and mechanical strengthening by precipitation of fine Laves phase particles is possible in novel ferritic HiperFer$^{\text{SCR (salt corrosion-resistant)}}$ steel. Microstructural examination unveiled the formation of a compact, continuous Al_2O_3 layer on the surface of the model alloys with Al contents of 5 wt. % and higher. Furthermore, a stable distribution of fine, strengthening Laves phase precipitates was achieved in the metal matrix, resulting in a combination of molten salt corrosion resistance and potentially high mechanical strength by a combination of solid solution and precipitation strengthening. These results show that high-strength ferritic alloys are suitable for use in CSP applications.

Keywords: thermal energy storage; concentrating solar power; protective Al_2O_3 scale; salt corrosion resistance; strengthening Laves phase precipitation

Citation: Aarab, F.; Kuhn, B. Development of Self-Passivating, High-Strength Ferritic Alloys for Concentrating Solar Power (CSP) and Thermal Energy Storage (TES) Applications. *Energies* **2023**, *16*, 4084. https://doi.org/10.3390/en16104084

Academic Editor: Jesús Polo

Received: 25 March 2023
Revised: 11 May 2023
Accepted: 12 May 2023
Published: 14 May 2023

1. Introduction

Currently, extensive research is being conducted in the field of renewable energy [1–3]. The continuous innovation and optimization of solar tower power plant technology is helping to ensure that it can play an important role in the transition to a sustainable and low-carbon energy future. Solar tower power plants constitute a promising renewable technology for large-scale power and process heat generation [4]. These power plants use a field of mirrors (heliostats) to concentrate sunlight onto a central receiver mounted on the top of a tower. The concentrated solar energy is then used to generate high-temperature heat, usually by heating a heat transfer fluid, such as molten salt or pressurized steam. This heat is then used to drive a turbine and generate electricity. Solar thermal power plants typically use molten salts for heat transfer and storage [5]. The so-called "solar salt" is a combination of 60 wt. % $NaNO_3$ and 40 wt. % KNO_3 [6]. In the case of nitrate salts, partial dissociation occurs at higher temperatures, before they finally completely decompose at further increasing temperatures [7]. The overall efficiency of commercially operated solar thermal power plants is currently limited by the thermal stability of the solar salt, which makes process temperatures above 600 °C difficult to achieve [8]. Even at these temperatures, increased corrosion of metallic components occurs, requiring careful material selection [7]. Although the cost of solar thermal power plants has decreased by about 68% over the last 10 years [9], further cost reductions are needed to make solar thermal power plants competitive with other regenerative (wind, PV) and fossil energy technologies. An important aspect of cost reduction is the materials' cost and durability. Ferritic alloys, which primarily consist of iron and chromium, offer a more cost-effective alternative for certain components in CSP, compared to commercially used nickel-based alloys. The abundance and widespread availability of iron contribute to the lower price of ferritic alloys. Additionally, chromium, the main alloying element in ferritic alloys, is more affordable compared to nickel, molybdenum, and other expensive alloying elements used in nickel-based alloys.

The development of Al_2O_3-forming stainless steels with high mechanical strength offers the potential for cost-effective structural material solutions. In the development of a cost-effective alternative, it is also important to ensure that it has good resistance to the heat transfer fluid used.

The aim of the presented alloy development was to combine the strengthening by precipitation of intermetallic Laves phase particles, known from ferritic "HiperFer" [10,11] alloys, with the formation of a salt corrosion-resistant oxide surface layer. Laves phases form a large class of intermetallic compounds that have tended to be considered detrimental phases to avoid [12]. However, with proper control of the precipitation process, Laves phases can serve as strengthening particles [10,11,13–15]. The main Laves phase-forming elements, W, Nb, and Si, strongly influence the mechanical properties by modifying the solid solution and precipitation strengthening effects.

By using model alloys, each weighing 100 g, alloying elements such as Al, Nb, and W were varied to investigate the influence of alloy chemistry on Al_2O_3 formation, and the precipitation of fine, long-term stable Laves phase particles in the ferrite matrix. Following the corrosion experiments, a detailed microstructure investigation was carried out with qualitative and quantitative evaluation of the particle size distribution (FE-SEM) and chemistry (EDX), as well as surface layer formation (FE-SEM).

2. Materials and Methods

2.1. Materials and Preparation

In this study, seven ferritic model alloys, each weighing 100 g, were investigated. The Al content was varied to investigate its effect on the precipitation of the Laves phase and the formation of the Al-oxide top layer. The W and Nb contents were also increased to examine the impact on the precipitation of the Laves phase. W, Nb, and Si were used for combined solid solution and precipitation hardening. In FeCrAl alloys, Nb is an essential component in the precipitation of the Laves phase [11,16,17]. Additionally, it hinders the occurrence of intergranular corrosion by preventing the formation of chromium carbides along grain boundaries [18]. The addition of W facilitates solid solution hardening and enhances the volume fraction of the Laves phase [17]. The intermetallic Laves phase particles, $(Fe,Cr,Si,Al)_2(Nb,W)$, significantly enhance the mechanical strength of these alloys [19]. Moreover, the addition of Si promotes the formation and stabilization [20,21] of the Laves phase, consequently increasing the service life of FeCrAl steel components. This is achieved by improving the adhesion of the protective oxide layers [22].

The model alloys were prepared under argon in a cold levitation crucible at normal pressure using high-purity raw materials. Investigations were then carried out to determine the homogeneous distribution of the alloying elements in the model alloys. The distribution of the alloying elements was largely homogeneous, and no segregations were detected after annealing for homogeneity. The chemical compositions of the model alloys (analyzed by inductively coupled plasma optical emission spectroscopy (ICP-OES); C, S, and N analyzed by infrared absorption) are presented in Table 1.

Table 1. Chemical compositions (wt. %) of the ferritic model alloys.

Alloy Designation	C	S	N	Cr	Al	W	Nb	Si	Fe
2 Al	0.0032	<0.0017	<0.0003	17.25	1.904	2.58	0.61	0.21	R
3.5 Al	0.002	<0.0022	<0.0003	17.25	3.41	2.93	0.7	0.259	R
5 Al	0.0011	<0.002	<0.0003	17.3	4.917	2.74	0.638	0.224	R
8 Al	0.0054	<0.002	<0.0003	17.1	7.84	2.79	0.671	0.255	R
14 Al	<0.001	<0.0022	<0.0003	17.0	13.9	2.78	0673	0.258	R
4W 1Nb 3.5Al	0.0026	<0.0018	<0.0003	17.1	3.56	4.04	1.05	0.236	R
4W 1Nb 5Al	0.006	<0.0022	<0.0003	17.2	5.04	4.22	1.09	0.247	R

After performing solution annealing, specimens with dimensions of 10 mm × 10 mm × 1 mm were extracted from the melts using electrical discharge machining. In accordance with ISO 17245:2015 [23], the surfaces and edges of these specimens were meticulously ground using 600-grit SiC paper to ensure consistent initial conditions in terms of surface roughness. Subsequently, the specimens were cleaned with ethanol and dried using hot air.

In air, temperatures of at least 1000 °C or higher are normally required for the formation of a protective α-Al_2O_3 layer on metal surfaces of FeCrAl alloys [24,25]. At temperatures above 1350 °C, significantly accelerated oxidation kinetics can occur, primarily driven by the formation of iron oxides [24]. To investigate the corrosion attack of the solar salt on a preformed Al_2O_3 layer, one sample per model alloy was annealed at 1100 °C for one hour in laboratory air. This temperature was selected to avoid uncontrolled precipitation of particles of the Laves phase, which can occur below 1050 °C [20]. Figure 1 shows an SEM image and a photograph of each model alloy before exposure to solar salt. In the pre-oxidized model alloys 4W 1Nb 3.5Al and 4W 1Nb 5Al, coarse Laves phase precipitates were still present. Due to the increased W and Nb contents, pre-oxidation temperatures of more than 1100 °C were apparently necessary to avoid uncontrolled precipitation of the Laves phase. After the heat treatment experiments, suitable pre-oxidation parameters (1175 °C/1 h) could be determined for these two model alloys, too. Samples treated with the optimized pre-oxidation parameters will be aged in solar salt in future work.

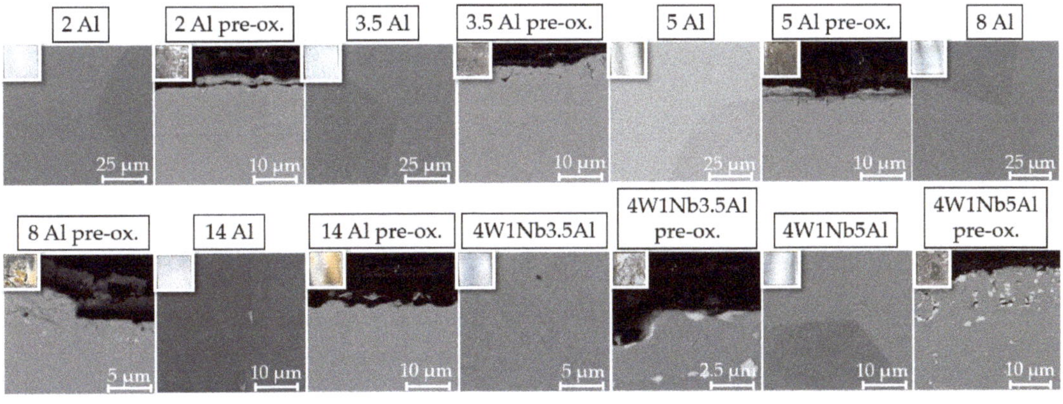

Figure 1. Scanning electron (SEM) micrographs (top left: inlay light optical photographs showing specimen surface color) of the cross-sections of the specimens before corrosion testing.

2.2. Experimental Methods

Discontinuous salt corrosion testing of alloy specimens was carried out by placing them inside salt-filled alumina crucibles for a total of 2000 h at 600 °C in a tube furnace under the flow of synthetic air, with a flow rate of 10 sl/h. For this study, a mixture consisting of 60 wt. % $NaNO_3$ (supplied by BASF SE, Ludwigshafen, Germany) and 40 wt. % KNO_3 (supplied by Haldor Topsoe, Lyngby, Denmark) was prepared. As the solar salt creeps out of the specimen containers with the aging time, fresh solar salt was replenished every 500 h to ensure continuous coverage of the alloy specimens. The specimens were weighed every 250 h to calculate their individual mass changes according to DIN 50905-1 [26]. Further details regarding the equipment and testing methodology for salt corrosion behavior can be found in [12].

High-resolution scanning electron images (SEM) with a resolution of 4096 × 3072 pixels were acquired to determine the particle size distribution and the widths of the particle-free zones at grain boundaries, which are characteristic features of HiperFer-type steels [20,21,27], as well as below the surface oxide layer. The images were quantitatively analyzed using the OLYMPUS Stream Desktop software package, and the method described by Lopez et al. [21], which has been successfully applied to Laves phase-strengthened alloys, was employed.

3. Results and Discussion

3.1. Discontinuous Corrosion Testing

For the model alloy with the lowest Al content of 2 wt. % (2 Al), a strong mass increase of about 0.875 mg/cm^2 was observed in the first 250 h (cf. Figure 2a). In the case of pre-oxidized 2 Al, the mass loss was even higher at about 1.625 mg/cm^2. From 500 h until the end of the experimental period of 2000 h of aging in solar salt, only net losses were observed, ranging from 0.167 mg/cm^2 to 0.792 mg/cm^2. These mass losses, which were even higher in the case of pre-oxidized 2 Al, result from spallation of the oxide layer and indicate insufficient adhesion to the base material. Obviously, the Al content of 2 wt. % is too low to form a protective top layer. The fact that Al is simultaneously consumed by the precipitation of Laves phase within the ferrite matrix may worsen the situation. In comparison to the 2 Al alloy, the mass gain of the 3.5 Al steel was relatively low, especially for the pre-oxidized 3.5 Al. The slight increase in mass gain for the 3.5 Al alloy in the first 500 h indicates uniform, slow oxide film growth. This could also be observed for the model alloys with an increased Al content up to 5 wt. % Al (cf. Figure 2b) and for the model alloys with increased W and Nb contents (cf. Figure 2c).

When considering the mass changes of the model alloys with varying Al content, it was demonstrated that pre-oxidation had a detrimental effect, as seen by the increased mass change in the case of the 5 Al alloy and the increased mass losses in the case of the 14 Al alloy.

The increase in W and Nb contents in combination with an increase in Al content from about 3.5 wt. % to about 5 wt. % seemed to result in a lower mass change (cf. Figure 2c). After the first 250 h, the mass increase of 4W 1Nb 3.5Al (1.083 mg/cm^2) was about three times higher than that of 4W 1Nb 5Al (0.417 mg/cm^2). Regarding 4W 1Nb 3.5Al and 4W 1Nb 5Al alloys, pre-oxidation resulted in similar mass changes, but compared to the other model alloys, 4W 1Nb 3.5Al exhibited a comparatively higher mass increase.

Comparing the weight changes of 4W 1Nb 3.5Al and 4W 1Nb 5Al, it can be concluded that the increased W and Nb contents, due to the interaction of oxide layer growth and Laves phase precipitation, also required an increased Al content to obtain a slow-growing oxide layer.

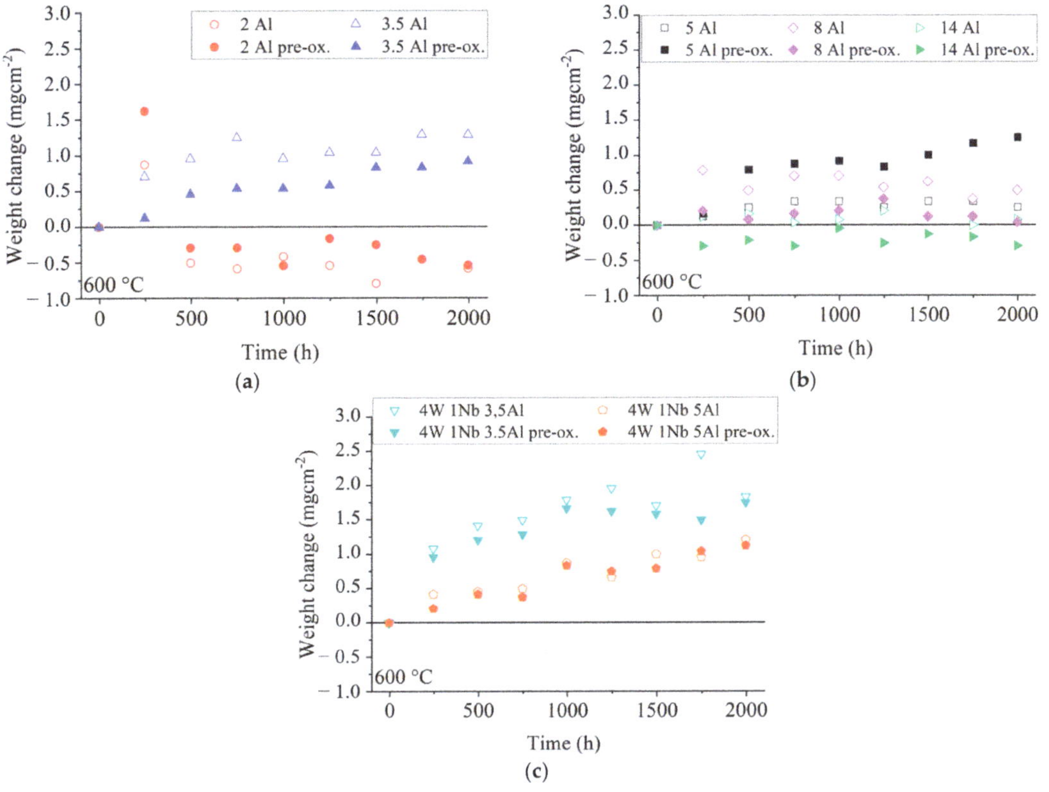

Figure 2. Weight changes of corrosion specimens during discontinuous annealing in solar salt at 600 °C: (**a**) low Al content, (**b**) high Al content, and (**c**) high W and Nb contents.

When considering the higher-grade model alloys, a clear influence of the Al content on the mass change was observed (cf. Figure 2a,b). An Al content of 3.5 wt. % and above ensured a consistently low mass change. With increased W and Nb contents, an Al content of 5 wt. % was necessary for this effect. Therefore, higher W and Nb contents require higher Al contents to ensure sufficient availability of Al for the formation of a stable Al_2O_3 top oxide layer, indicating further potential for optimization of strength (due to higher W and Nb contents [24]), while maintaining high salt corrosion resistance. At an Al content of 14 wt. %, the mass change behaved significantly differently compared to the other model alloys. Over the entire test period of 2000 h, there were only small mass increases of up to 0.208 mg/cm^2. In the case of the pre-oxidized variant, there were small mass losses of up to 0.291 mg/cm^2 over the entire testing period.

SEM surface examination results, including associated light photographic images of the specimen surfaces, are presented in Figure 3. In the photographs of the specimen surfaces, the oxide layer appears reddish-brown for specimens with Al contents up to 3.5 wt. % and increased W and Nb contents, indicating an Fe-rich oxide layer. For Al contents from 5 wt. % to 14 wt. %, the sample surfaces appear grayish-beige, indicating an Al-containing oxide layer. For model alloys with Al contents up to 3.5 wt. %, on which bulky multi-layer Na-Fe-oxides formed, SEM exhibited indications of spallation, but no superficial cracks. These findings are consistent with the weight change curves presented in Figure 2. With a higher Al content, the oxide layer had a fine-grained morphology with fewer bulges, consisting of mixed Na-Fe-oxide forming on top of a mixed Al-Cr-Fe-oxide layer. The pre-oxidized specimen exhibited a coarser oxide layer, which was marked by the presence of surface bulges. Only mixed Al-oxide without bulges of Na-Fe-oxides was detected on the 14 Al alloy. In the sample cross-section of the 14 Al alloy (cf. Figure 4), a protective, continuous Al_2O_3 layer with a thickness of about 40 nm formed after 2000 h at 600 °C in solar salt. In contrast to the other model alloys, the 14 Al specimen did not form a uniform distribution of small Laves phase particles in the base metal during aging in solar salt. The Al_2O_3 layer of the pre-oxidized 14 Al was about 30 nm-thick, with comparatively large Laves phase precipitations underneath.

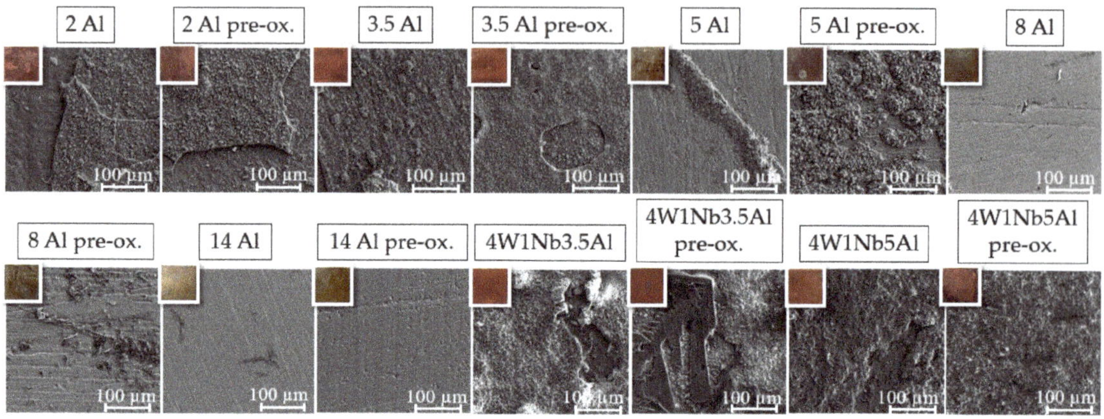

Figure 3. SEM micrographs of the specimen surfaces after discontinuous corrosion tests in solar salt for 2000 h at 600 °C (top left: inlay light optical photographs showing specimen surface color).

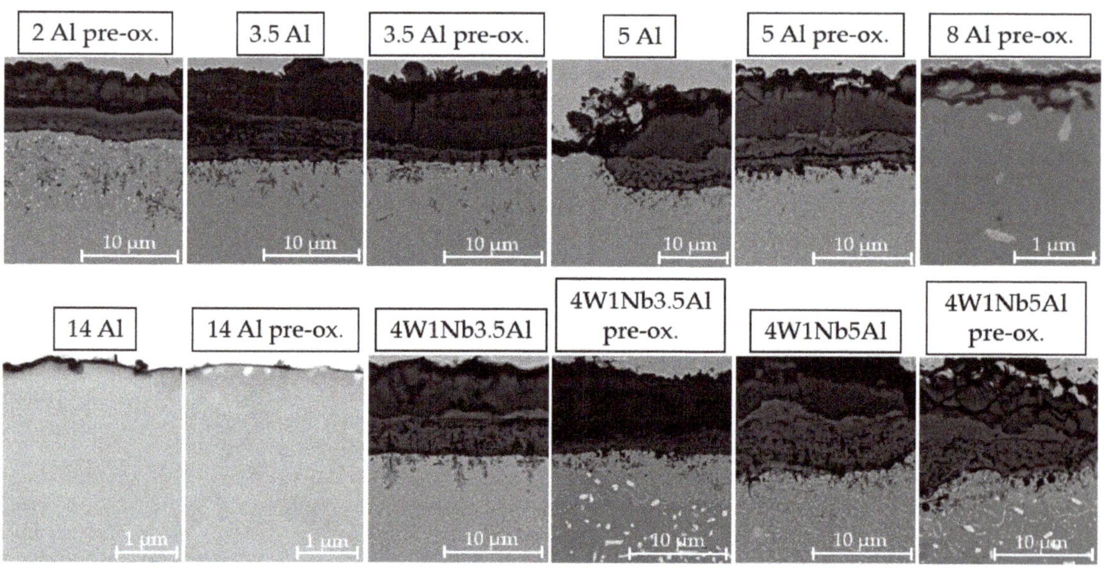

Figure 4. SEM micrographs of the specimen cross-sections after discontinuous corrosion tests in solar salt for 2000 h at 600 °C.

The EDX element mappings of the cross-sections of expressive model alloys are displayed in Figures 5–7. For the model alloys with lower Al contents of up to 3.5 wt. % (cf. Figure 5), the oxide formed in three layers: an intermediate layer composed of Cr-Na-Fe-oxide was formed between an upper layer consisting of Na-Fe-mixed oxide and a lower layer of Cr-Al-oxide on the steel base. This type of three-layer oxide film formation is already known from previous salt corrosion experiments with ferritic FeCrAl alloys [12]. Coarse AlN precipitates could be found below the oxide layer. As expected, intermetallic phase precipitates composed of Fe, Cr, Al, and W were observed in the base material. For the pre-oxidized 5 Al alloy, the oxide film formation was similar to the lower Al content variants (cf. Figure 6). For 5 Al, the multilayer oxide formation was only observed in the few bulky areas. In the flat areas, a thin and compact Al_2O_3 layer was found in the case of the 5 Al alloy.

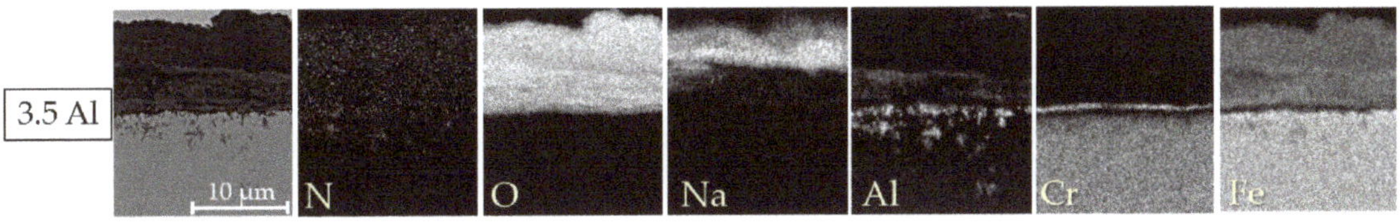

Figure 5. Energy-dispersive X-ray element mappings of the cross-sections of the specimens with a low Al content of 3.5 wt. % after discontinuous corrosion tests in solar salt for 2000 h at 600 °C.

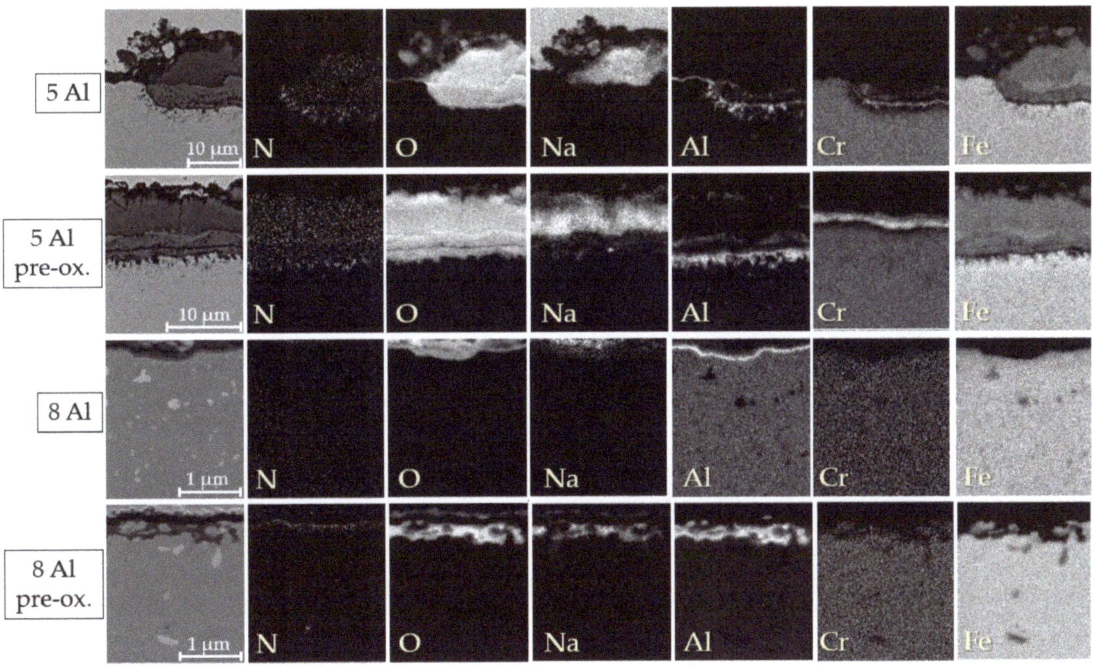

Figure 6. EDX element mappings of the cross-sections of the specimens with high Al contents of 5 wt. % and 8 wt. % after 2000 h of discontinuous corrosion testing in solar salt at 600 °C.

Furthermore, AlN precipitates were detected below the multilayer oxide areas only. In contrast, below the thin, flat Al_2O_3 layers of the 8 Al alloys, no AlN precipitates were found. However, the disadvantage of pre-oxidation became evident when examining the 8 Al alloy with and without pre-oxidation. The oxide layer formed by pre-oxidation appeared to be incomplete, allowing the solar salt to permeate through. Consequently, internal oxidation by Na occurred. Pure Na-oxide, visible on top of the Al_2O_3 layer of the 8 Al alloys, could also be salt residues left behind during sample preparation. For the model alloys with increased W and Nb contents, the surface oxide also formed in layers, similar to the alloys with a low Al content (cf. Figure 7).

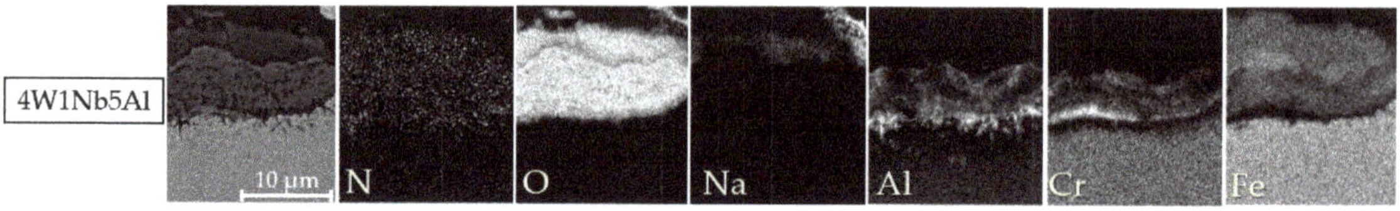

Figure 7. EDX element mappings of the cross-sections of the specimens with high W and Nb content after discontinuous corrosion tests in solar salt for 2000 h at 600 °C.

It is remarkable that a protective Al_2O_3 layer formed on top of the model steels during annealing in solar salt at the intermediate temperature of 600 °C, even in the absence of pre-oxidation in air, as previously observed in corrosion experiments in solar salt with a similar alloy composition and aging temperature [9]. The formation of Al_2O_3 at lower temperatures can be attributed to the interplay between the alloy composition and the corrosive environment containing oxygen. Increasing the Al content provides a greater availability of Al for oxide formation. Moreover, the presence of high Cr and Al contents in the alloy triggers the so-called third element effect [28]. As Cr exhibits a lower affinity for oxygen compared to Al, it facilitates the reduction of oxygen influx into the base metal, allowing Al to diffuse to the metal surface without precipitating as an internal oxide. Subsequently, once Al has diffused to the metal surface, it rapidly undergoes oxidation due to the presence of an oxygenated, corrosive environment that effectively wets the material surface.

These results imply that the alloy concept with all the tested variations of Al, W, and Nb has the potential for self-passivation upon exposure to solar salt. The reason for the Al_2O_3 formation at lower temperatures could be an interplay between the alloy composition and the oxygen-containing, corrosive environment.

3.2. *Laves Phase Precipitation*

Exemplary SEM micrographs of Laves phase precipitation in the 2 Al, 5 Al, 8 Al, 4W 1Nb 3.5Al, and 4W 1Nb 5Al model alloys after discontinuous corrosion tests in solar salt for 2000 h at 600 °C are presented in Figure 8. To assess the stability of the Laves phase particles during operation, the mean particle size and the widths of particle-free zones along high-angle grain boundaries were determined after exposure to solar salt for 2000 h at 600 °C (cf. Table 2). Associated size distribution histograms (particle frequency, i.e., fraction of particles within corresponding equivalent circle diameter ranges in relation to the overall number of particles counted) are depicted in Figures 9 and 10.

Near-surface particle-free zones were found exclusively in the 5 Al and 8 Al pre-oxidized variants. In the case of the 8 Al alloy, on the other hand, a narrow region of much finer Laves phase precipitates below the oxide layer was encountered (cf. Figure 8). In the other samples, the particle size remained stable along the depth of the material. Due to pre-oxidation and with the increasing Al content, the particle-free zones in the bulk material expanded along the high-angle grain boundaries, which is typical [29] for HiperFer-type steels. Since Al_2O_3 is more stable than Cr_2O_3, there was reduced consumption of Al and Cr through surface oxidation. As a result, a larger amount of Al and Cr were available for the growth of the Laves phase in the pre-oxidized model alloys. In the untreated model alloys, the oxide layer formed during the corrosion experiment and consumed Al and Cr from the surface near the alloy matrix. This inhibited the nucleation and growth of Laves particles near the surface, as seen in 5 Al, 8 Al, and 8 Al pre-oxidized specimens. However, as the depth within the material increased, then consumption of Al and Cr at the surface no longer significantly affected the Laves phase formation, which consequently led to equalization of the particle sizes.

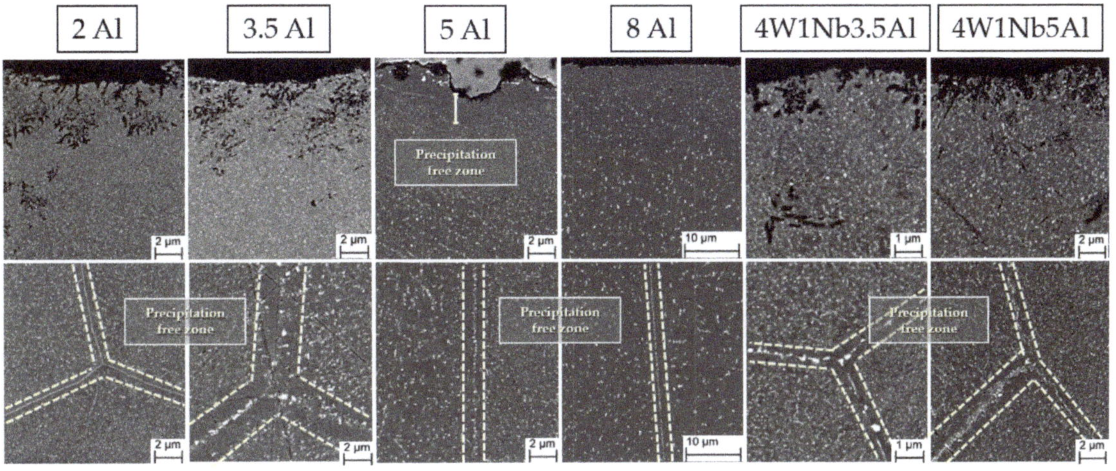

Figure 8. SEM micrographs of Laves phase particle population after discontinuous corrosion tests in solar salt for 2000 h at 600 °C. The dashed lines mark the areas of the particle-free zones at the grain boundaries.

Table 2. Mean equivalent circle diameters with standard deviations of Laves phase particles (investigated area depth: 100 to 600 μm), width of the particle-free zone near the surface (down to 6 μm depth), and at high-angle grain boundaries within the bulk material (depth of the investigated area: 100 to 600 μm) in the model alloys after discontinuous corrosion tests in solar salt for 2000 h at 600 °C.

Alloy Designation	Mean Equivalent Circle Diameter (nm)	Precipitation-Free Zone (Near Surface Area) (nm)	Precipitation-Free Zone (Bulk Material) (nm)
2 Al	64.46 ± 30.18	-	484.66 ± 71.83
2 Al pre-ox.	81.01 ± 34.24	-	856.24 ± 355.28
3.5 Al	65.74 ± 27.11	-	552.20 ± 85.87
3.5 Al pre-ox.	83.38 ± 46.65	-	1056.31 ± 257.09
5 Al	90.12 ± 47.78	1792.88 ± 264.80	757.44 ± 180.37
5 Al pre-ox.	93.99 ± 40.95	-	3223.66 ± 171.27
8 Al	226.06 ± 101.42	-	1856.02 ± 475.22

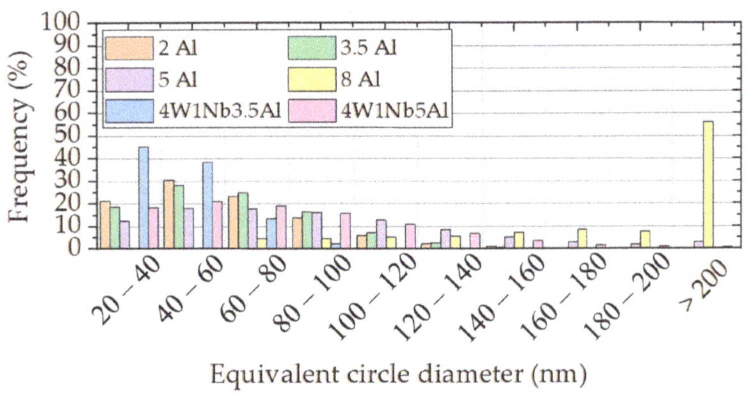

Figure 9. Equivalent circle diameter distributions of precipitates in the model alloys (determined over an area of approximately 1450 μm^2 each) after 2000 h of discontinuous corrosion testing in solar salt at 600 °C.

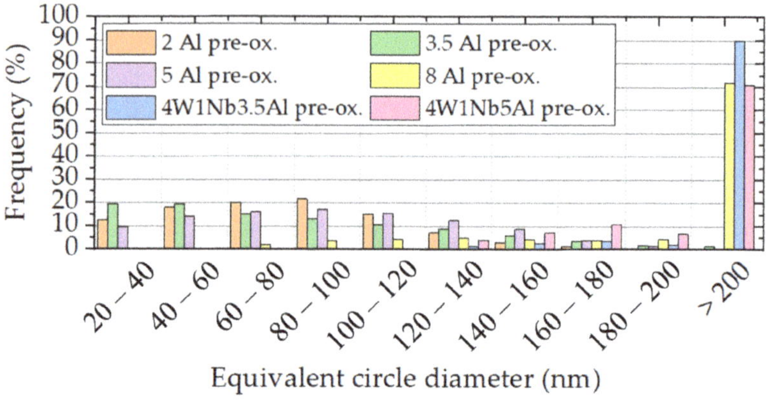

Figure 10. Equivalent circle diameter distributions of precipitates in the pre-oxidized model alloys (determined over an area of approximately 1450 μm^2 each) after discontinuous corrosion tests in solar salt for 2000 h at 600 °C.

Looking at Figure 9 and Table 2, the Laves phase precipitates also tended to coarsen with the rising Al content. Pre-oxidation of the specimens before corrosion testing slightly enhanced this effect (cf. Figure 10). The pre-oxidation of the alloys with increased W and Nb contents led to particularly strong coarsening, because of the obviously unsuitable pre-oxidation parameters, which led to nucleation, growth, and/or incomplete dissolution of the Laves phase during pre-oxidation. The increase in W and Nb contents per se caused a decrease in the equivalent circle diameter. However, it is striking that at an elevated content of 14 wt. % Al, no Laves phase precipitates were detected after corrosion testing in solar salt. In principle, increasing the Al content seems to lead to coarsening and reduction of the number of Laves phase precipitation. Since the Laves phase precipitation can be influenced by heat treatment parameters [30], it can be assumed that higher temperatures are necessary for particle precipitation at an Al content of 14 wt. %.

4. Conclusions

Salt corrosion experiments were carried out at 600 °C for 2000 h on seven ferritic model alloys with different Al, W, and Nb contents. The key conclusions of the practical research presented here can be summarized as follows:

- For a low Al content of up to 3.5 wt. %, spallation of the formed surface oxide occurred. The oxide formed in three layers: an intermediate layer of Cr-Na-Fe-oxide between an upper layer of Na-Fe-mixed oxide and a bottom layer of Cr-Al-oxide adjacent to the steel substrate. Coarse AlN precipitates could be found below the oxide layer for an Al content of up to 5 wt. %. As expected, strengthening intermetallic Laves phase precipitates of Cr, Fe, W, Nb, and Al formed in the base material. Pre-oxidation proved detrimental due to increased mass changes and coarse, discontinuous oxide layers.
- The 14 Al formed a protective, continuous Al_2O_3 layer with a thickness of about 40 nm, but did not form small Laves phase particles in the base metal during aging in solar salt. For this reason, it is not considered a good candidate for structural applications, despite its superior salt corrosion resistance.
- Increased W and Nb contents necessitated a higher Al content to obtain slow-growing oxide layers, because of the interaction of oxide layer growth and Laves phase precipitation.
- Protective Al_2O_3 layers formed in solar salt at a moderate temperature of 600 °C, even without prior oxidation in air at a high(er) temperature. The alloying concept (in all the tested variations of Al, W, and Nb contents) exhibited potential self-passivation upon exposure to solar salt.

- The amount of strengthening Laves precipitates in the size range from 20 to 60 nm clearly correlated with increased W and Nb contents. Changes in the Al content in the range from 2 to 5 wt. % had a minor impact on the number of particles in the size range from 20 to 100 nm and are thus considered advantageous in terms of mechanical strength. In contrast, an Al content of 8 wt. % favored increased particle growth and should be avoided.

As a final summary, it can be stated that the presented HiperFer^{SCR} steel type is a promising candidate for a significant reduction of investment costs in molten salt technology. It combines self-passivation in solar salt, mechanical strength, and low cost. For this reason, it could provide an enabling material solution for improving the competitiveness of CSP and TES equipment. However, it should be noted that despite extensive research efforts on the corrosion resistance of molten salts for solar energy applications, the underlying corrosion mechanisms are still not fully understood. Long-term tests or tests in practical operation are still necessary to ensure the reliability of the entire CSP-TES system.

Author Contributions: Conceptualization, F.A. and B.K.; methodology, F.A. and B.K.; investigation, F.A.; resources, B.K.; data curation, F.A.; writing—original draft preparation, F.A.; writing—review and editing, B.K. and F.A.; visualization, F.A.; supervision, B.K.; project administration, B.K.; funding acquisition, B.K. All authors have read and agreed to the published version of the manuscript.

Funding: This research was funded by the German Federal Ministry of Education and Research (BMBF) under grant number 03EE5048A, which is greatly appreciated.

Data Availability Statement: Data sharing is not relevant/appropriate.

Acknowledgments: The authors would like to extend their gratitude to D. Grüner and E. Wessel from Forschungszentrum Jülich GmbH for their assistance with the microstructural examination.

Conflicts of Interest: The authors have no conflict of interest to declare.

References

1. Du, Z.; Liu, G.; Huang, X.; Xiao, T.; Yang, X.; He, Y. Numerical studies on a fin-foam composite structure towards improving melting phase change. *Int. J. Heat Mass Transf.* **2023**, *208*, 124076. [CrossRef]
2. Malkawi, D.S.; Rabady, R.I.; Malkawi, M.S.; Al Rabadi, S.J. Application of Paraffin-Based Phase Change Materials for the Amelioration of Thermal Energy Storage in Hydronic Systems. *Energies* **2023**, *16*, 126. [CrossRef]
3. Fabrykiewicz, M.; Cieśliński, J.T. Effect of Tube Bundle Arrangement on the Performance of PCM Heat Storage Units. *Energies* **2022**, *15*, 9343. [CrossRef]
4. Guédez, R.; Topel, M.; Spelling, J.; Laumert, B. Enhancing the Profitability of Solar Tower Power Plants through Thermoeconomic Analysis Based on Multi-objective Optimization. *Energy Procedia* **2015**, *69*, 1277–1286. [CrossRef]
5. Zaversky, F.; García-Barberena, J.; Sánchez, M.; Astrain, D. Transient molten salt two-tank thermal storage modeling for CSP performance simulations. *Sol. Energy* **2013**, *93*, 294–311. [CrossRef]
6. Bonk, B.; Martin, C.; Braun, M.; Bauer, T. Material investigations on the thermal stability of solar salt and potential filler materials for molten salt storage. *AIP Conf. Proc.* **2019**, *1850*, 080008. [CrossRef]
7. Aarab, F.; Kuhn, B.; Bonk, A.; Bauer, T. A New Approach to Low-cost, Solar Salt Resistant Structural Materials for Concentrating Solar Power (CSP) and Thermal Energy Storage (TES). *Metals* **2021**, *11*, 1970. [CrossRef]
8. Kuizenga, A.M.; Cordaro, J.G. *Preliminary Development of Thermal Stability Criterion for Alkali Nitrates*; Sandia Technical Report SAND2011-5837C; Sandia National Laboratories: Albuquerque, NM, USA, 2011.
9. IRENA. *Latest Cost Trends in Renewable Power Generation Costs in 2020*; International Renewable Energy Agency: Abu Dhabi, United Arab Emirates, 2021.
10. Kuhn, B.; Fischer, T.; Fan, X.; Talik, M.; Aarab, F.; Yamamoto, Y. HiperFer-Weiterentwicklungs-Und Anwendungspotenziale. In Proceedings of the 43th FVWHT Vortragsveranstaltung Langzeitverhalten Warmfester Stähle und Hochtemperaturwerkstoffe, Online, 27 November 2020; p. 43.
11. Kuhn, B.; Talik, M.; Fischer, T.; Fan, X.; Yamamoto, Y.; Lopez Barrilao, J. Science and Technology of High Performance Ferritic (HiperFer) Stainless Steels. *Metals* **2020**, *10*, 463. [CrossRef]
12. Stein, F.; Leineweber, A. Laves phases: A review of their functional and structural applications and an improved fundamental understanding of stability and properties. *J. Mater. Sci.* **2021**, *56*, 5321–5427. [CrossRef]
13. Hald, J. Microstructure and long-term creep properties of 9–12% Cr steels. *Int. J. Press. Vessel. Pip.* **2008**, *85*, 30–37. [CrossRef]
14. Kobayashi, S.; Kimura, K.; Tsuzaki, K. Interphase precipitation of Fe2Hf Laves phase in a Fe–9Cr/Fe–9Cr–Hf diffusion couple. *Intermetallics* **2014**, *46*, 80–84. [CrossRef]

15. Kobayashi, S. Formation of the Fe2Hf Laves Phase through Eutectoid Type Reaction of δ→γ+Fe2Hf in Ferritic Heat Resistant Steels. *ISIJ Int.* **2014**, *55*, 2. [CrossRef]
16. Kuhn, B.; Talik, M.; Niewolak, L.; Zureka, J.; Hattendorf, H.; Ennis, P.J.; Quadakkers, W.J.; Beck, T.; Singheiser, L. Development of high chromium ferritic steels strengthened by intermetallic phases. *Mater. Sci. Eng. A* **2014**, *594*, 372–380. [CrossRef]
17. Fan, X.; Kuhn, B.; Pöpperlová, J.; Bleck, W.; Krupp, U. Compositional Optimization of High-Performance Ferritic (HiperFer) Steels—Effect of Niobium and Tungsten Content. *Metals* **2020**, *10*, 1300. [CrossRef]
18. Żuk, M.; Czupryński, A.; Czarnecki, D.; Poloczek, T. The effect of niobium and titanium in base metal and filler metal on intergranular corrosion of stainless steels. *Weld. Technol. Rev.* **2019**, *91*, 30–38. [CrossRef]
19. Beck, T.; Kuhn, B.; Eckardt, T.; Singheiser, L. Microstructure, Creep and Thermomechanical Fatigue of Novel Solid Solution and Laves Phase Strengthened Cr_2O_3 and Al_2O_3 Forming Ferrites for Car Engine Exhaust and Heat Exchanger Systems. *Trans. Indian Inst. Met.* **2016**, *69*, 379–385. [CrossRef]
20. Lopez Barrilao, J.; Kuhn, B.; Wessel, E. Microstructure and intermetallic particle evolution in fully ferritic steels. In Proceedings of the 8th International Conference on Advances in Materials Technology for Fossil Power Plants, Albufeira, Portugal, 11–14 October 2016; pp. 1029–1037.
21. Lopez Barrilao, J. *Microstructure Evolution of Laves Phase Strengthened Ferritic Steels for High Temperature Applications*; Energie & Umwelt; RWTH Aachen University: Aachen, Germany, 2017; Volume 375, ISBN 978-3-95806-231-3.
22. Klöwer, J. Factors affecting the oxidation behaviour of thin Fe-Cr-Al foils Part II: The effect of alloying elements: Overdoping. *Mater. Corros.* **2000**, *51*, 373–385. [CrossRef]
23. *ISO 17245:2015*; Corrosion of Metals and Alloys—Test Method for High Temperature Corrosion Testing of Metallic Materials by Immersing in Molten Salt or Other Liquids under Static Conditions. ISO: Geneva, Switzerland, 2015.
24. Kim, C.; Tang, C.; Grosse, M.; Steinbrueck, M.; Jang, C.; Maeng, Y. Oxidation Kinetics of Nuclear Grade FeCrAl Alloys in Steam in the Temperature Range 600–1500 °C. In Proceedings of the TOPFUEL 2021 Conference, Santander, Spain, 24–28 October 2021. [CrossRef]
25. Gunduz, K.O.; Visibile, A.; Sattari, M.; Fedorova, I.; Saleem, S.; Stiller, K.; Halvarsson, M.; Froitzheim, J. The effect of additive manufacturing on the initial High temperature oxidation properties of RE-containing FeCrAl alloys. *Corros. Sci.* **2021**, *188*, 109553. [CrossRef]
26. *DIN 50905-1*; Korrosion der Metalle—Korrosionsuntersuchungen: Teil 1: Grundsätze. Beuth Verlag GmbH: Berlin, Germany, 2009. [CrossRef]
27. Lopez Barrilao, J.; Kuhn, B.; Wessel, E. Identification, size classification and evolution of Laves phase precipitates in high chromium, fully ferritic steels. *Micron* **2017**, *101*, 221–231. [CrossRef]
28. Wagner, C. Passivity and inhibition during the oxidation of metals at elevated temperatures. *Corros. Sci.* **1965**, *5*, 751–764. [CrossRef]
29. Lopez Barrilao, J.; Kuhn, B.; Wessel, E. Microstructure evolution and dislocation behaviour in high chromium, fully ferritic steels strengthened by intermetallic Laves phases. *Micron* **2018**, *108*, 11–18. [CrossRef] [PubMed]
30. Kuhn, B.; Talik, M. Impact of Processing on the Creep Properties of High Performance Ferritic (HiperFer) Steels. *Metals* **2022**, *12*, 1459. [CrossRef]

energies

MDPI

Article

Energy Performance Evaluation of a Solar PVT Thermal Energy Storage System Based on Small Size Borefield

Evangelos I. Sakellariou [1], Petros J. Axaopoulos [1,*], Bill Vaneck Bot [1,2] and Ioannis E. Sarris [1]

[1] Department of Mechanical Engineering, Campus Ancient Olive Grove, University of West Attica, 250, Thivon & P. Ralli Str., 12241 Athens, Greece

[2] Laboratory of Energy, Materials, Modelling and Methods, Higher National Polytechnic School, University of Douala, Douala P.O. Box 2701, Cameroon

* Correspondence: pax@uniwa.gr

Abstract: In this study, a PVT-based solar-assisted ground source heat pump (SAGSHP) system with a small size borefield as the long-term heat storage component was energetically evaluated. The mathematical model of the system was formulated in TRNSYS and three cities with distinctive climates were chosen: Athens (Greece); Melbourne (Australia); and Ottawa (Canada). The parametric analyses were carried out for 10 years by varying the number of the PVT collectors and the size of the earth energy bank (EEB). The evaluation of the systems was made via two energy indicators, and the heat flow across the EEB was analyzed. The under-consideration system was found capable of establishing self-sufficiency as regards the energy consumption (renewable power fraction RPF > 1) for all locations. Namely, for Athens, any system with more than four PVT collectors, and for Melbourne, any system with more than eight PVTs was found with an RPF higher than 1, regardless of the EEB size. For Ottawa, self-sufficiency can be achieved with PVT arrays larger than 12 collectors for small EEBs, and with eight collectors for larger EEBs. The storage capacity was found to be an important parameter for the energy performance of the system. In particular, it was determined that, as the storage capacity enlarges the RPF and the seasonal performance factor (SPF) of the system improves, mainly due to the reduction of the electricity consumed by the heat pump and the auxiliary heating. Moreover, a larger storage capacity facilitates solar heat production by enlarging the available heat storage volume and by maintaining the EEB at relatively low temperatures.

Keywords: PVT; GHE; SAGSHP; GSHP; PVT-SAGSHP; EEB

Citation: Sakellariou, E.I.; Axaopoulos, P.J.; Bot, B.V.; Sarris, I.E. Energy Performance Evaluation of a Solar PVT Thermal Energy Storage System Based on Small Size Borefield. *Energies* **2022**, *15*, 7906. https://doi.org/10.3390/en15217906

Academic Editor: Ann Lee

Received: 7 October 2022
Accepted: 21 October 2022
Published: 25 October 2022

1. Introduction

Manmade pollution and climate change are the greatest challenges of today. Combustion of conventional fuels releases greenhouse gases into the atmosphere, and these are considered to raise the average global temperature. A solution to this issue may be to use more renewable energy systems (RESs) and to displace fossil fuel-based systems. RESs are implemented in the household sector by providing thermal and electrical energy. Solar assisted heat pumps (SAHPs) [1,2] and shallow ground source heat pumps (GSHPs) [3–5] are both promising space heating technologies. The combination of the aforementioned systems is called solar-assisted ground source heat pump (SAGSHP) [6,7]. The first attempt to assess the SAGSHP systems dates back to the 1970s, with the innovative work conducted by Metz [8], from that time, the attention on this technology has increased.

The combination of the solar and geothermal systems can promote long-term, or interseasonal, heat storage. Despite the fact that SAGSHP systems dominate the market at the current stage, the idea of interseasonal solar heat storage goes back to 1939. Thus, the first attempt to store summer solar heat was made by MIT's solar house [9]. The storage container was a horizontal insulated cylindrical steel water tank of 68 m^3. A house of 45 m^2 was built, and the storage water temperature was 90 °C at the end of the charging period and dropped to 55 °C in February. The area of the solar collectors was 33 m^2, thus the storage capacity of the system was about 1.6 m^3 m^{-2}. Since then, many seasonal thermal energy methods have been invented [10–15]. The majority of the utilized interseasonal thermal energy storage (TES) methods store the heat sensibly via water tanks or lakes [16] and into the soil via borefields as well. TES systems based on soil as the sensible heat storage material are becoming popular because they can be used on systems with heat pumps, thus the storage temperature can be relatively low (no excessive heat loss) and the construction cost is lower than that of a water tank. For the soil-based TES, a geothermal heat exchanger (GHE) is required, this type of system is called the borehole thermal energy storage (BTES) system [15,17,18] and the majority of the SAGSHP systems are based on this method for long-term-heat storage.

Up to now, the majority of the experimental [19–22] or theoretically [23–26] studies on SAGSHP systems are made with a variety of components, layouts, operation-control, and for different weather conditions. The solar collectors used for the studies on SAGSHP systems vary from flat plate collectors (FPC) [27] to PVTs [28], and the GHE includes BHEs [29] and very shallow borefields [22,30,31]. All these types of components can be combined in many configurations in addition to numerus layouts (hydraulic connection of systems) [32–34]. In conclusion, SAGSHP systems are multiple-aspect projects, and their design is unique for every case.

A special configuration of SAGSHP systems is the one with PVT collectors, which has the ability to coproduce heat and power. Bertram et al. [28] investigated a PVT-based SAGSHP system with a concentrical BHE built in Munich. PVT collectors and one photovoltaic panel were placed alongside with the goal to assess their energy production. With a two-year trial data referring to their performance, the PVT collector was more productive by 4% than the photovoltaic as regards to the power, and the system's seasonal performance factor (SPF) was found to be 4.2. A trial based on PVT utilization was carried out by Wright et al. [35], the study shows a newly built dwelling designed in accordance with zero carbon emission regulations. A similar study with a PVT-based SAGSHP system was coupled with a novel borefield and presented by Mendoza et al. [36]. The borefield was very shallow and it was installed beneath the building with the goal of restricting the surface heat losses. A pilot system was designed and implemented by De Montfort University with the aim of assessing its performance energetically for the UK Midlands climate [22]. The seven PVT collectors and a borefield of 16 BHEs with only 1.5 m of depth were its main components. With the experimental data, the SPF of the system was found to be 2.51 out of 20 months of operation.

Computer-based work has been carried out for the Netherlands' climate, aiming to evaluate the contribution of PVTs paired with a conventional GSHP system [37]. The study was carried out with the TRNSYS simulation platform [38]. The outcomes were that: 96% of the electricity demanded by the SAGSHP system was offered by the PVTs; and the heat energy requirements were covered by the GHE to 83% of its total demand. A theoretical study with a low exergy concept was conducted to appraise the performance of a PVT-based SAGSHP system [39]. Via simulations carried out with TRNSYS, the SPF was estimated to be 6, which is a significantly high value.

A study was carried out to assess the contribution of PVTs added to a conventional GSHP system by considering many parameters [40]. In particular, the evaluation was made of the system's energy and financial performance for a block of flats in Stockholm. The primary results were a reduction of 18% in the borefield's size and a 50% decrease in the distance among BHEs. Both above-mentioned reductions assist the economy of the system. It is important that the system was benefited by the PVTs' addition and performed with an SPF equal to that of the conventional GSHP system (without PVTs). The reduction in the GHE's size, paired with PVTs, against the GSHP system is translated into a reduced initial capital investment. A trial was conducted, seeking to find out the advantages of having flat plate solar collectors paired with a GSHP system [41]. A low-rise building built in Milton (Canada) was monitored as regards to its energy demand. Data from the building was utilized and, along with the SAGSHP system, were modeled in TRNSYS. The storage capacity of the systems was one of the important parameters. The storage capacity with the best energy performance of the system was that with 4.7 m of BHE per m^2 of FPC. With the optimum storage capacity, the required total length of the BHE was determined to be shorter by 32 m. This reduction can be achieved by installing 6.81 m^2 of FPC. The economy of the systems was appraised via the net present value (NPV) index for a 20-year period. It can be concluded that the SAGSHP has a somewhat higher NPV compared to the conventional GSHP system.

SAGSHP may be an alternative solution for space heating systems by being based on renewable energy sources and by facilitating interseasonal solar heat storage. Thus, in this paper, a novel PVT-based SAGSHP system was studied energetically as a space heating system solution for low-rise dwellings in three locations: Athens (Greece); Melbourne (Australia); and Ottawa (Canada). The novelty of the system is based on the very shallow borefield (2 m deep), which may potentially reduce the construction cost of the GHE by avoiding costly deep drilling. The investigated borefield is the long-term solar heat storage component of the system. The system's mathematical model was built in TRNSYS by having the main parts of the system validated via experimental data. Based on the built model, parametric analyses were conducted with the variation of the PVT collectors and the size of the GHE (Earth Energy Bank, EEB). Through the simulation results, the systems were evaluated by two energy indexes. The evaluation of the SAGSHP system is an essential task, with the aim of identifying the contribution of the long-term heat storage component to the energy performance of the system.

2. Methodology

In the current study, a PVT-based SAGSHP system is evaluated energetically for three cities with distinctive climates. The three cities are: Athens, Greece (37.54° N, 23.44° E), with a Mediterranean climate Csa; Melbourne, Australia (37.40° S, 144.50° E), with a temperate oceanic climate Cfb; and Ottawa, ON, Canada (43.40° N, 79.19° W), with a warm summer continental climate Dfb (Köppen–Geiger [42,43]). For all three cities, the SAGSHP system was installed to provide thermal energy for a low-rise single-family dwelling for space heating and domestic hot water (DHW). Figures 1 and 2 show the mean ambient db air temperature and the monthly global horizontal solar energy, respectively, for all locations. Data from Meteonorm for the typical meteorological year (TMY) were utilized for these figures.

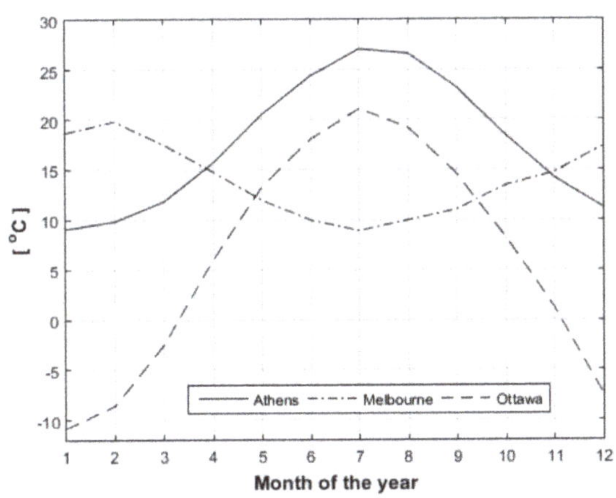

Figure 1. Monthly mean dry bulb ambient air temperature for all three cities. (Derived from TMY2, Meteonorm).

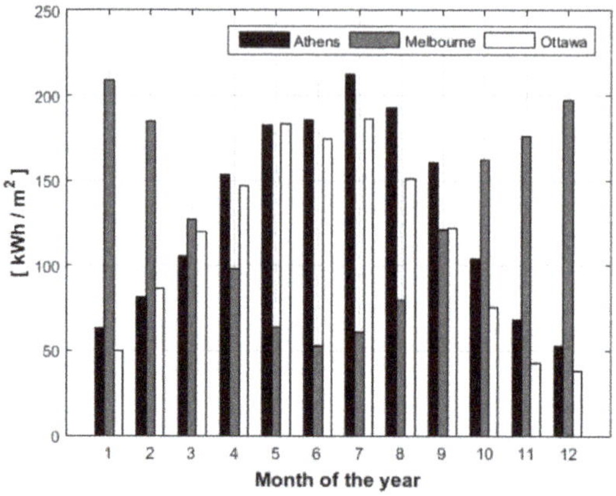

Figure 2. Monthly global horizontal solar energy for all three cities. (Derived from TMY2, Meteonorm).

The main components of a SAGSHP system are the PVT array, the earth energy bank (EEB), and the heat pump. The EEB is comprised of the GHE and the adjacent soil volume, in other words, this is the long-term solar heat storage element of the system. The analysis of the system along with the dwellings is illustrated in Section 2.1, while the topology of the system is shown in Figure 3. A combined mathematical model of the SAGSHP system and of the dwelling was formulated in the TRNSYS simulation platform [38]. With the developed model, simulation-based parametric analyses were conducted by varying the size of the PVT array and of the EEB. Based on that, twenty-five simulations were carried out for each city, with five different sizes of the PVT array and five different sizes of the EEB, Section 2.1.

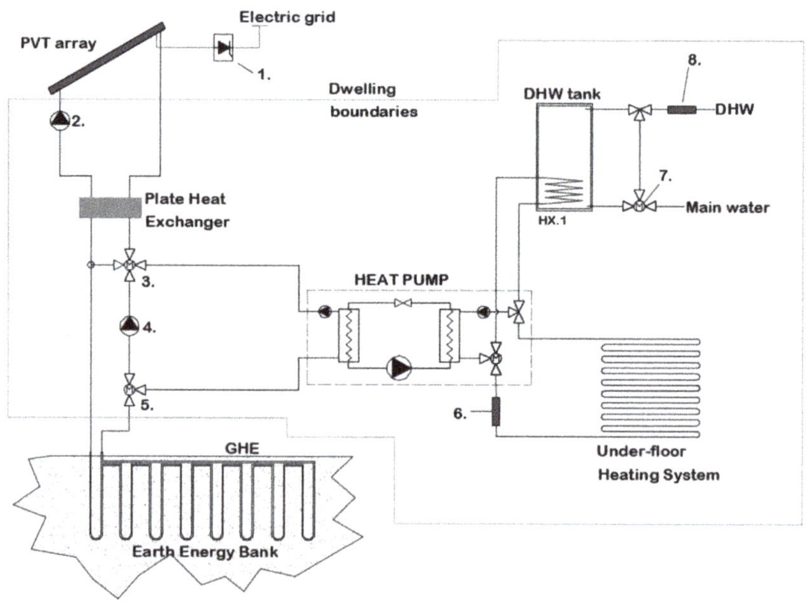

Figure 3. SAGSHP system's layout: 1. DC to AC inverters; 2. Solar system pump; 3. Four-way deviator; 4. Solar-soil charging pump; 5. Three-way deviator; 6. Space heating auxiliary heater; 7. Temperature control valve; 8. DHW auxiliary heater.

For the evaluation of the systems, a 20-year period was considered by executing simulations with hourly steps. The considered period over which the analysis takes place is substantial to get accurate results, especially about the transient behavior of the soil during the initial years of operation. The meteorological data provided by Meteonorm formulated according to the Typical Meteorological Year (TMY) format was inserted into the model. The results from simulations were utilized to energetically assess the SAGSHP system. Two energy indexes were found to be adequate for the analysis (Section 2.2), and these can provide a holistic view regarding the performance of the system.

The operation of the proposed system is briefly described in Table 1. The implementation of the control was made via Type 40 of TRNSYS, which provides the ability to program a microcontroller as inputs to Type 40, where: the collectors' exiting brine temperature; the adjacent to the BHEs' mean ground temperature; and the space hating (SH) and DHW thermostats. It is important to note that the system was responsible for covering the SH demand first and then DHW. Moreover, the additional thermal energy was provided (if needed) via two elements installed at each consumption point, SH and DHW (Figure 3).

Table 1. The system's main operation modes. Where: D.V is deviation valve; PHE is plate heat exchanger; and HP is heat pump.

Operation Mode	Description	Control
PVTs-GHE-Heat Pump-Load	Both PVTs and GHE are the thermal energy sources of HP. The PVT outlet temperature is 6 K higher than the soil temperature (EEB)	D.V 3 and 5 (Figure 3) are engaged to drive the brine via the GHE and PHE. Pump 2 operates for the delivery of solar heat. Pump 4 is bypassed
GHE-Heat Pump-Load	The system operates as a conventional GSHP. Low solar irradiance or solar heat cannot be utilized. The ΔT between the PVT outlet temperature and that of the soil is lower than 6 K	D.V 3 and 5 (Figure 3) are responsible for delivering heat to HP by bypassing the PHE and pump 4
PVTs-GHE	Solar heat is delivered to the EEB for storage. SH or DHW are not required. The PVT outlet temperature is 6 K higher than the soil temperature (EEB)	D.V 3, 5 (Figure 3) and pump 4 deliver the solar heat to the EEB

2.1. The Energy System and the Dwelling

The studied system is inspired by pilot experimental work carried out by DMU [22,36]. The model of the pilot project with minor modifications was implemented in TRNSYS [44]. Moreover, the major components of the system, such as the PVTs and the borefield, are validated via experimental data from DMU's project and additional experimentation [44,45]. Due to the already made assessment of the system's mathematical model, the analysis of the system is overviewed. However, basic information is presented below regarding the mathematical model of each component, with the aim of facilitating the comprehension of the presented study.

The mathematical model of the PVT collector is based on Sakellariou and Axaopoulos [45] transient model, which is an evolution of the steady state energy balance model introduced by the Hottel and Whillier [46]. The useful heat rate produced by the PVT collectors can be estimated by the general solution of Equation (1), which is the transient energy balance equation of the collector [45]. The transient model has been validated through experimental data [45] by achieving RMSD of 0.66% to 4.22% for the temperature of the fluid exiting the collector and from 4.15% to 14.9% for the power production. Moreover, the model was formulated as a new TYPE (component) in TRNSYS for the needs of the parametric analysis. In Table 2, the main parameters of the PVT collectors used for the current study are listed.

$$(Mc_p)_C \frac{dT_{pm}}{dt} = SA_C - \dot{Q}_u - \dot{P}_e - \dot{Q}_{loss} \quad (1)$$

where, in Equation (1): T_{pm} is the collector's absorber mean temperature (K); t is time (s); $(MC_P)_C$ is the heat capacity of the collector; S is the absorbed irradiance (W m^{-2}); Ac is the collector's area (m^2); Qu is the useful heat rate produced by the collector (W); P_e is the power produced by the PV cells (W); and Q_{loss} are the heat loss rate of the collector (W).

Table 2. Main system parameters.

Subsystem	Details
PVTs	PVT arrays: 4, 8, 12, 16, and 20 collectors. Peak power 235W_p. Absorber's area 1.58 m^2 [45]. Inclination 30 degrees. South or north (Melbourne) facing fixed.
Borefield	4, 8, 12, 16, and 20 BHEs with a very short length of 2 m. Soil thermal conductivity 1.5 W m^{-1} K^{-1} Soil heat capacity (clay) 2400 kJ m^{-3} K^{-1} Spacing between BHEs at 2 m.
Heat pump	Nominal heat capacity 6 kW_{TH} for Athens and Melbourne, 8 kW_{TH} for Ottawa. Operation envelope is −10 °C to 25 °C for the evaporator and 30 °C to 55 °C for the condenser. Nominal power 1.5 kW_e for Athens and Melbourne, 2 kW_e for Ottawa. Flowrate evaporator side 1200 kg h^{-1} for Athens and Melbourne, 1720 kg h^{-1} for Ottawa. Flowrate condenser side 700 kg h^{-1} for Athens and Melbourne, 1400 kg h^{-1} for Ottawa.

With regard to the mathematical model of the geothermal heat exchanger, this was formulated via Hellström's Duct Ground Heat Storage Model (DST) [47]. The solution provided by the DST model is estimated by superimposing two numerical calculations and one analytical calculation of the heat conduction equation in the soil (TYPE 557 from TRNSYS). The proposed heat exchanger has been validated via four months of experimental data. The RMSD of the model against the data was found to be 4.43% for the GHE outlet temperature. Concerning the GHE used at the current study, it is a very shallow borefield, 2 m deep, with a distance among the BHEs of 2 m. It has to be noted that the size (volume, V_{EEB}) of the EEB is related to the number (N_{bor}), the length of the borehole heat exchangers (L_{bor}) and the distance between boreholes (D_{bor}) according to Equation (2). The borefield was 1 m underneath the dwelling without any thermal insulation, and the main parameters of the GHE are listed in Table 2.

$$V_{EEB} = L_{bor} \cdot N_{bor} \cdot \pi \cdot (0.525D_{bor})^2 \quad (2)$$

For the heat pump, a new TYPE was formulated in TRNSYS. The heat pump's new type was developed to account for the performance data according to EN 14511 and the delivered heat, forming the GHE and PVTs. The performance data of the heat pump are sourced from a well-known German manufacturer. The operation envelop of the heat pump is from −10 °C to 25 °C for the evaporator and from 30 °C to 55 °C for the condenser.

The hydraulic connection between the PVTs and the GHE was implemented to be in-series, via the plate heat exchanger, as is shown in Figure 3. The heat from PVTs was used by the system when the PVT's exit bribe temperature was 6 K higher than that of the EEB (soil temperature adjacent to BHEs). The evaporator of the heat pump supplied heat by solar and/or the ground (EEB) loop. If no heat was demanded by the dwelling, the heat from PVTs was stored in the EEB (Table 1). The dwelling's heat transfer system was an under-floor heating system, with heat pump providing water at 31 °C. As regards the DHW, this was based on an immersed heat exchanger in the DHW tank supplying water from the heat pump at 45 °C. As regards the electric energy yield of PVTs, this was delivered totally to the power grid by accepting an overall loss of 10%. The aforementioned loss accounts for joule cable losses and the efficiency of the inverters. Moreover, an annual drop of 1% in the PVTs' electrical efficiency was considered. The above operation modes of the system are listed in Table 1.

As regards the simulation scenarios, these were carried out with five sizes of PVT arrays paired with five sizes of EEB. For PVT arrays of 4, 8, 12, 16, and 20 collectors (strings of four PVTs connected hydraulically) and with EEB of 4, 8, 12, 16, and 20 BHE; in total, 25 simulations for each city. For each string of PVTs, a flowrate of 100 kg h^{-1} was applied (25 kg h^{-1} PVT^{-1}) and 8 m of thermally insulated piping was assumed for every string. Table 3 illustrates all the PVT arrays along with the borefields and the volume of the EEB and the corresponding storage capacity (SC) for each pair. The storage capacity is the ratio between the heat storage volume of the EEB and the PVT collectors' area. In other words, here the SC depicts the ratio between the storage volume–capacity and the apparatuses of the heat production.

Table 3. Storage capacity [m^3/m^2) as product of the heat storage volume allocated to the EEB and the utilized PVT arrays.

	4 BHE V_{EEB} (28 m^3)	8 BHE V_{EEB} (56 m^3)	12 BHE V_{EEB} (83 m^3)	16 BHE V_{EEB} (111 m^3)	20 BHE V_{EEB} (139 m^3)
4 (PVTs)	4.43	8.86	13.13	17.56	21.99
8 (PVTs)	2.22	4.43	6.57	8.78	11.00
12 (PVTs)	1.48	2.95	4.38	5.85	7.33
16 (PVTs)	1.11	2.22	3.28	4.39	5.50
20 (PVTs)	0.89	1.77	2.63	3.51	4.40

As regards the dwelling, this was assumed to be a two-story building with a total occupation area of 100 m^2 for a single family of four. It was hypothesized that the dwelling was built according to the national energy-saving regulations for residential buildings applied in each country. The dwelling's model was developed in TRNSYS via TYPE 56, and the layout was common for all cities. Moreover, it was assumed that both stories were maintained at a mean air db temperature of 19 °C all day round (during the heating season). The DHW needs of the dwelling were set to be 35 L per day per person (140 L day^{-1} for a family of four) at 45 °C. The volume of the DHW tank was set to be 200 L and the consumption profile of DHW was based on the well-known f-Chart method [48]. The main parameters of the heating load (space heating and DHW) are listed in Table 4 for all locations. Concerning the space heating load (Table 4), this was estimated through the TYPE 56, which can provide the user with the hourly thermal energy demand of the building. Similarly, the DHW load was estimated by the difference between the water main temperature at the given time and the DHW set temperature of 45 °C multiplied by the hourly water demand. The water main temperature was provided by the weather TYPE 15 of TRNSYS. The auxiliary space heating and DHW systems were set to consume electricity when necessary. This is an acceptable solution for a relatively small-sized system, especially one in which electricity is already available.

Table 4. Dwellings and DHW main parameters values for all three cities.

Part	Athens		Melbourne		Ottawa	
Exterior walls U-value	0.45 W m^{-2} K^{-1}	Regulation for energy efficiency on buildings [49]	0.375 W m^{-2} K^{-1}	Australian construction code volume two [50]	0.32 W m^{-2} K^{-1}	National Energy Code of Canada for Buildings [51]
Ground floor U-value	0.80 W m^{-2} K^{-1}		0.444 W m^{-2} K^{-1}		0.45 W m^{-2} K^{-1}	
Exterior roof U-value	0.40 W m^{-2} K^{-1}		0.244 W m^{-2} K^{-1}		0.21 W m^{-2} K^{-1}	
Windows U-value	2.60 W m^{-2} K^{-1}		2.60 W m^{-2} K^{-1}		1.60 W m^{-2} K^{-1}	
Windows to walls ratio	0.15		0.15		0.15	
Infiltration	1 ACH		1 ACH		0.5 ACH	
Consumption of DHW	140 L day^{-1} at 45 °C		140 L day^{-1} at 45 °C		140 L day^{-1} at 45 °C	
Annual heating load	3393 kWh (19 °C)		7150 kWh (19 °C)		5240 kWh (19 °C)	
Annual DHW load	1445 kWh (45 °C)		1667 kWh (45 °C)		2185 kWh (45 °C)	

2.2. Energy Metrics

The energy performance evaluation of the system was carried out through two energy indicators. The first index is the ratio between the delivered electricity from PVTs and the electricity consumed (in total) by the system. In other words, the renewable power fraction (RPF) [52] depicts the level of the energy self-sufficiency of the system (Equation (3)). It is worth mentioning that, in the current study, household consumption is not considered (kitchen, fridge, etc.). The RPF is a ratio which gets on its numerator the electricity delivered to the power grid by the PVTs (E_{PVT_u}), while the denominator receives all the system's consumption. In more detail, the power consumption of the system is the heat pump (E_{HP}), the consumption of the circulation pumps ($E_{parasitic}$), and the auxiliary heat (E_{aux}). In this study, Q_{aux} and the E_{aux} represent the same amount of energy, by assuming the conversion factor from electricity to heat equal to 1. Lastly, it is important note that an RPF higher than one indicates an energetically self-sufficient system.

$$\mathrm{RPF} = \sum_{i=\mathrm{year}}^{20} \left[\frac{E_{PVT\_u}}{E_{HP} + E_{parasitic} + E_{aux}}\right]_i \quad (3)$$

As the second energy indicator, the seasonal performance factor (SPF) is chosen. In the present work, the 4th boundary for the estimation of the SPF is implemented according to the [53]. By that, all the consumption up to the heat to be delivered from the sources to the dwelling are considered in the denominator of Equation (4). The SPF_{H4} states the ratio between the delivered heat (Q_{cond}) by the system and the system's consumed electricity, Equation (4).

$$\mathrm{SPF} = \frac{\sum_{i=\mathrm{year}}^{20} (Q_{cond})_i}{\sum_{i=\mathrm{year}}^{20} (E_{HP} + E_{parasitic} + E_{aux})_i} \quad (4)$$

3. Results and Discussion

The simulation results are presented and discussed in three sections, which correspond to three different aspects and energy indicators: the renewable power fraction (Section 3.1); the seasonal performance factor (Section 3.2); and the heat flow on the EEB (Section 3.3).

3.1. Renewable Power Fraction

The investigated PVT-based SAGSHP system has the capability to coproduce heat and power. In this regard, power production is interrelated to heat production and, particularly, to the temperature at which the heat is stored in the EEB. The EEB is the long-term thermal energy storage component of the system. As the storage temperature increases, the soil temperatrue negatively influences the power production of the PVT array by increasing the PV-cell temperature [54]. A valid method to evaluate the overall energy performance of the system equipped with PVTs is to utilize the RPF (Section 2.2), this indicator has been used successfully in many cases [52].

In Figure 4, the RPF for Athens is illustrated as a function of the storage capacity. Based on the results, the system, out of all its size variations, obtains an RPF of about 0.4 to 4.7. All RPF values higher than 1 indicate a self-sufficient system, as regards its annual electricity anergy consumption. Thus, all systems with a PVT array of 8 PVTs and larger are self-sufficient for Athens, regardless of the size of the EEB. That is justified by the tabulated results in Table 5, which show the consumption and the generated electricity for the systems with all PVT array variations paired with a borefield of 12 BHEs. The results of Table 5 refer to the midline of Figure 4 (the third from the bottom). As it can be seen in Table 5, the electricity produced by PVTs is proportional to the size of the array. As regards the consumption, that of the heat pump (E_{HP}) is larger across all scenarios and drops as the PVT array enlarges. The reduction is caused by the larger amount of augmented solar heat production, which charges the EEB with more heat and, due to that, raises the average temperature of the soil. The higher temperature of the soil positively influences the performance of the heat pump by elevating the temperature of the brine entering the evaporator. As is well known, the higher the evaporator temperature, the higher the performance of the heat pump, and this lowers the consumption of electricity by the compressor. The second significant consumption of the system is that of the auxiliary heat (E_{aux_SH}, E_{aux_DHW}), the consumption also reduces as the PVT array enlarges. This improvement is made by the system's capability to have more available heat, thus the requirement for auxiliary heat is reduced.

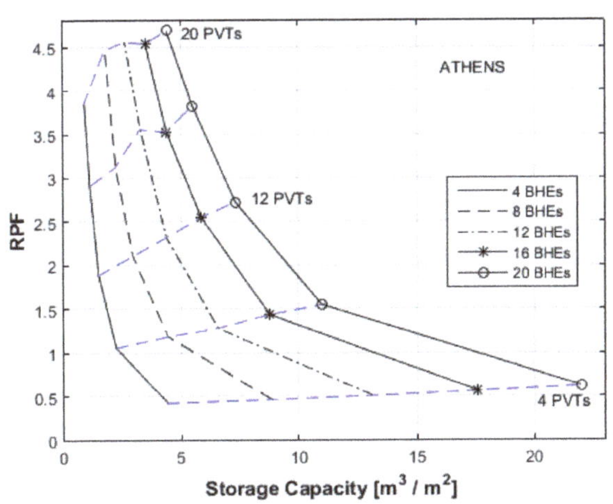

Figure 4. RPF for Athens as function of the storage capacity.

Table 5. Annual electric energy (E) consumed on the SAGSHP system and the annual electricity and heat produced by the PVTs, along with the heat provided by the heat pump for Athens.

	4 PVTs X 12 BHEs	8 PVTs X 12 BHEs	12 PVTs X 12 BHEs	16 PVTs X 12 BHEs	20 PVTs X 12 BHEs
E_{HP} (kWh/year)	1797.3	1699.3	1572.1	1550.9	1456.7
$E_{parasitic}$ (kWh/year)	5.3	8.5	10.6	12.6	13.5
E_{aux_SH} (kWh/year)	1188.0	668.4	376.6	122.6	105.3
E_{aux_DHW} (kWh/year)	119.6	85.9	78.3	81.5	68.1
E_{PVT} (kWh/year)	1576.6	3149.5	4722.1	6293.4	7860.7
Q_{PVT} $(kWh/year)_{heat}$	880.2	1604.8	2150.1	2648.2	2885.6
Q_{HP_cond} $(kWh/year)_{heat}$	3145.1	3665.3	3957.3	4211.3	4148.6

A trend that can be identified in Figure 4 is that the enlargement of the storage capacity improves the RPF of the system for all PVT arrays. Moreover, it can be seen that the larger storage capacity enhances the RPF as the PVT array enlarges. Therefore, for the system with four PVTs, the RPF increases by about 0.2 from the smallest EEB (four BHEs) to the largest one (20 BHEs), versus the system with 20 PVTs, where the largest EEB improves the RPF by 0.9 from the smallest one. The asset of having large EEBs lies in the fact that the available solar heat from the PVT collectors can be stored and used later, on the evaporator of the heat pump or by reducing the electricity consumed by the systems, as is discussed in the previous paragraph.

Figure 5 illustrates the RPF for Melbourne for all the investigated scenarios. The obtained RPF for Melbourne varies from about 0.2 to 3 for the smallest to the largest system, respectively. The RPF for Melbourne exhibits similar trends to those described for Athens, the RPF increases as the PVT array and the EEB enlarge. As regards the self-sustainability of the system, this can be obtained with the PVT array having more than eight collectors. The storage capacity for all scenarios with 20 collectors remained lower than 5 m^3 m^{-2}, but in this range, the system increases the RPF by about 0.8. In other words, the systems with 20 PVTs can produce 80% of their annual electric energy consumption by installing the largest EEB against the smallest one. The contribution of the storage capacity to the RPF diminishes as the number of collectors reduces.

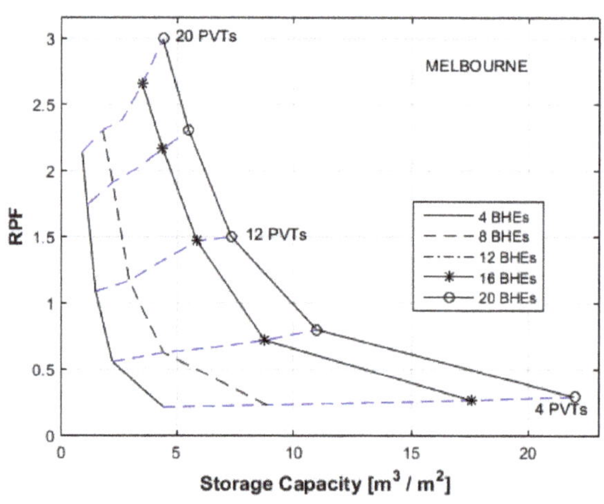

Figure 5. RPF for Melbourne as function of the storage capacity.

In Table 6, the electric energy consumption and the electricity produced by the PVTs for the systems with 12 BHEs and all PVT arrays for Melbourne are listed. As can be noted, the demand for electricity by the heat pump (E_{HP}) is the larger consumption for almost all scenarios. The only exception is for the array of four PVTs, where the auxiliary energy (E_{aux_SH}, E_{aux_DHW}) is larger than that of the heat pump. It is worth mentioning that auxiliary consumption is significantly high, and it is assumed that this is provided by electricity. It turned out that this amount of auxiliary energy should be considered during the design of the system, with the aim of achieving higher self-sustainability. Other than that, the auxiliary heat is reduced rabidly as the solar heat from PVTs (Q_{PVT}) contributes more to the system by elevating the average temperature of the soil and increasing the available heat in the EEB.

Table 6. Annual electric energy (E) consumed on the SAGSHP system and the annual electricity and heat produced by the PVTs, along with the heat provided by the heat pump for Melbourne.

	4 PVTs X 12 BHEs	8 PVTs X 12 BHEs	12 PVTs X 12 BHEs	16 PVTs X 12 BHEs	20 PVTs X 12 BHEs
E_{HP} (kWh/year)	3351.5	3277.5	2987.1	2735.4	2600.9
$E_{parasitic}$ (kWh/year)	12.3	21.4	28.0	30.5	32.8
E_{aux_SH} (kWh/year)	3366.9	1618.7	691.6	458.7	405.5
E_{aux_DHW} (kWh/year)	319.4	252.0	175.7	164.5	156.5
E_{PVT} (kWh/year)	1629.5	3257.6	4878.6	6495.0	8088.7
Q_{PVT} (kWh/year)$_{heat}$	2234.3	4437.2	6116.3	6322.5	8072.7
Q_{HP_cond} (kWh/year)$_{heat}$	6365.2	8113.4	9040.5	9273.4	8966.6

In Figure 6, the RPF for Ottawa is illustrated as a function of the storage capacity. The largest value for Ottawa is recorded as 2.1 for the largest system and the lowest as 0.25 for the smallest one, respectively. The obtained RPF for Ottawa can be marked as significantly lower than that of Athens (Figure 4), with the main cause being the hard winter (Figure 1). Low temperatures implicitly influence the space heating load and the temperature of the soil, thus, consequently, heat loss from the EEB. The self-sufficiency for Ottawa can be distinguished into three sections: (a) for small EEBs of four and eight BHEs with PVT arrays larger than 12 collectors are required; (b) the case of 12 BHEs, where the system gets RPF larger than 1 from the array of 12 PVTs; and (c) for the larger EEBs of 16 and 20 BHEs, all PVT arrays equal and above 12 collectors are capable of RPF larger than 1.

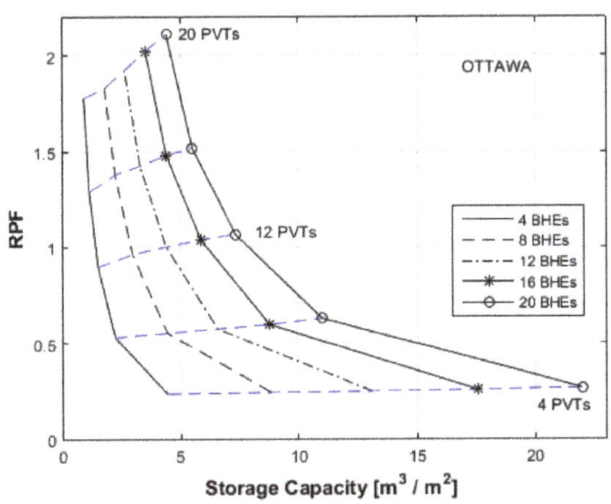

Figure 6. RPF for Ottawa as function of the storage capacity.

The annual electric energy consumption for Ottawa is listed in Table 7, along with the generated electricity from PVTs for the system with 12 BHEs and all PVT arrays. Similarly to the results for Athens and Melbourne, the annual consumption of the heat pump is the largest consumption of the system throughout all the scenarios, followed by the auxiliary heat, which reduces significantly as the PVT array enlarges.

Table 7. Annual electric energy (E) consumed on the SAGSHP system and the annual electricity and heat produced by the PVTs, along with the heat provided by the heat pump for Ottawa.

	4 PVTs X 12 BHEs	8 PVTs X 12 BHEs	12 PVTs X 12 BHEs	16 PVTs X 12 BHEs	20 PVTs X 12 BHEs
E_{HP} (kWh/year)	3469.7	3438.9	3469.2	3354.0	3192.0
$E_{parasitic}$ (kWh/year)	4.1	7.6	11.4	13.9	16.1
E_{aux_SH} (kWh/year)	2694.1	1908.5	1129.3	928.9	778.3
E_{aux_DHW} (kWh/year)	442.1	328.9	277.1	278.3	248.2
E_{PVT} (kWh/year)	1628.2	3263.1	4897.2	6519.4	8144.7
Q_{PVT} (kWh/year)$_{heat}$	768.1	1596.4	2509.9	2972.5	3473.6
Q_{HP_cond} (kWh/year)$_{heat}$	4618.3	5403.9	6183.2	6383.5	6534.2

3.2. *Seasonal Performance Factor*

The SPF depicts the ratio between the heat delivered by the heat pump and the electricity consumed by the system (annually $kWh_{heat}/kWh_{electricity}$). In the current analysis, it is assumed that the electricity from PVTs does not contribute to the SPF (Section 2.2). This is not exact regarding the performance of the system, but with this index, useful information can be gained regarding the heat production by the heat pump. In addition, by subtracting the E_{PVT} in the denominator from the consumption, the SPF receives an infinity value for RPF larger than 1, and that does not provide any useful information. In this regard, Figure 7 illustrates the SPF for Athens and for all PVT and BHE arrays. The obtained SPF for Athens was found to be 0.65 for the smallest system and up to about 2.6 for the largest one. The SPF is influenced significantly by the storage capacity, with the higher values improving the energy performance of the system with specific PVT arrays. Table 5 illustrates many energy aspects—the results for Athens and for the system with 12 BHEs and all available PVT arrays. These results stand for the mid line of Figure 7 and, as it can be seen, the larger the PVT array, the higher the SPF. By reading Figure 7 in conjunction with Table 5, for the EEB of 12 BHEs, the heat provided by the heat pump (Q_{HP_cond}) increases as the PVT array enlarges. Concurrently, the consumption of all these decreases as the number of collectors increases, with the only exception being the parasitic energy ($E_{parasitic}$), which is negligible compared to the others. Additionally, for a specific EEB, as the PVT array enlarges, the improvement in the SPF is due to the additional solar heat. In more detail, from the array of 4 PVT to this of 8, the SPF increases by 0.5, but for the larger number of collectors, the incrementation of the SPF drops.

Figure 8 illustrates the SPF for all investigated scenarios of PVTs and BHEs for Melbourne. The SPF for Melbourne was found to vary from about 0.6 to 3.4 for the smallest and the largest systems, accordingly. Overall, the SPF calculated for Melbourne is higher than that for Athens, this is justified by the substantially higher Q_{HP_heat} (Table 6) delivered for Melbourne compared to that for Athens (Table 5). The high heating load of Melbourne drives the heating system (heat pump) to deliver more heat, which depletes the stored heat in the EEB and reduces the soil temperature. Consequently, this facilitates the capability of the system to deliver more solar heat and to store this in the ground. Given this, it can be stated that the ability of the system to provide heat is restricted by the demanded heating load: no or low load, the soil temperature stagnates and the solar energy cannot be stored further.

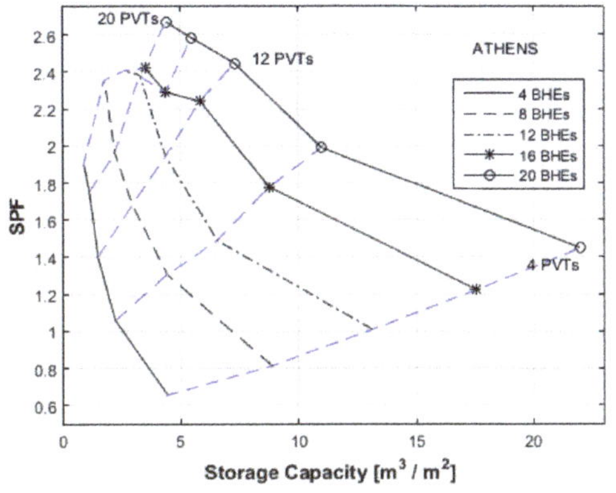

Figure 7. Annual seasonal performance factor as function of the storage capacity for Athens.

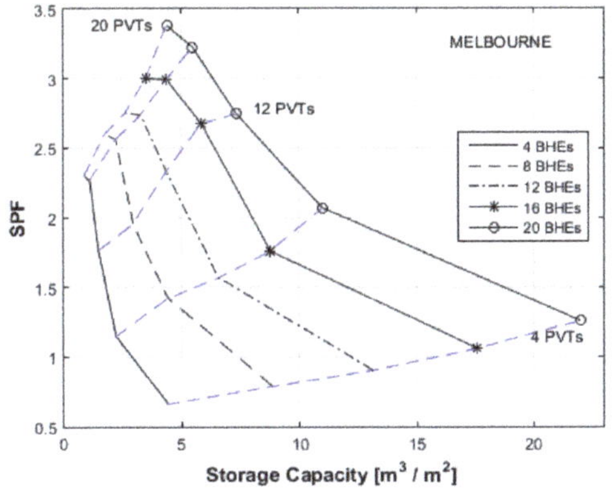

Figure 8. Annual seasonal performance factor as function of the storage capacity for Melbourne.

As regards the SPF for Ottawa, this is illustrated by Figure 9 in the simulation scenarios. Based on the simulation results, the SPF was estimated to be from 0.5 to about 1.8 for the system with the smallest PVT array and EEB to the largest PVT array paired with the largest EEB. The high E_{HP} along with the significant large E_{aux} (Table 7) are the major factors for the relatively low SPF.

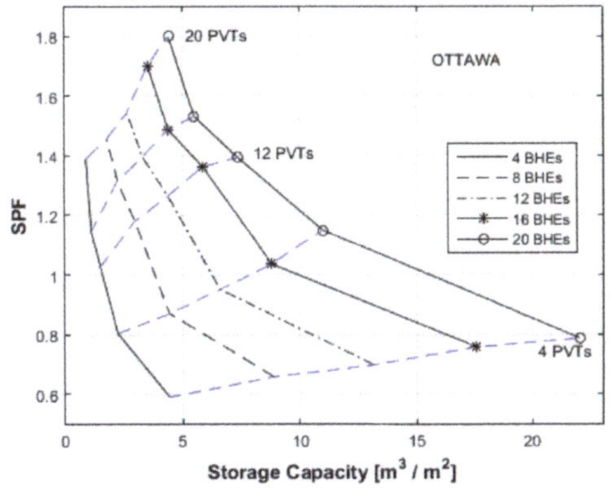

Figure 9. Annual seasonal performance factor as function of the storage capacity for Ottawa.

3.3. Heat Flow on the EEB

In Figures 10–12, the monthly heat flow across the EEB for the 6th year of simulation for all cities and for the system with 12 PVTs paired with 12 BHEs (EEB = 83 m^3) is illustrated. The 6th year is chosen with the need to get results referring to the EEB after the initial transient period as regards the soil temperature, which, for the current analyses, was the post-three period after initiation. These figures show three monthly thermal energy values related to the EEB: the heat provided by the solar heat exchanger (Q_{SHE}), the heat transferred from the solar loop to the ground loop via the plate heat exchanger (Figure 3); the heat transferred to (+) and from (−) the soil (Q_{GHE}) via the GHE (borefield); and the heat loss from the EEB (Q_{GHE_loss}) to the adjacent soil mass (which can be loss and gains as well).

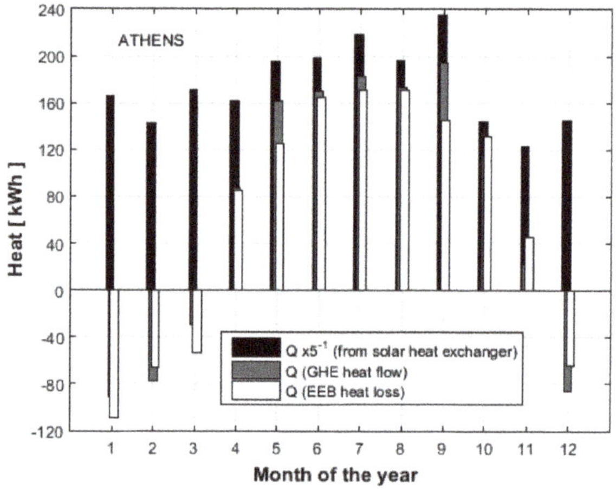

Figure 10. Monthly heat flow across the EEB for Athens, for the scenario of 12 PVTs paired with 12 BHEs and for the 6th year of simulation.

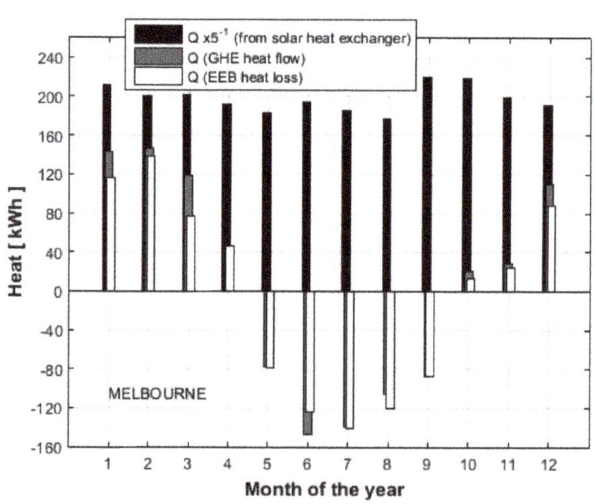

Figure 11. Monthly heat flow across the EEB for Melbourne, for the scenario of 12 PVTs paired with 12 BHEs and for the 6th year of simulation.

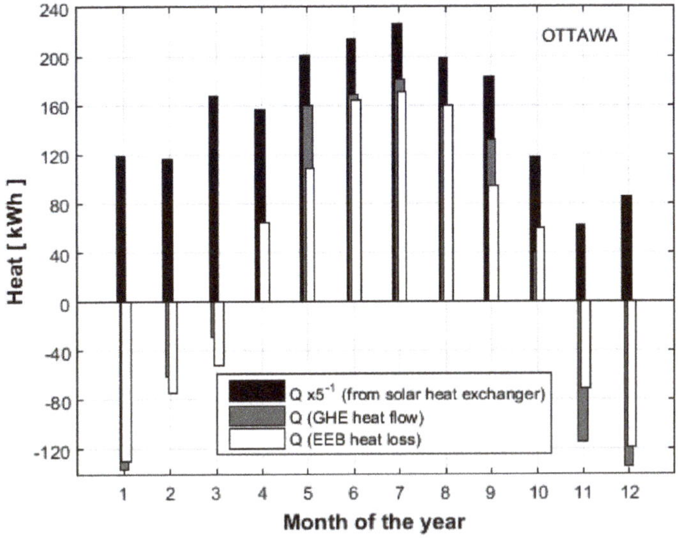

Figure 12. Monthly heat flow across the EEB for Ottawa, for the scenario of 12 PVTs paired with 12 BHEs and for the 6th year of simulation.

As illustrated in Figure 10, for Athens, the solar heat Q_{SHE} is provided to the systems all year long, with the most productive month being September. The solar heat production remains high during the heating season as well due to low soil temperatures, which augments the operation of the PVTs. As regards the solar heat stored in the EEB, this is the positive part of the Q_{GHE} and takes place from April to November, with the highest value being September. From December to March, the heat stored in the ground is provided as additional heat to the solar on the heat pump, this is the negative part of the Q_{GHE}. Moreover, the stored heat in the ground is utilized disproportionally during the heating season, from the about 1000 kWh transferred into the ground during the charging period, only about 200 kWh is used during the heating period. This discrepancy is due to two major causes: the EEB suffers from heat loss Q_{GHE_loss}, which follows the storage production pattern and is proportional to the soil temperature of the EEB; and secondly, the offered solar heat Q_{SHE} is substantial during winter, thus it can be used directly with no need to extract the stored amount. It is worth noting that, during the heating season, the EEB receives heat from the adjacent soil mass (negative loss), which contributes to the available heat.

Figure 11 shows the heat flow across the EEB for Melbourne and for the systems of 12 PVTs installed with 12 BHEs. Similarly to the results for Athens, the solar heat production can be considered all year round for Melbourne. The most productive months are September and October, which are spring months for Melbourne. The high levels of solar heat (Table 6) production all year round are caused by the high heat demand, which discharges the EEB and, by that, reduces the soil temperature. With a low temperature regime in the soil, the solar heat production augments (more frequently reaching the ΔT of 6 K between the PVTs and the soil). The storage period of the solar heat for Melbourne can be seen to be from October until April; similarly, the discharging period (heating season) extends from May to September. The stored solar heat ($+Q_{GHE}$) and the heat discharged from the EEB ($-Q_{GHE}$) are balanced, with about 500 kWh of heat to be stored and the same amount to be discharged. The heat loss of the EEB follows the pattern of the heat from the GHE, with positive and negative values to be balanced annually.

Ottawa has the coldest climate among the study locations (Figures 1 and 2). The relative heat flows across the EEB for the system in Ottawa are graphed by Figure 12. The solar heat Q_{SHE} is produced throughout the year, by having the highest value in July and significantly lower production during the winter months. The heat storage period is from April to October, and from November to March, the heat is discharged from the EEB. The heat storage is uneven for the heat drug from the EEB, which indicates excessive heat loss during the charging period, which reduces the available heat of the EEB.

4. Conclusions

In the present work, a PVT-SAGSHP system is investigated regarding its energy performance for three cities with distinctive climates: Athens (Greece); Melbourne (Australia); and Ottawa (Canada). The mathematical model of the system along with the three buildings was formulated in TRNSYS and parametric analyses were carried out. The major aim was to get quantitative information regarding the influence of the system's performance by the size of its long-term storage component (EEB) and the PVT array.

The self-sufficiency of the system as regards the annual energy was evaluated via the RPF; values larger than 1 indicate self-sufficiency. The highest RPF was found for all locations with the largest system (20 PVTs and 20 BHEs). The highest RPF was found at 4.7, 3, and 2.1 for Athens, Melbourne, and Ottawa, respectively. Athens and Melbourne were found to be self-sufficient with all PVT arrays larger than four collectors and four collectors, respectively, regardless of the size of the EEB. As regards Ottawa, the self-sufficiency was found to be related to the size of the EEB, thus, for small EEBs (four and eight BHEs), more than 12 PVTs are needed, for the larger EEB, any PVT array larger than eight collectors is substantial. Moreover, it was found that as the storage capacity enlarges, the RPF improves, and this augmentation is amplified as the number of PVT collectors grows.

Similarly to the RPF, the SPF was found to be improved as the storage capacity of the system increased. For both indices, the improvement was caused by the reduction of the electricity consumed by the heat pump; the reduction of the auxiliary heat; the increase of the solar heat produced by the PVTs; and the generally higher heat availability in the EEB. Melbourne was found to have the highest values of SPF across all locations. The systems with the highest SPF were found to be for all the cities, with 20 PVTs and 20 BHEs, namely, 2.6, 3.7, and 1.8 for Athens, Melbourne, and Ottawa, accordingly. The greater SPF for Melbourne than the other two locations was due to a larger heating load, which drove the heat pump to deliver more heat.

Solar heat was found to be delivered throughout the year for all cities, with the largest amount being delivered during summer for Athens and Ottawa, while for Melbourne, this was at the end of spring. For the systems with 12 PVTs and 12 BHEs located in Athens and Ottawa, the solar heat stored in the EEB was found to be unbalanced with the heat being discarded during the heating season. For Athens, the excessive heat was dissipated into the adjacent soil mass as heat loss, whereas, for Ottawa, it was all utilized within the heating period. For Melbourne, the charging and discharging of heat from the EEB was found to be about balanced annually. The heat loss and gains were found to follow the same pattern as that of the charging and discharging of the EEB. Heat loss is taking place during the charging months where the ground temperature is higher than that of the adjacent soil mass, whereas heat gains are entering the EEB during the heating months when the soil temperature in the EEB drops.

Author Contributions: Conceptualization, E.I.S.; Formal analysis, I.E.S.; Investigation, P.J.A.; Methodology, E.I.S. and B.V.B.; Software, E.I.S. and B.V.B.; Supervision, P.J.A.; Writing—original draft, E.I.S., P.J.A. and I.E.S. All authors have read and agreed to the published version of the manuscript.

Funding: This research received no external funding.

Conflicts of Interest: The authors declare no conflict of interest.

Nomenclature

Ac	PVT collector area, m^2
E	electricity, kWh
PVT_i	electricity generated by PVT, kWh
PVT_u	electricity delivered to the power grid, kWh
Q	heat, kWh
RPF	renewable power fraction, -
SPF	seasonal performance factor, -
Abbreviations	
BHE	borehole heat exchanger
DHW	domestic hot water
EEB	earth energy bank
FPC	flat plate collector
GHE	geothermal heat exchanger
GSHP	ground source heat pump
PV	photovoltaic panel
PVT	photovoltaic and thermal collector
SAGSHP	solar assisted ground source heat pump
SAHP	solar assisted heat pump
TES	thermal energy storage
Subscripts	
aux	auxiliary
cond	condenser
ev	evaporator
HP	heat pump
parasitic	system's parasitic electricity

References

1. Ozgener, O.; Hepbasli, A. A review on the energy and exergy analysis of solar assisted heat pump systems. *Renew. Sustain. Energy Rev.* **2007**, *11*, 482–496. [CrossRef]
2. Buker, M.S.; Riffat, S.B. Solar assisted heat pump systems for low temperature water heating applications: A systematic review. *Renew. Sustain. Energy Rev.* **2016**, *55*, 399–413. [CrossRef]
3. Sarbu, I.; Sebarchievici, C. General review of ground-source heat pump systems for heating and cooling of buildings. *Energy Build.* **2014**, *70*, 441–454. [CrossRef]
4. Atam, E.; Helsen, L. Ground-coupled heat pumps: Part 1—Literature review and research challenges in modeling and optimal control. *Renew. Sustain. Energy Rev.* **2016**, *54*, 1653–1667. [CrossRef]
5. Shukla, S.; Bayomy, A.M.; Antoun, S.; Mwesigye, A.; Leong, W.H.; Dworkin, S.B. Performance characterization of novel caisson-based thermal storage for ground source heat pumps. *Renew. Energy* **2021**, *174*, 43–54. [CrossRef]
6. Nouri, G.; Noorollahi, Y.; Yousefi, H. Solar assisted ground source heat pump systems—A review. *Appl. Therm. Eng.* **2019**, *163*, 114351. [CrossRef]
7. Naili, N.; Kooli, S. Solar-assisted ground source heat pump system operated in heating mode: A case study in Tunisia. *Renew. Sustain. Energy Rev.* **2021**, *145*, 111144. [CrossRef]
8. Metz, P.D. Development of a validated model of ground coupling. In Proceedings of the American Section of the International Solar Energy Society Conference, Phoenix, AZ, USA, 2–6 June 1980.
9. Givoni, B. Underground longterm storage of solar energy—An overview. *Sol. Energy* **1977**, *19*, 617–623. [CrossRef]
10. Pinel, P.; Cruickshank, C.A.; Beausoleil-Morrison, I.; Wills, A. A review of available methods for seasonal storage of solar thermal energy in residential applications. *Renew. Sustain. Energy Rev.* **2011**, *15*, 3341–3359. [CrossRef]
11. Xu, J.; Wang, R.Z.; Li, Y. A review of available technologies for seasonal thermal energy storage. *Sol. Energy* **2014**, *103*, 610–638. [CrossRef]
12. Braun, J.E.; Klein, S.A.; Mitchell, J.W. Seasonal storage of energy in solar heating. *Sol. Energy* **1981**, *26*, 403–411. [CrossRef]
13. Yang, T.; Liu, W.; Kramer, G.J.; Sun, Q. Seasonal thermal energy storage: A techno-economic literature review. *Renew. Sustain. Energy Rev.* **2021**, *139*, 110732. [CrossRef]
14. Dahash, A.; Ochs, F.; Janetti, M.B.; Streicher, W. Advances in seasonal thermal energy storage for solar district heating applications: A critical review on large-scale hot-water tank and pit thermal energy storage systems. *Appl. Energy* **2019**, *239*, 296–315. [CrossRef]
15. Mahon, H.; O'Connor, D.; Friedrich, D.; Hughes, B. A review of thermal energy storage technologies for seasonal loops. *Energy* **2022**, *239*, 122207. [CrossRef]
16. Bott, C.; Dressel, I.; Bayer, P. State-of-technology review of water-based closed seasonal thermal energy storage systems. Renew. Sustain. *Energy Rev.* **2019**, *113*, 109241. [CrossRef]

17. Chapuis, S.; Bernier, M. Seasonal storage of solar energy in borehole heat exchangers. In Proceedings of the 11th International IBPSA Conference, Glasgow, UK, 27–30 July 2009; pp. 599–606.
18. Lanahan, M.; Tabares-Velasco, P.C. Seasonal thermal-energy storage: A critical review on BTES systems, modeling, and system design for higher system efficiency. *Energies* **2017**, *10*, 743. [CrossRef]
19. Dai, L.; Li, S.; DuanMu, L.; Li, X.; Shang, Y.; Dong, M. Experimental performance analysis of a solar assisted ground source heat pump system under different heating operation modes. *Appl. Therm. Eng.* **2015**, *75*, 325–333. [CrossRef]
20. Stojanović, B.; Akander, J. Build-up and long-term performance test of a full-scale solar-assisted heat pump system for residential heating in Nordic climatic conditions. *Appl. Therm. Eng.* **2010**, *30*, 188–195. [CrossRef]
21. Bakirci, K.; Ozyurt, O.; Comakli, K.; Comakli, O. Energy analy sis of a solar-ground source heat pump system with vertical closed-loop for heating applications. *Energy* **2011**, *36*, 3224–3232. [CrossRef]
22. Naranjo-Mendoza, C.; Oyinlola, M.A.; Wright, A.J.; Greenough, R.M. Experimental study of a domestic solar-assisted ground source heat pump with seasonal underground thermal energy storage through shallow boreholes. *Appl. Therm. Eng.* **2019**, *162*, 114218. [CrossRef]
23. Eslami-nejad, P.; Langlois, A.; Chapuis, S.; Bernier, M.; Faraj, W. Solar heat injection into boreholes. In Proceedings of the 4th Canadian Solar Buildings Conference, Toronto, ON, Canada, 25–27 June 2009; pp. 237–246.
24. Han, Z.; Zheng, M.; Kong, F.; Wang, F.; Li, Z.; Bai, T. Numerical simulation of solar assisted ground-source heat pump heating system with latent heat energy storage in severely cold area. *Appl. Therm. Eng.* **2008**, *28*, 1427–1436. [CrossRef]
25. Kjellsson, E.; Hellström, G.; Perers, B. Optimization of systems with the combination of ground-source heat pump and solar collectors in dwellings. *Energy* **2010**, *35*, 2667–2673. [CrossRef]
26. Carbonell, D.; Haller, M.Y.; Frank, E. Potential Benefit of Combining Heat Pumps with Solar Thermal for Heating and Domestic Hot Water Preparation. *Energy Procedia* **2014**, *57*, 2656–2665. [CrossRef]
27. Trillat-Berdal, V.; Souyri, B.; Fraisse, G. Experimental study of a ground-coupled heat pump combined with thermal solar collectors. *Energy Build.* **2006**, *38*, 1477–1484. [CrossRef]
28. Bertram, E.; Glembin, J.; Rockendorf, G. Unglazed PVT collectors as additional heat source in heat pump systems with borehole heat exchanger. *Energy Procedia* **2012**, *30*, 414–423. [CrossRef]
29. Yang, W.; Sun, L.; Chen, Y. Experimental investigations of the performance of a solar-ground source heat pump system operated in heating modes. *Energy Build.* **2015**, *89*, 97–111. [CrossRef]
30. Sakellariou, E.I.; Wright, A.J.; Axaopoulos, P.J. Energy, economic and emission assessment of a solar assisted shallow earth borehole field heat pump system for domestic space heating in a north European climate. *Geothermics* **2021**, *95*, 102159. [CrossRef]
31. Sakellariou, E.I.; Axaopoulos, P.J.; Wright, A.J. Energy and economic evaluation of a solar assisted ground source heat pump system for a north Mediterranean city. *Energy Build.* **2020**, *231*, 110640. [CrossRef]
32. Nicholson-Cole, D. Domestic Solar Earth Charging: Modelling the process for augmentation of heat pump. In Proceedings of the CIBSE ASHRAE Technical Symposium, Imperial College, London, UK, 18–19 April 2012; pp. 1–14.
33. Verma, V.; Murugesan, K. Experimental study of solar energy storage and space heating using solar assisted ground source heat pump system for Indian climatic conditions. *Energy Build.* **2017**, *139*, 569–577. [CrossRef]
34. Loose, A.; Drück, H. Field test of an advanced solar thermal and heat pump system with solar roof tile collectors and geothermal heat source. *Energy Procedia* **2014**, *48*, 904–913. [CrossRef]
35. Wright, A.J.; Talbot, R.; Goddard, M. The solar house—A true low carbon solution for 2016? In Proceedings of the CIBSE ASHRAE Technical Symposium, Dublin, Ireland, 3–4 April 2014; pp. 3–4.
36. Naranjo-Mendoza, C.; Greenough, R.M.; Wright, A.J. Are shallow boreholes a suitable option for inter-seasonal ground heat storage for the small housing sector? *Conf. IGSHPA* **2018**, 1–10. [CrossRef]
37. Bakker, M.; Zondag, H.A.; Elswijk, M.J.; Strootman, K.J.; Jong, M.J.M. Performance and costs of a roof-sized PV/thermal array combined with a ground coupled heat pump. *Sol. Energy* **2005**, *78*, 331–339. [CrossRef]
38. Solar Energy Laboratory University of Wisconsin-Madison. GmbH—TRANSSOLAR Energietechnik, CSTB—Centre Scientifique et Technique du Bâtiment, TESS—Thermal Energy Systems Specialists. 2009. TRNSYS 17—A TRaNsientSYstem. Simulation program. Simulation.
39. Baetschmann, M.; Leibundgut, H. LowEx Solar Building System: Integration of PV/T Collectors into Low Exergy Building Systems. *Energy Procedia* **2012**, *30*, 1052–1059. [CrossRef]
40. Sommerfeldt, N.; Madani, H. In-depth techno-economic analysis of PV / Thermal plus ground source heat pump systems for multi-family houses in a heating dominated climate. *Sol. Energy* **2019**, *190*, 44–62. [CrossRef]
41. Rad, F.M.; Fung, A.S.; Leong, W.H. Combined solar thermal and ground source heat pump system. In Proceedings of the IBPSA 2009—International Building Performance Simulation Association, Glasgow, Scotland, 27–30 July 2009; pp. 2297–2305.
42. Beck, H.E.; Zimmermann, N.E.; McVicar, T.R.; Vergopolan, N.; Berg, A.; Wood, E.F. Present and future köppen-geiger climate classification maps at 1-km resolution. *Sci. Data* **2018**, *5*, 180214. [CrossRef]
43. Kottek, M.; Grieser, J.; Beck, C.; Rudolf, B.; Rubel, F. World map of the Köppen-Geiger climate classification updated. *Meteorol. Zeitschrift* **2006**. [CrossRef]
44. Sakellariou, E.I.; Wright, A.J.; Axaopoulos, P.; Oyinlola, M.A. PVT based solar assisted ground source heat pump system: Modelling approach and sensitivity analyses. *Sol. Energy* **2019**, *193*, 37–50. [CrossRef]

45. Sakellariou, E.; Axaopoulos, P. An experimentally validated, transient model for sheet and tube PVT collector. *Sol. Energy* **2018**, *174*, 709–718. [CrossRef]
46. Hottel, H.C.; Whillier, A. Evaluation of flat-plate solar collector performance. *Transcr. Conf. Use Sol. Energy* **1955**, *2*, 74–104.
47. Hellström, G. Duct Ground Heat Storage Model, Manual for Computer Code. *Energy* **1989**, *915*.
48. Duffie, J.A.; Beckman, W.A. *Solar Engineering of Thermal Processes*, 4th ed.; John Wiley & Sons: Hoboken, NJ, USA, 2013. [CrossRef]
49. Technical Chamber of Greece. *Technical Directive 20701-1: National Specifications of Parameters for Calculating the Energy Performance of Buildings and the Issue of the Energy Performance Certificate*; Technical Chamber of Greece: Athens, Greece, 2017.
50. Australian Building Codes Board. 2022. Available online: https://ncc.abcb.gov.au/ (accessed on 1 October 2022).
51. National Research Council of Canada. *National Energy Code of Canada for Buildings*; National Research Council of Canada: Ottawa, ON, Canada, 2017.
52. Sakellariou, E.I.; Axaopoulos, P.J. Energy performance indexes for solar assisted ground source heat pump systems with photovoltaic-thermal collectors. *Appl. Energy* **2020**, *272*, 115241. [CrossRef]
53. European Parliament. *Decision (2013/114/EU) Establishing the Guidelines for Member States on Calculating Renewable Energy from Heat Pumps from Different Heat Pump Technologies pursuant to Article 5 of Directive 2009/28/EC of the European Parliament an dof the Council*; Official Journal of the European Union: Maastricht, The Netherlands, 2016; Union 27–35.
54. Sakellariou, E.I.; Axaopoulos, P.J.; Sarris, I.E.; Abdullaev, N. Improving the Electrical Efficiency of the PV Panel via Geothermal Heat Exchanger: Mathematical Model, Validation and Parametric Analysis. *Energies* **2021**, *14*, 6415. [CrossRef]

energies

MDPI

Article

Simulation Study of Solidification in the Shell-And-Tube Energy Storage System with a Novel Dual-PCM Configuration

Moslem Mozafari, Ann Lee * and Shaokoon Cheng

School of Engineering, Macquarie University, Sydney, NSW 2109, Australia;
ryan.mozafari@hdr.mq.edu.au (M.M.); shaokoon.cheng@mq.edu.au (S.C.)
* Correspondence: ann.lee@mq.edu.au

Abstract: This study proposes a novel dual-PCM configuration with outstanding solidification response in a horizontal shell-and-tube energy storage system. To demonstrate that the proposed PCM configuration is superior in its thermal responses, results from a range of numerical simulations are presented and compared between different configurations of dual-PCM. As the melting/solidus point is a crucial factor for the solidification rate, dual PCMs are chosen such that the average of their melting point is equal to the melting point of the single-PCM in the reference case. Additionally, equal-area sectors are considered for all cases to ensure the same quantities of PCMs are compared. The temporal liquid fraction and temperature contours reveal that solidification is delayed in the upper half of the system due to strong natural convection motions. Therefore, a dual-PCM configuration is offered to improve the solidification rate in this region and accelerate the full solidification process. Results show that placing a PCM with a lower solidus point in the lower half or an annulus-shaped zone around the cold tube can save the full recovery time up to 8.51% and 9.36%, respectively. The integration of these two strategies results in a novel and optimum design that saves the solidification time up to 15.09%.

Keywords: phase change material; energy storage; dual-PCM; solidification; heat exchanger; numerical

check for updates

Citation: Mozafari, M.; Lee, A.; Cheng, S. Simulation Study of Solidification in the Shell-And-Tube Energy Storage System with a Novel Dual-PCM Configuration. *Energies* **2022**, *15*, 832. https://doi.org/10.3390/en15030832

Academic Editor: Adrián Mota Babiloni

Received: 1 December 2021
Accepted: 21 January 2022
Published: 24 January 2022

1. Introduction

There is a growing environmental concern regarding the increasing usage of fossil fuels, and solar energy has gained much attention as an important substitute [1]. An impending problem with the application of solar energy is that its supply fluctuates with weather conditions. Thermal energy storage (*TES*) systems have been introduced as an attractive way to store solar energy [2,3]. Three methods of energy storage could be applied in *TES* systems: sensible (change in the temperature of the storage material), latent (change in the phase of storage material), and chemical (by reversible chemical reactions). Phase change materials (PCMs) are highly used in latent thermal energy storage (LTES) systems due to their high latent energy capacity [3–5]. A *TES* system with PCM can store 5–14 times more energy than a system using suitable storage materials of the same volume [6]. The usage of PCMs has received widespread interest, with clear evidence demonstrating the important roles they have in the storage and recovery of solar energy [7,8], energy savings in buildings [9–11], and electronics cooling [12,13].

Although PCMs have high energy storage capacity, their poor thermal conductivity is a major challenge, especially in solidification, where the conduction heat transfer is mainly dominated, and natural convection is less effective. Therefore, novel techniques could be used for an accelerated solidification or melting of the PCMs. Eccentric configuration of the tubes in horizontal *TES* systems has been identified as an efficient, economical, practical, yet simple technique to improve the rate of phase change process [14]. Senthil [15] experimentally investigated the effect of heat transfer fluid (HTF) tube orientation in a horizontal *TES* unit filled with PCM. Three types of HTF tube (concentric, eccentric, and inclined) were examined, while water was selected as the HTF with the temperature of 60 °C. The inclined configuration was found to be more efficient due to revealing 21.42% lesser charging duration compared with the regular concentric configuration. In a numerical study, Kadivar et al. [16] investigated the effect of tube eccentricity in a horizontal PCM-based *TES* unit. They conducted an optimization to find the radial and tangential eccentricities of the inner tube (HTF tube). The optimized geometry improved the charging time up to 7.1 time faster but worsened the discharging time up to 3 times slower than the reference concentric case. Alnakeeb et al. [17] investigated the effects of the inner flat-tube eccentricity of a *TES* unit on the melting performance of the PCM. They carried out numerical simulations to examine different eccentricities and aspect ratios of the inner flat-tube and discovered that the best melting performance resulted from the circular tube with the smallest eccentricity value. Although some studies have been conducted to improve the melting performance of single PCM in *TES* systems, to the best of authors' knowledge, this technique (eccentric configuration of tubes) has not been attempted to improve the energy recovery in a *TES* unit (which is associated with the solidification rate of PCM).

Furthermore, the use of multiple PCMs with different melting temperatures has been shown to significantly improve the energy storage and recovery process, which has gained increasing attention in recent years [3,12,18–20]. However, when a single-PCM is used, the rate of melting or solidification in different regions of the *TES* is different due to the non-uniform distribution of the natural convection induced vortexes. Thus, a non-uniform distribution of solid–liquid interface and temperature results. Consequently, a weak storage/recovery performance results. Therefore, an optimum configuration of multiple PCMs with different melting temperatures could alleviate this problem. For example, the solidification process is delayed in regions with strong convective motions. Therefore, placing a PCM with a higher solidus point in these regions can help to compensate for the negative effect of natural convection. The benefit of applying multiple PCMs with varying melting points is studied by Farid et al. [21,22]. They discovered an improved thermal response in a *TES* system when several PCMs with different melting temperatures are housed in cylindrical capsules, compared to a conventional system with a single PCM. In our latest work [12], the advantage of employing multiple PCMs was also reported. It was found that employing a pair of PCMs (n-Eicosane/RT44) results in a better thermal performance of the heat sink than employing only one PCM (n-Eicosane or RT44). Siyabi et al. [23] employed experimental and numerical approaches to investigate the benefit of multiple PCMs for a better heat transfer rate in a shell-and-tube heat exchanger. They reported that the case with multiple PCMs is the best arrangement due to the shorter melting time. Mozafari et al. [3] investigated the effect of different arrangements of dual PCMs in a triplex tube heat exchanger, when subjected to simultaneous charging and discharging. They developed a new dual-PCM design which improved the thermal energy storage and recovery as 37.93%, and 21.06%, respectively, compared with the reference case with single PCM. This performance was also found to be further improved to 76.9% in storage and 32.9% in recovery by adding 3% nanoparticles to the PCMs. Although some researchers have studied multiple PCMs in the *TES* system, to the best of the authors' knowledge, there is no reported study to offer an optimum configuration of PCMs for the energy recovery in a horizontal shell-and-tube heat exchanger.

Despite the poor thermal conductivity of PCMs, a common problem in the solidification of some phase change materials (such as salt hydrates, sugar alcohols, and alkanes) is supercooling [24,25]. In supercooling, the PCM solidifies below its typical solidus temperature, limiting its thermal stability in long-term applications. In the present work, paraffin-based PCMs were selected to avoid the supercooling effect during the solidification.

In this study, PCM solidification in a horizontal shell-and-tube heat exchanger was numerically investigated for different dual-PCM configurations for the first time. An annular geometry was preferred for the heat exchanger due to the high surface-area-to-volume ratio. The full solidification time was comparatively investigated for different dual-PCM configurations. This study aimed to develop a new dual-PCM design for improved heat recovery of the *TES* system. Numerical simulations were carried out to investigate the thermo-fluidic characteristics of the PCMs during solidification. Dual-PCM configurations that yielded faster solidifications were further optimized through parametric studies. This novel design would benefit wide *TES* applications in industrial waste heat recovery, solar energy utilization, energy saving in buildings, and cold energy storage systems.

2. Problem Description

Figure 1 shows the schematic of the current storage unit, which includes three concentric copper tubes, where $r_i = 6.35$ mm and $r_o = 24.13$ mm. The HTF passes through the inner tube while the outer tube is insulated. Single PCM or different configurations of dual-PCMs are housed in the middle annulus. The inner tube is assumed to be isothermally cooled with a constant temperature of 300 K. Therefore, the solidification grows initially from a layer around the inner tube. The thermophysical properties of PCMs and the copper tube are summarized in Table 1.

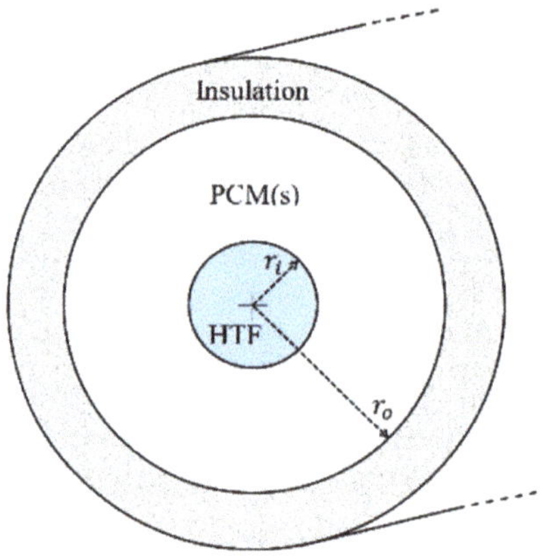

Figure 1. Schematic of the *TES* unit in the current study.

Table 1. Properties of materials employed in this study [26].

Material	k ($\frac{W}{m}$·K)	C_p ($\frac{kJ}{kg}$·K)	L ($\frac{kJ}{kg}$)	T_s (K)	T_l (K)	ρ (kg/m^3)	β ($\frac{1}{K}$).	μ (N·$\frac{s}{m^2}$)
PCM-1 (RT-55)	0.2	2	170	324	330	770	0.0005	0.0264
PCM-2 (RT-60)	0.2	2	160	328	334	770	0.0005	0.0288
PCM-3 (RT-65)	0.2	2	150	331	338	770	0.0005	0.03
Copper	400	0.38	—	—	—	8920	—	—

Firstly, simulations were conducted to examine the solidification rate for a different arrangement of dual PCMs (Figure 2). To have a fair comparison, the PCM sectors in all dual-PCM configurations had equal areas. In addition, the selection of PCMs was done such that the effect of other parameters (properties of PCMs) on the comparison results were minimized. For example, the average values of melting temperature and latent heat in dual-PCM cases (PCM-1 and PCM-3) was equal to the melting temperature and latent heat of single-PCM (PCM-2) in the reference case (case-1), as could be seen in Table 1. Furthermore, all the PCMs had the same density of 770 kg/m^3, the same thermal conductivity of 0.2 W/m·K, and the same specific heat of 2 kJ/kg·K.

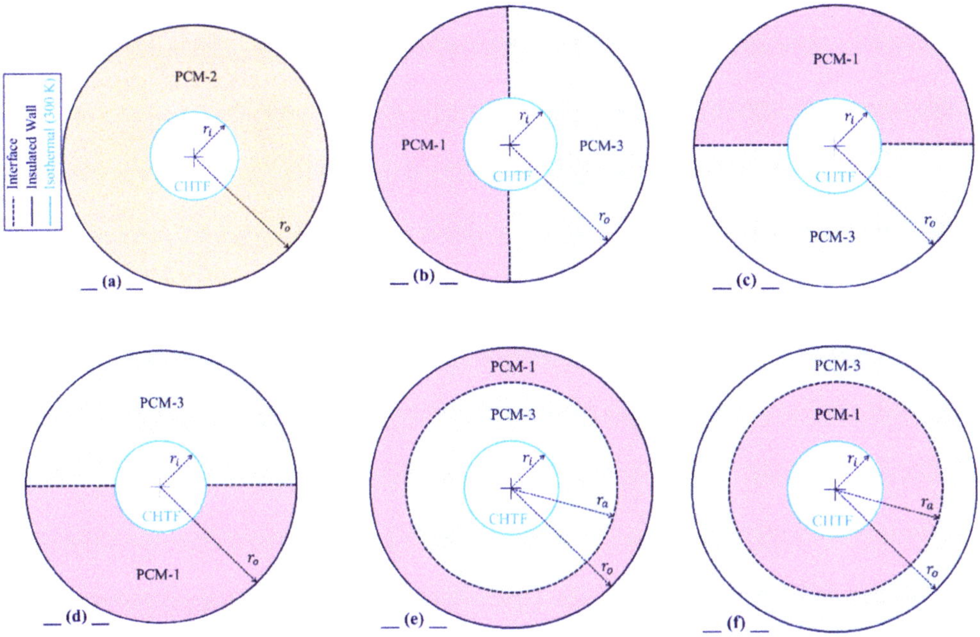

Figure 2. Computational domains in the current study; (**a**) case-1, (**b**) case-2, (**c**) case-3, (**d**) case-4, (**e**) case-5, (**f**) case-6.

3. Numerical Method

3.1. Initial Boundary Conditions and Assumptions

In this work, a transient 2D model was developed to study the PCM solidification in a shell-and-tube heat exchanger. The natural convection effect inside the liquid PCM was taken into account. The storage unit was considered to be initially fully melted ($T_{initial} = 353$ K), which could be expressed as:

$$t = 0 \rightarrow \quad u_r = 0,\ u_\theta = 0,\ T = T_{init} = 353\ \text{K} \tag{1}$$

For the symmetric cases (all cases except case-2), only the right-half of the unit was considered in the computational domain. A constant temperature of 300 K was considered for the inner tube, while the outer tube was insulated (zero heat flux). The boundary conditions can be expressed as follows:

$$\begin{gathered} r = r_i \rightarrow \quad u_r = 0,\ u_\theta = 0,\ T = T_{CHFT} = 300\ \text{K} \\ r = r_o \rightarrow \quad u_r = 0,\ u_\theta = 0,\ \frac{\partial T}{\partial r} = 0 \\ \theta = \pm\pi/2 \rightarrow \quad \frac{\partial u_r}{\partial \theta} = 0,\ u_\theta = 0,\ \frac{\partial T}{\partial \theta} = 0 \quad (cases 1,\ 3,\ 4, 5\ \&\ 6) \end{gathered} \tag{2}$$

For the borders between the PCM zones (illustrated with thick dashed lines in Figure 2), the thin wall approximation was applied to allow the heat exchange from one zone to the other. At these boundaries, the temperature was directly mapped by coupling the two sides of each border together as follows:

$$border\ walls \begin{cases} T_{PCM(i)} = T_{PCM(i+1)} \\ u_r = 0 \\ u_\theta = 0 \end{cases} \tag{3}$$

The above equation means that the rate of heat leaving the $(i)_{th}$ PCM layer across the interface is equal to the rate of heat entering the $(i+1)_{th}$ layer along the same direction.

In this numerical work, natural convection and heat conduction coexisted during the PCM solidification. The following assumptions were made:

- The liquid phase was laminar, incompressible, and Newtonian.
- The radiation heat transfer was not considered in the system.
- Volume change of PCM during the solidification was neglected.
- Isotropic and homogeneous materials were considered for the tubes.
- The HTF was assumed to keep a constant temperature.
- The thickness of tubes was ignored due to the relatively higher thermal conductivity of the material.
- Viscous dissipation and slip velocity on the boundaries were ignored.

3.2. Governing Equations

In the present study, continuity, momentum, and energy equations were employed as:
(I) Continuity equation

$$\nabla . V = 0 \tag{4}$$

(II) Momentum equation

$$\rho(\frac{\partial u_r}{\partial t} + V.\nabla u_r) = -\nabla P + \mu \nabla^2 u_r + \rho g \sin\theta + S_{u_r} \tag{5}$$

$$\rho(\frac{\partial u_\theta}{\partial t} + V.\nabla u_\theta) = -\nabla P + \mu \nabla^2 u_\theta + \rho g \cos\theta + S_{u_\theta} \tag{6}$$

(III) Energy equation

$$\frac{\partial}{\partial t}(\rho H) + \nabla.(\rho V H) = \nabla.(k \nabla T) \tag{7}$$

where h and ΔH represent sensible enthalpy and latent heat, respectively $(H = h + \Delta H)$. The term $\nabla.(\rho V H)$ is the transport of latent-heat evolution. Momentum sink was defined as:

$$S_{u_r} = -C(1-\lambda)^2 \frac{u_r}{\lambda^3 + \delta} \tag{8}$$

$$S_{u_\theta} = -C(1-\lambda)^2 \frac{u_\theta}{\lambda^3 + \delta} \tag{9}$$

where C is the mushy zone constant (set to 10^5 in this study), led to the best match with predictions of Khodadadi et al. [27]. δ = 0.001 is a small constant to avoid zero in the denominator. The enthalpy h was defined as:

$$h = h_{ref} + \int_{T_{ref}}^{T} C_p dT \tag{10}$$

where h_{ref} is the reference enthalpy at $(T_{ref} = 273\text{ K})$. Melting heat was calculated based on latent heat L, i.e.:

$$\Delta H = \lambda L \tag{11}$$

λ denotes the liquid fraction of PCM during the solidification process, between the solidus and liquidus temperatures ($T_s < T < T_l$.):

$$\lambda = \begin{cases} 0, & T \leq T_s \\ \frac{T-T_s}{T_l-T_s} & T_s < T < T_l \\ 1, & T \geq T_l \end{cases} \tag{12}$$

The Boussinesq approximation was used to account for the natural convection motions inside the liquid PCM, considering the variation in density. Thus, the density of liquid PCM could be expressed as:

$$\rho = \frac{\rho_m}{\beta(T - T_m) + 1} \quad (13)$$

where ρ_m indicates the original density of liquid PCM, β is the thermal expansion coefficient, and $T_m = (T_s + T_l)/2$. T_s and T_l are the solidus and liquidus temperatures, respectively.

3.3. Numerical Procedure

The current study aimed to predict the solidification of PCM in a shell-and-tube heat exchanger. ANSYS FLUENT 21.1 software was used to conduct numerical simulations. The enthalpy-porosity approach [28] was employed to compute the solidification of PCM, wherein the porosity in each cell was set equal to the liquid fraction in that cell. Finite Volume Method (FVM) was used with high-order quadratic upstream interpolation for QUICK scheme [29] to discretize the governing equations. The pressure-based method was applied to solve the heat and fluid flow equations. The pressure staggering option (PRESTO) scheme was employed for pressure correction during the iterative solution process. The semi-implicit method for pressure linked equations (SIMPLE) algorithm recommended by Patankar [30] was also applied for pressure-velocity coupling. The under-relaxation factors for the momentum, pressure, and energy were set as 0.5, 0.3, and 1, respectively. Convergence criteria for continuity and momentum equations were set as 10^{-4}, while it was set as 10^{-6} for the energy.

3.4. Validation and Verification

In this section, the simulation method was validated against the experimental reports of Al-Abidi et al. [31]. For this aim, the initial and boundary conditions were set based on the reported data in the reference study [31]. For this validation, the phase change process of paraffin (RT82) was numerically examined in a triplex tube heat exchanger. The comparison plot is illustrated in Figure 3, where a good prediction of temperature during a period of 65 min could be seen.

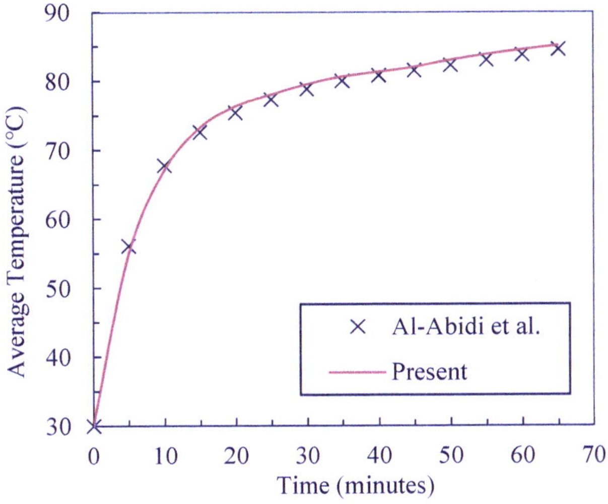

Figure 3. Validation of the numerical method against the experimental results reported by Al-Abidi et al. [31].

The effects of mesh size and time-step on the numerical results were thoroughly examined. Figure 4 shows the result of grid independence examination, where transient liquid fraction was predicted for dual-PCM configuration of case-6. Various number of cells (from 6000 to 22,800) were examined in the numerical simulations, and it was observed that increasing the number of cells to higher than 11,900 did not make any significant change in predicted values of liquid fraction. Therefore, computational domains with 11,900 cells were selected for the predictions. The time step was also found to be set to 0.1 s for an accurate prediction.

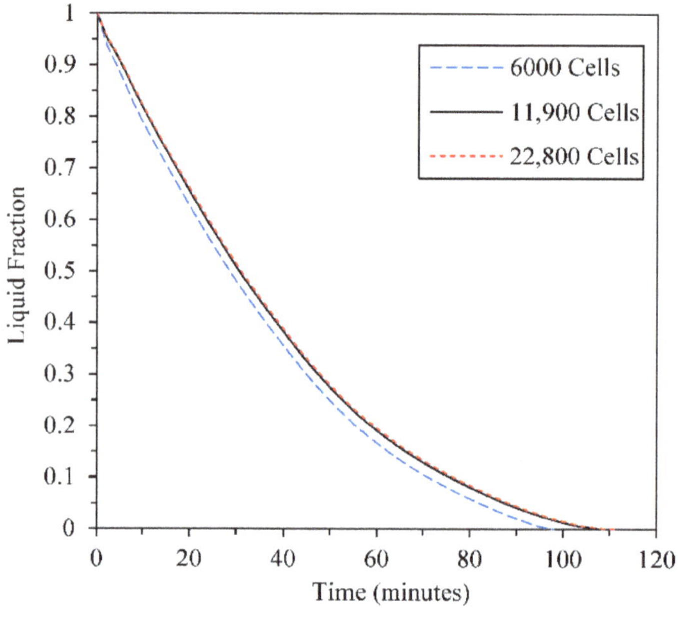

Figure 4. Effect of grid size on the transient liquid fraction in *TES* with configuration of case-6.

4. Results and Discussion

Computational Fluid Dynamics (CFD) was employed to examine the heat transfer and fluid flow behavior of PCMs during a heat recovery process in the shell-and-tube heat exchanger.

4.1. Comparison of the Configurations

In this section, six cases (refer to Figure 2) were investigated through the discussion on their liquid-fraction and temperature analysis. Next, the configuration with the fastest solidification response was refined with a parametric optimization.

4.1.1. Liquid-Fraction and Temperature Analysis

Figure 5 presents the liquid-fraction contours in three time-stages of discharging of the *TES*. The evolution of solid–liquid interfaces for all cases indicated that solidification started from the neighbor of the cold inner tube and grew in a radial direction but mostly toward the lower zones. This is due to the strong natural convection induced motions in the upper zone. A buoyancy force drove the hot liquid toward the upper regions since it had a lower density than the cold liquid. Therefore, strong natural convection movements in the upper zone cause delayed solidification in this region. Moreover, a lower range of temperature could be seen in the lower zones of temperature contours illustrated in Figure 6. However, this solidification non-uniformity could be reduced or increased by arranging two different PCMs with different melting-solidification points. Due to higher solidus temperature, PCM-3 showed a faster solidification response than PCM-1 and PCM-2. This could be seen in contour plots of case-2 at Figure 5, where PCMs were housed in left and right sections with the same condition against gravity. Figure 6 reveals that the temperature was mostly changed in layers along the radial direction, indicating the dominance of heat conduction over natural convection, especially in the lower zone.

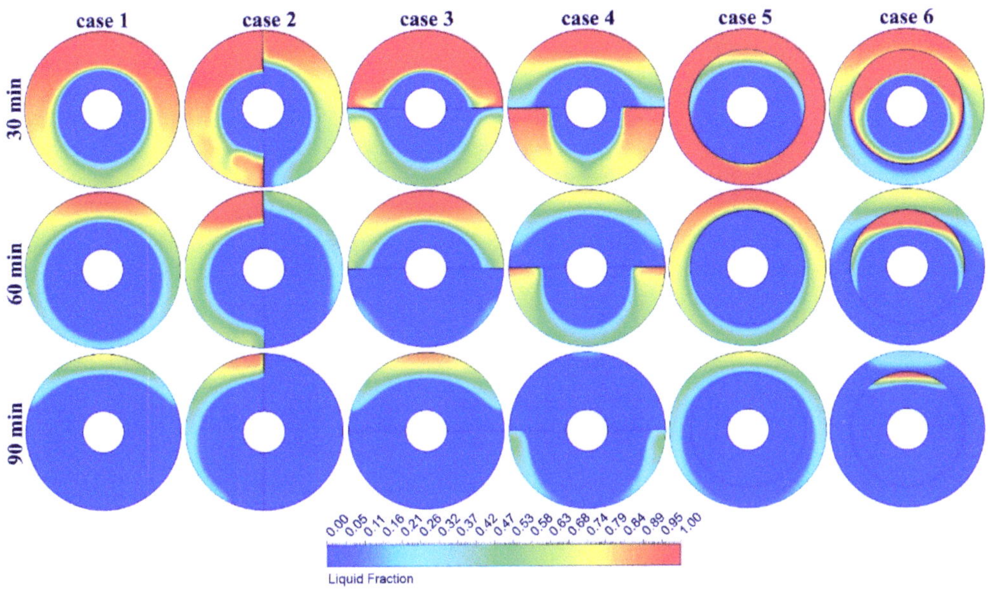

Figure 5. Liquid-fraction contours of the studied cases at different time-stages.

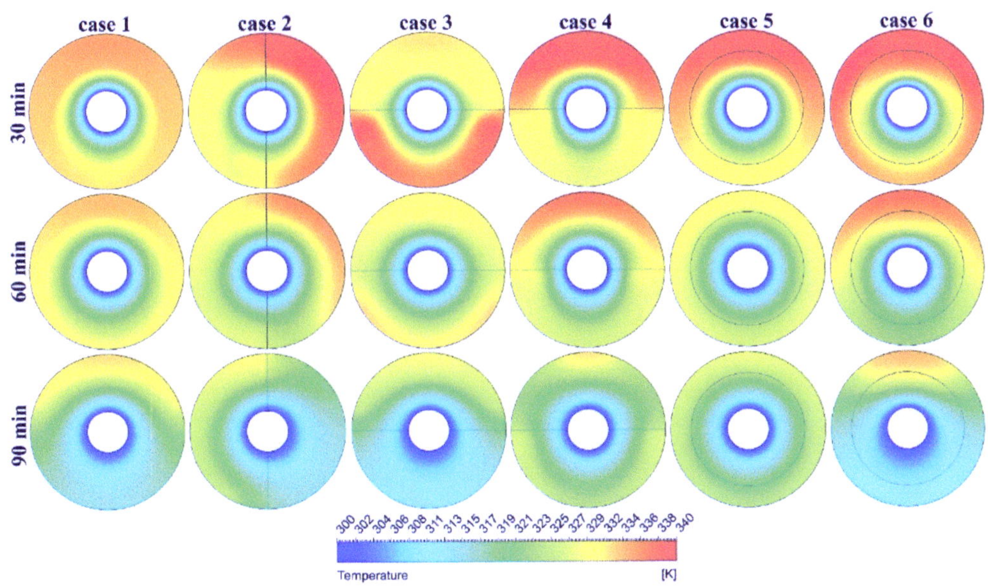

Figure 6. Temperature contours of the studied cases at different time-stages.

At the time-stage of 90 min, the liquid fractions of 0.65, 0.079, 0.108, 0.0437, 0.114, and 0.0421 are reported for cases 1–6, respectively. Case-4 and case-6 reveal better solidification responses according to the liquid-fraction evolutions (refer to Figure 5). PCM-3 (which had the quickest solidification response) was housed in the upper half of the unit in case-4. Therefore, the strong natural convection in the upper half was compensated with the quick response of PCM-3 at this zone. As a result, both PCMs fully solidified almost at the same time. Case-6 favored another strategy with the same goal, such that PCM-1 (which had a relatively delayed solidification) was placed around the cold tube and PCM-3 was placed in the outer zone. Therefore, the solidification process finished almost at the same time in both PCMs.

4.1.2. Heat Transfer Analysis

Transient evolution of liquid fraction is presented in Figure 7 for all studied cases so far. For dual-PCM cases, the average liquid fraction of two PCMs were counted. The required time for a full solidification process was found to be 116, 126.5, 134.5, 108, 128, and 109 min for cases 1–6, respectively. Although case-4 revealed the quickest response, Figure 7 shows that case-6 revealed the lowest values of the liquid fraction among the cases most time during the discharging. For example, the liquid fraction of case-6 at time-stages of 30 and 60 min was 0.509 and 0.191, respectively, while case-4 showed higher liquid fractions of 0.553 and 0.247. Figure 7 reveals that the solidification process could be accelerated up to 6.89% or 6.03% when effective configurations of dual-PCM (case-4 or case-6) were preferred over a single-PCM in a shell-and-tube heat exchanger.

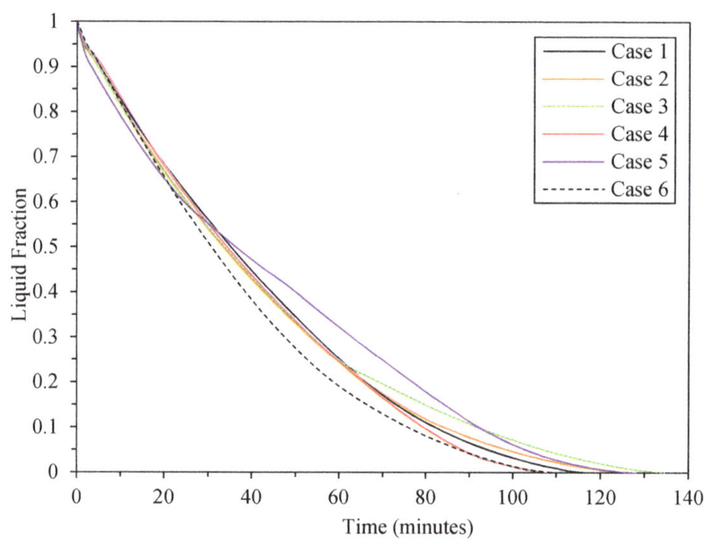

Figure 7. Liquid-fraction evolution for studied cases.

Equation (14) expresses the storage heat contribution (*SHR*) in the *TES* system, which is the ratio of the latent stored energy to the maximum amount of latent energy that could be stored in the unit:

$$SHR = \frac{Latent\ heat\ stored\ in\ the\ TES}{Maximum\ heat\ storage\ capacity\ of\ TES} \quad (14)$$

Figure 8 shows temporal plots of *SHR* for all of the studied cases so far. It could be seen that quick, full solidifications could be achieved when both PCMs were almost equally involved in the energy recovery during the whole discharging process (case-4 and case-6). Case-6 revealed the most consistent *SHR* share of PCMs during the discharging process so that both PCMs went through the solidification process almost equally and were fully solidified with only 4 min time difference. On the other hand, in dual-PCM cases with slow solidification responses (cases 2, 3, and 5), inconsistent solidifications were observed according to the data provided in Figure 8. For example, PCM-1 was fully solidified after 46 min, while PCM-3 was fully solidified after 128 min discharging in case-5.

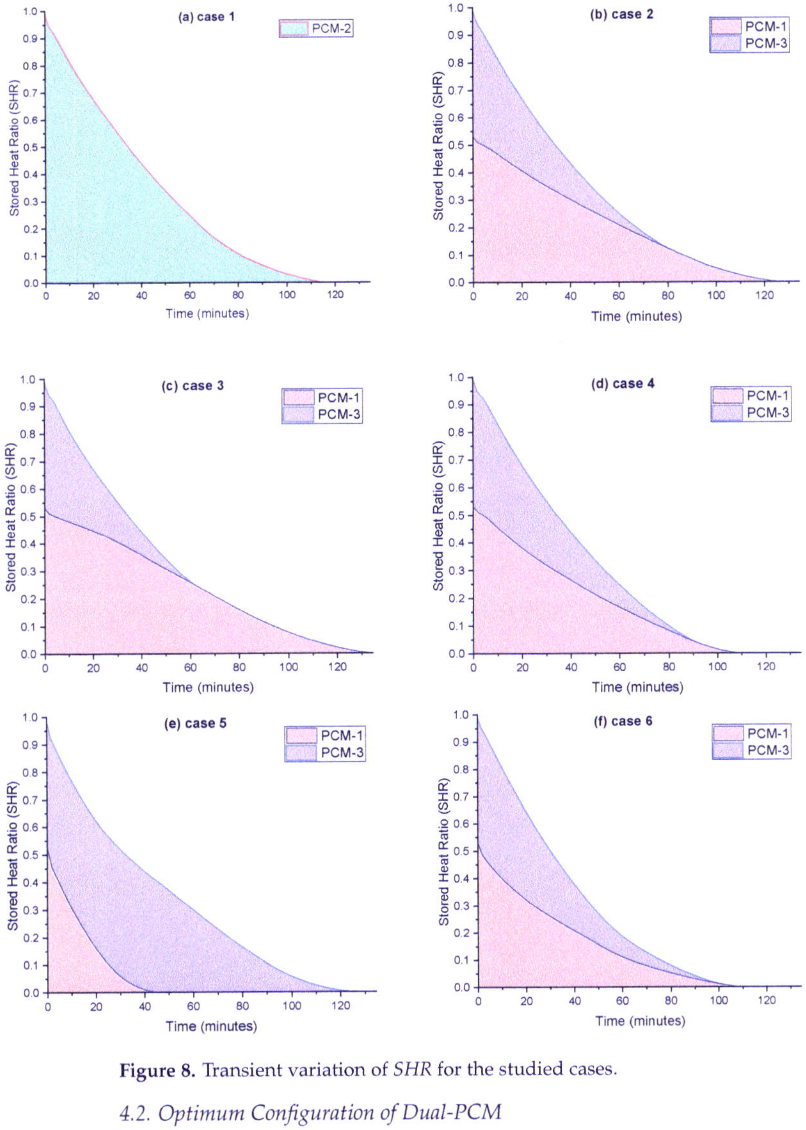

Figure 8. Transient variation of *SHR* for the studied cases.

4.2. Optimum Configuration of Dual-PCM

A good *TES* system should reveal a quick response in storage and recovery of the thermal energy. As discussed in Section 4.1.2, case-4 and case-6 showed better responses in recovering latent heat in the storage unit. Case-6 had the radial distribution advantage where the PCM with delayed solidification (PCM-1) was placed around the cold tube, leading to an almost uniform rate of solidification in both PCMs. On the other hand, housing a quick-solidifying PCM (PCM-3) in the upper half of the unit (where strong natural convection motions prevent a quick solidification) provided an almost consistent recovery in both upper and lower halves. Therefore, two important points must be taken into account for an optimum design of a shell-and-tube system in energy recovery:

1. Arrangement of PCMs in the radial direction, such that PCM with lower solidus point is placed around the cold tube.
2. Filling the upper half of the unit with the PCM with a higher solidus point.

Considering these points, the configuration of case-6 could be improved if a downward eccentricity is applied for the sector tube (Figure 9). Thus, the optimum eccentricity of the sector tube is suggested in a parametrical study for accelerated energy recovery in the shell-and-tube storage system.

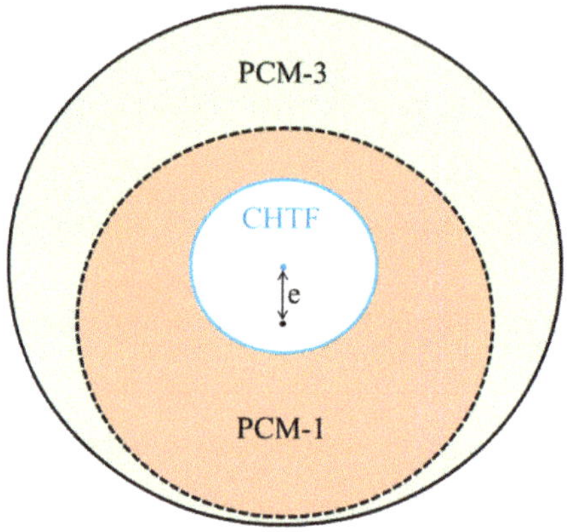

Figure 9. Eccentric configuration of the sector tube.

Figure 10 presents the transient reduction in liquid fraction for eccentricities of 3, 4, 5, and 6 mm. A unit with e = 6 mm needs the shortest time (98.5 min) for a full solidification process. It indicates that the new design (case-7: e = 6 mm) accelerates the full solidification process by 15.09% compared with the reference case with a single PCM. The liquid fraction evolution is also compared between cases 1, 4, 6, and 7 in Figure 11. It could be seen that case-7 shows the fastest solidification. The novel design accelerated the solidification time up to 8.51% and 9.36% compared with case-4 and case-6, respectively.

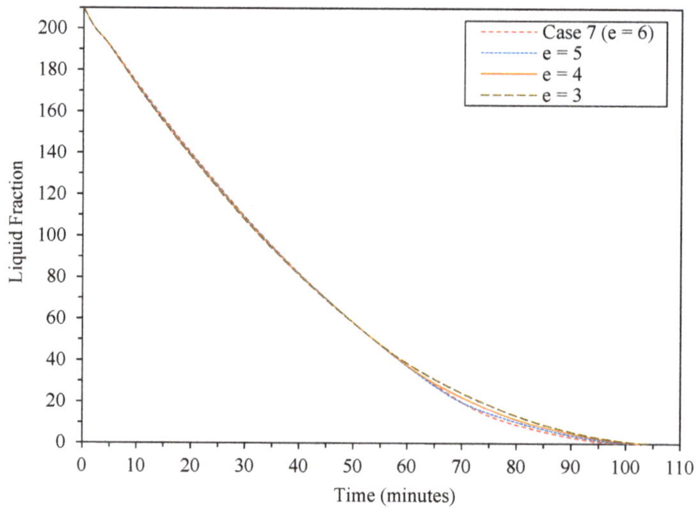

Figure 10. Transient liquid fraction for different eccentricities of the sector tube.

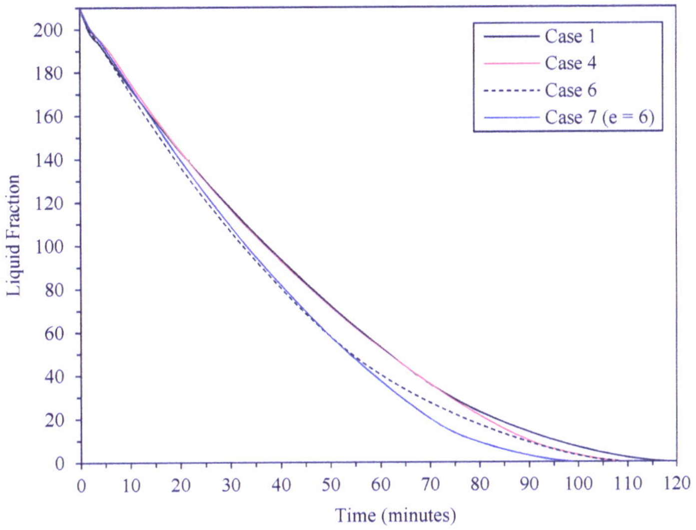

Figure 11. Comparison of transient liquid fraction between some selected cases.

Liquid-fraction and temperature contours at different time-stages are illustrated in Figure 12 for different eccentricities. Only three eccentricities of 0, 3, and 6 mm were selected due to the close similarity of cases. The evolution of contour plots shows that by increasing the eccentricity, a higher portion of PCM-1 (the PCM with delayed solidification) was placed in the lower region, then its delayed solidification was compensated with the advantage of being in the area with weak natural convection. Temporal contour plots clearly show how an eccentricity of 6 mm could improve the solidification rate in both PCMs during the heat recovery process.

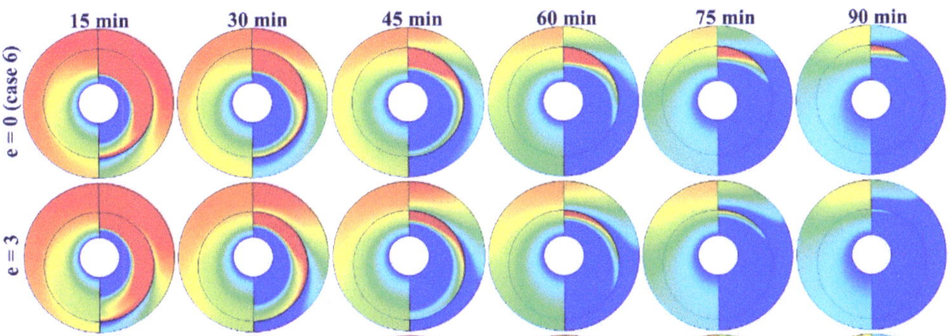

Figure 12. Histories of liquid fraction (**right halves**) and temperature distribution (**left halves**) for different eccentricities.

4.3. Melting Performance of the New Design

To complement the *TES* system explored in this study, the new design that shows great reduction in the full solidification time was investigated for melting processes as well. The optimum design was compared with the melting performance of the reference case in order to have a brief evaluation of the heat storage process. For this aim, the HTF with hot constant temperature of 353 K was passed through the inner tube, while the outer tube was thermally insulated. The system was initially kept at a temperature of 300 K. Figure 13 presents the liquid-fraction variation of case-1 and case-7 (optimum design) during the melting process. The full melting times for case-1 and case-7 were reported as 128 and 127 min, respectively, indicating a slight improvement (of 0.8%) in the melting performance of the optimum design.

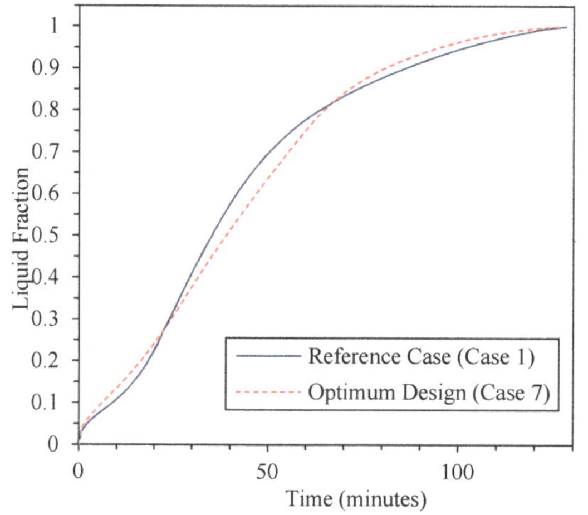

Figure 13. Melting evolution comparison between the reference case and the new design.

5. Conclusions

The heat recovery was studied in the solidification of PCMs in a shell-and-tube storage system. Different configurations of dual-PCM were comparatively investigated against a single-PCM configuration. In each dual-PCM configuration, the storage unit was divided into equal-area sections including: left/right sections, lower/upper sections, and annulus-shaped sections. The numerical results show that good arrangements of dual-PCM can improve the energy recovery compared with a traditional unit with a single-PCM. The results can be summarized as follows:

- Applying dual-PCM configurations instead of a single-PCM configuration does not necessarily improve the solidification performance inside a horizontal *TES* system, but effective arrangements of PCMs with different solidus points could improve the energy recovery performance.
- Placement of a PCM with a higher solidus point at the upper half and a PCM with a lower solidus point at the lower half of the system could significantly accelerate the solidification time (case-4), up to 6.89% compared with the reference case (case-1 with a single-PCM).
- When dual-PCMs were radially arranged in the storage unit (annulus-shaped sections) such that the PCM with a lower solidus point is placed around the cold tube (case-6), the energy recovery time improved up to 6.03% compared with the reference case.
- A better recovery response in the *TES* system could be achieved when both design advantages of case-4 and case-6 are taken into account by applying a downward eccentricity to the sector tube of the radial configuration. Therefore, a novel design

with 6 mm eccentricity was introduced to accelerate the solidification time up to 15.09%, 8.51%, and 9.36% compared with cases 1, 4, and 6, respectively.
- Although this paper was focused only on solidification performance of the PCMs in a shell-and-tube storage unit, evaluation of the melting process showed a slight storage improvement for the new design compared with the reference case.

The numerical results in this study are limited to the horizontal shell-and-tube storage units and PCMs used. Such study should be extended to other types of storage containers and further conclusions could be discovered by adding finned structures.

Author Contributions: Conceptualization, M.M.; methodology, M.M.; software, M.M.; validation, M.M.; formal analysis, M.M.; investigation, M.M.; resources, M.M.; data curation, M.M.; writing—original draft preparation, M.M.; writing—review and editing, M.M, S.C. and A.L.; supervision, A.L. and S.C. All authors have read and agreed to the published version of the manuscript.

Funding: This research received no external funding.

Data Availability Statement: Data are contained within the article.

Conflicts of Interest: The authors declare no conflict of interest.

Nomenclature

C	Mushy zone constant
C_p	Specific heat (J/kg·K)
d	Diameter
g	Gravity acceleration (m/s^2)
h	Specific enthalpy (J/kg)
k	Thermal conductivity (W/m·K)
L	Latent heat of fusion (J/kg)
P	Pressure (Pa)
S	Momentum source term (N/m^3)
t	Time (s)
T	Temperature (K)
T_l	Liquidus temperature (K)
T_s	Solidus temperature (K)
u	Velocity in r direction (m/s)
v	Velocity in θ direction (m/s)
r	Radial coordinate
Greek symbols	
ρ	Density (kg/m^3)
β	Thermal expansion coefficient (K^{-1})
μ	Dynamic viscosity (kg/ms)
θ	Angular coordinate
δ	Constant small number
λ	Liquid fraction
Φ	Volume fraction
Subscripts	
i	Inner tube
$init$	Initial
l	Liquid PCM
o	Outer shell
ref	Reference
s	Solid PCM

Acronyms

CFD	Computational fluid dynamics
FVM	Finite volume method
HTF	Heat transfer fluid
LTES	Latent thermal energy storage
PCM	Phase change materials
PRESTO	Pressure staggering option
SIMPLE	Semi-implicit method for pressure linked equations
TES	Thermal energy storage

References

1. Kolosok, S.; Bilan, Y.; Vasylieva, T.; Wojciechowski, A.; Morawski, M. A Scoping Review of Renewable Energy, Sustainability and the Environment. *Energies* **2021**, *14*, 4490. [CrossRef]
2. Alva, G.; Liu, L.; Huang, X.; Fang, G. Thermal energy storage materials and systems for solar energy applications. *Renew. Sustain. Energy Rev.* **2017**, *68*, 693–706. [CrossRef]
3. Mozafari, M.; Lee, A.; Cheng, S. A novel dual-PCM configuration to improve simultaneous energy storage and recovery in triplex-tube heat exchanger. *Int. J. Heat Mass Transf.* **2022**, *186*, 122420. [CrossRef]
4. Chaturvedi, R.; Islam, A.; Sharma, K. A review on the applications of PCM in thermal storage of solar energy. *Mater. Today Proc.* **2021**, *43*, 293–297. [CrossRef]
5. Tao, Y.; He, Y.-L. A review of phase change material and performance enhancement method for latent heat storage system. *Renew. Sustain. Energy Rev.* **2018**, *93*, 245–259. [CrossRef]
6. Ghoneim, A. Comparison of theoretical models of phase-change and sensible heat storage for air and water-based solar heating systems. *Sol. Energy* **1989**, *42*, 209–220. [CrossRef]
7. Qiu, L.; Ouyang, Y.; Feng, Y.; Zhang, X. Review on micro/nano phase change materials for solar thermal applications. *Renew. Energy* **2019**, *140*, 513–538. [CrossRef]
8. Kumar, A.; Tiwari, A.K.; Said, Z. A comprehensive review analysis on advances of evacuated tube solar collector using nanofluids and PCM. *Sustain. Energy Technol. Assess.* **2021**, *47*, 101417. [CrossRef]
9. Saffari, M.; De Gracia, A.; Fernández, C.; Cabeza, L.F. Simulation-based optimization of PCM melting temperature to improve the energy performance in buildings. *Appl. Energy* **2017**, *202*, 420–434. [CrossRef]
10. Jia, J.; Liu, B.; Ma, L.; Wang, H.; Li, D.; Wang, Y. Energy saving performance optimization and regional adaptability of prefabricated buildings with PCM in different climates. *Case Stud. Therm. Eng.* **2021**, *26*, 101164. [CrossRef]
11. Park, J.H.; Berardi, U.; Chang, S.J.; Wi, S.; Kang, Y.; Kim, S. Energy retrofit of PCM-applied apartment buildings considering building orientation and height. *Energy* **2021**, *222*, 119877. [CrossRef]
12. Mozafari, M.; Lee, A.; Mohammadpour, J. Thermal Management of Single and Multiple PCMs based Heat Sinks for Electronics Cooling. *Therm. Sci. Eng. Prog.* **2021**, *23*, 100919. [CrossRef]
13. Mozafari, M.; Lee, A.; Cheng, S. Improvement on the cyclic thermal shock resistance of the electronics heat sinks using two-objective optimization. *J. Energy Storage* **2022**, *46*, 103923. [CrossRef]
14. Mozafari, M.; Akhavan-Behabadi, M.; Qobadi-Arfaee, H.; Hanafizadeh, P.; Fakoor-Pakdaman, M. Experimental study on condensation flow patterns inside inclined U-bend tubes. *Exp. Therm. Fluid Sci.* **2015**, *68*, 276–287. [CrossRef]
15. Senthil, R. Effect of position of heat transfer fluid tube on the melting of phase change material in cylindrical thermal energy storage. *Energy Sources Part A Recovery Util. Environ. Eff.* **2019**, 1–11. [CrossRef]
16. Kadivar, M.; Moghimi, M.; Sapin, P.; Markides, C. Annulus eccentricity optimisation of a phase-change material (PCM) horizontal double-pipe thermal energy store. *J. Energy Storage* **2019**, *26*, 101030. [CrossRef]
17. Alnakeeb, M.A.; Salam, M.A.A.; Hassab, M.A. Eccentricity optimization of an inner flat-tube double-pipe latent-heat thermal energy storage unit. *Case Stud. Therm. Eng.* **2021**, *25*, 100969. [CrossRef]
18. Gong, Z.-X.; Mujumdar, A. A new solar receiver thermal store for space-based activities using multiple composite phase-change materials. *J. Sol. Energy Eng.* **1995**, *117*, 215–220. [CrossRef]
19. Wang, J.; Ouyang, Y.; Chen, G. Experimental study on charging processes of a cylindrical heat storage capsule employing multiple-phase-change materials. *Int. J. Energy Res.* **2001**, *25*, 439–447. [CrossRef]
20. Al Siyabi, I.; Khanna, S.; Mallick, T.; Sundaram, S. Multiple phase change material (PCM) configuration for PCM-based heat sinks—an experimental study. *Energies* **2018**, *11*, 1629. [CrossRef]
21. Farid, M.M.; Kanzawa, A. Thermal performance of a heat storage module using PCM's with different melting temperatures: Mathematical modeling. *J. Sol. Energy Eng.* **1989**, *111*, 152–157. [CrossRef]
22. Farid, M.M.; Kim, Y.; Kansawa, A. Thermal performance of a heat storage module using PCM's with different melting temperature: Experimental. *J. Sol. Energy Eng.* **1990**, *112*, 125–131. [CrossRef]
23. Al Siyabi, I.; Khanna, S.; Mallick, T.; Sundaram, S. Experimental and numerical study on the effect of multiple phase change materials thermal energy storage system. *J. Energy Storage* **2021**, *36*, 102226. [CrossRef]
24. Thonon, M.; Fraisse, G.; Zalewski, L.; Pailha, M. Analytical modelling of PCM supercooling including recalescence for complete and partial heating/cooling cycles. *Appl. Therm. Eng.* **2021**, *190*, 116751. [CrossRef]

25. Tan, P.; Lindberg, P.; Eichler, K.; Löveryd, P.; Johansson, P.; Kalagasidis, A.S. Effect of phase separation and supercooling on the storage capacity in a commercial latent heat thermal energy storage: Experimental cycling of a salt hydrate PCM. *J. Energy Storage* **2020**, *29*, 101266. [CrossRef]
26. Rubitherm GmbH. Available online: https://www.rubitherm.eu. (accessed on 28 November 2021).
27. Khodadadi, J.M.; Hosseinizadeh, S.F. Nanoparticle-enhanced phase change materials (NEPCM) with great potential for improved thermal energy storage. *Int. Commun. Heat Mass Transf.* **2007**, *34*, 534–543. [CrossRef]
28. Brent, A.; Voller, V.R.; Reid, K. Enthalpy-porosity technique for modeling convection-diffusion phase change: Application to the melting of a pure metal. *Numer. Heat Transf. Part A Appl.* **1988**, *13*, 297–318.
29. Leonard, B.P. A stable and accurate convective modelling procedure based on quadratic upstream interpolation. *Comput. Methods Appl. Mech. Eng.* **1979**, *19*, 59–98. [CrossRef]
30. Patankar, S. *Numerical Heat Transfer and Fluid Flow*; Taylor & Francis: Abingdon, UK, 2018.
31. Al-Abidi, A.A.; Mat, S.; Sopian, K.; Sulaiman, M.; Mohammad, A.T. Experimental study of melting and solidification of PCM in a triplex tube heat exchanger with fins. *Energy Build.* **2014**, *68*, 33–41. [CrossRef]

energies

Article

Molecular Dynamics Simulation of Thermophysical Properties and the Microstructure of Na_2CO_3 Heat Storage Materials

Haiming Long [1], Yunkun Lu [2], Liang Chang [3], Haifeng Zhang [1], Jingcen Zhang [1], Gaoqun Zhang [3] and Junjie Hao [1,3,*]

[1] Institute for Advanced Materials and Technology, University of Science and Technology Beijing, Beijing 100083, China
[2] Science and Technology Innovation Center, China General Nuclear Powder New Energy Holdings Co., Ltd., Beijing 100160, China
[3] Beijing Institute of Smart Energy, Future Science and Technology City, Changping District, Beijing 102211, China
* Correspondence: ustbhaojunjie@126.com; Tel.: +86-10-8237-1528

Abstract: In recent years, heat storage technology has attracted wide attention in the fields of renewable energy storage for its relatively high melting point, high heat storage capacity and economy, Na_2CO_3 and eutectic salt mixtures containing Na_2CO_3 are promising candidates in the field of solar energy storage. In this paper, a molecular dynamics (MD) simulation of Na_2CO_3 was conducted with the Born–Mayer potential function. The simulated solid–liquid phase change temperature is 1200 K, and the error is 5.4%. The heat capacity at constant pressure (C_p) is higher in liquid than in solid, the average C_p of solid is 1.45 J/g and that of liquid is 1.79 J/g, and the minimum error is 2.8%. The simulation results revealed the change rules of density and thermal expansion coefficient of Na_2CO_3 in the process of heating up, and these changes were analyzed by radial distribution functions (RDF) and angular distribution functions (ADF). Moreover, the RDF and ADF results show that the atomic spacing of Na_2CO_3 increases, the coordination number decreases, and the angle distribution between atoms becomes wider as the temperature rises. Finally, this paper examined the microscopic changes of ions during the phase transition of Na_2CO_3 from solid to liquid. It is concluded that the angle change of CO_3^{2-} in the liquid state is more sharply. This study improves the understanding of the thermodynamic properties and local structure of Na_2CO_3 and provides theoretical support for Na_2CO_3 heat storage materials.

Keywords: molecular dynamics; Na_2CO_3; local structures; thermodynamic properties; phase change

Citation: Long, H.; Lu, Y.; Chang, L.; Zhang, H.; Zhang, J.; Zhang, G.; Hao, J. Molecular Dynamics Simulation of Thermophysical Properties and the Microstructure of Na_2CO_3 Heat Storage Materials. *Energies* **2022**, *15*, 7080. https://doi.org/10.3390/en15197080

Academic Editor: Ann Lee

Received: 1 September 2022
Accepted: 21 September 2022
Published: 27 September 2022

1. Introduction

With the development of society and the progress of science and technology, the energy demand is increasing year by year, and the waste of energy is becoming more and more serious. The overexploitation of fossil energy (coal, oil, etc.) has not only led to the energy crisis but also caused serious pollution problems [1,2]. Solving the energy crisis has become a global consensus. Therefore, it is urgent to seek renewable energy [3–5]. Heat storage technology is an important part of renewable energy, and heat storage materials play a key role in heat storage technology [6]. Due to the intermittency and instability of solar and wind power generation, as well as the peak and trough of power consumption during the day and night, it is urgent to develop efficient energy storage technologies [7–10].

Phase change materials (PCMs) are the core of energy storage technology. For PCMs, the absorption and release of heat energy are carried out during the charging and discharging process, which is accompanied by melting [11]. In the aspect of solar energy utilization, PCMs can directly convert solar energy into sensible heat and phase change latent heat of materials. There is a wide range of applications for organic, inorganic salts and their hydrates in low temperature applications of solar energy such as building space insulation and cooling [12]. In the aspect of medium and high temperature applications of solar energy, except for parabolic trough systems with relatively low working temperatures (use nitrate, 673.15 K to 773.15 K, thermal decomposed above 773.15 K), there are two other major methods used to use solar energy, which both have higher maximum working temperatures. One method is a solar tower surrounded by thousands of small mirrors. The upper limit temperature of the solar tower is 773.15 K to 1223.15 K. The other method, a solar dish, is a large reflective parabolic dish which has a receiver set on its focal point. The upper limit temperature of the solar dish is 973.15 K to 1473.15 K. However, chloride salts, which corrode pipes, heat exchangers, tanks and other thermal components at high temperatures, are difficult to apply despite their high melting point [13]. Due to its relatively high melting point and high heat storage capacity and economy, Na_2CO_3 and eutectic salt mixtures containing Na_2CO_3 are promising candidates in the field of solar energy storage [14,15].

Good thermal characteristics such as wide working temperature range, large heat capacity and excellent thermal stability make the molten salt an ideal material for thermal energy storage [16–19]. Some salts used in the heat storage industry are displayed in Figure 1 according to their thermophysical properties. It shows that the phase change temperature of nitrate is low, and the low phase change temperature, which limits its application range; the latent heat of NaCl and $MgCl_2$ is high, but chloride is very corrosive. When used for a long time, it will cause great pollution to the system and environment. Therefore, carbonate with low corrosivity has obvious advantages. Among various carbonates, Li_2CO_3 has high latent heat, but its cost is too high, and it is difficult to be widely used, and low-cost heat storage materials such as Na_2CO_3 have received extensive attention from researchers. Nobuyuki [20] and Feng [21] studied phase change materials (PCMs) with Na_2CO_3 as the heat storage medium. However, the macroscopic thermophysical properties of molten salt are determined by its microstructure. By studying the microstructure of molten salt, you can gain a deeper understanding of its thermophysical properties as well as explore its molecular motion state at high temperatures [22], which has important guiding significance for further studying the variation law between thermophysical properties and temperature of molten salt.

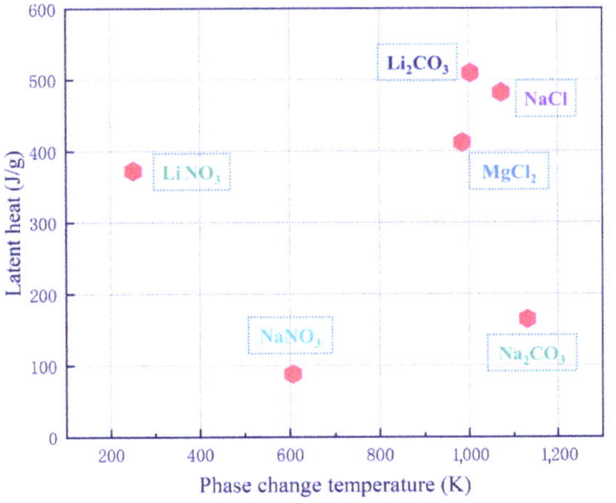

Figure 1. Comparison of thermophysical properties of some salts [23–26].

The change law of local structure and thermodynamic properties of salts can be explored by the Molecular Dynamics (MD) simulation [27]. Wang [22] studied the transport characteristics of liquid chloride and explained the local structure and transport characteristics of chloride at high temperatures. Du [28] and Ding [29] simulated the thermophysical properties of Na_2CO_3 in a liquid environment by the MD method and described the changing trends of microstructure and thermophysical properties of carbonate in the liquid state. However, the local microstructure and thermophysical properties of Na_2CO_3 in solid and during the transformation from solid to liquid have not been studied, which will help us to deeply understand the thermophysical properties of Na_2CO_3.

In this study, we used MD simulation to study the thermophysical properties (phase change temperature, heat capacity at constant pressure (C_p), and density, etc.) of Na_2CO_3 in solid-state and the process of solid-state transformation into the liquid state and verified the reliability of the model by comparing with the experimental data and literature results. Through the simulation of local microstructure (radial distribution functions (RDF) and angular distribution functions (ADF)), the law of local structure transformation of Na_2CO_3 during heating and the relationship between local transformation and thermophysical properties were clarified.

2. MD Simulation Methods and Experimental Details

2.1. Interaction Potential and Parameter Setting

In the MD simulations of this study, the effective pair potential was a Born–Mayer type combined with a Coulomb term, which was proposed by M.P. Tosi [30,31] in 1964. The thermophysical properties of molten salts are often calculated using this effective potential [32,33]:

$$U(r_{ij}) = \frac{z_i z_j e^2}{r_{ij}} + b\left(1 + \frac{z_i}{n_i} + \frac{z_j}{n_j}\right) exp[\alpha(\sigma_i + \sigma_j - r_{ij})] \tag{1}$$

where, r_{ij} is the distance between the centers of two particles i and j; α (3.45 Å) and b (4.865 kcal/mol) are constants, representing the softness parameter and the pre-exponential term, respectively. Furthermore, z_i is the valence of ion i, n_i is the number of ion's electrons in the outer shell, e is the electronic charge unit and σ_i is the ion's effective radius. In this study, the parameters of the relevant potential model for Na_2CO_3 simulation are listed in Table 1, which were obtained by Janssen [34].

Table 1. Parameters of Na_2CO_3 in MD simulation.

Atom	Na	C	O
z(e)	1.00	1.54	−1.18
n	8.00	2.46	7.18
σ(Å)	1.07	1.10	1.33

In this paper, the open-source program LAMMPS is used as the MD simulation tool, and the model Na_2CO_3 refers to the work of Arakcheeva [35]. The lattice structure constants of Na_2CO_3 are shown in Table 2, and a cubic box containing 1500 ions is used as the MD simulation system. O, C and Na are distributed in a ratio of 3:1:2 in the box.

Table 2. The lattice constant of Na_2CO_3.

Salt	a, b, c (Å)	α, β, γ (°)
Na_2CO_3	8.851, 5.240, 6.021	90.000, 101.080, 90.000

A Nose–Hoover thermostat was used to balance the simulation system under the NVT ensemble, followed by a range of temperatures from 300 K to 1600 K. The short-range interaction cutoff distance was set at 15 Å, which is half the simulation box length, to maintain a constant number of particles. The particle numbers were maintained by a periodic boundary condition. The method of eliminating truncation error in K-space was based on the Ewald summation, an accuracy of 1.0×10^{-4} was achieved in the calculation of force. Initial velocity follows a Gaussian distribution and is randomly distributed. The

Verlet algorithm was used to solve the Newton equation of motion and the time step was set to 1 fs. To ensure a good statistical representation of the required characteristics, it has been run 5 million times in the NPT ensemble environment.

2.2. Evaluation of Properties

2.2.1. Phase Change Temperature and Heat Capacity at Constant Pressure

In this study, the self-diffusion coefficient (*D*) was used to verify the phase change temperature of Na_2CO_3 [36,37]. The *D* increases as temperature rises, which means that the fluidity of particles increases gradually [38,39]. Einstein's law describes *D* as Equation (2), where *M(t)* is the mean square displacement (MSD) which was shown in Equation (3), where r_i is the position of ion *i* at different temperatures. The sign of phase change is marked by a significant change of *D*, and with the change of temperature, the *D* curve appears a turning point [40,41].

$$D = \frac{1}{6}\frac{dM(t)}{dt} \quad (2)$$

$$M(t) = \langle |r_i(t + t_0) - r_i(t)|^2 \rangle \quad (3)$$

Heat capacity at constant pressure (C_p) is determined by the change of enthalpy with temperature, and the C_p of Na_2CO_3 can be calculated as shown in Equation (4), where *H* is the enthalpy, *T* is the temperature.

$$C_p = (\frac{\partial H}{\partial T})_p \quad (4)$$

2.2.2. Density and Thermal Expansion Coefficient

From initial NPT simulations, the densities ρ of Na_2CO_3 could be determined by the following Equation (5), where *n* is the particle number, *N* is the total number of atoms in the simulation box, N_A denotes the Avogadro constant, *M* is the molar mass of the simulated molten salt, and V_E is the equilibrated volume of the simulation cell at the given temperature in the NPT ensemble simulations. The thermal expansion coefficient depends on the volume changes with temperature, and Equation (6) is used to evaluate it, where *T* is the temperature, *V* is the volume of the simulation box, ρ is the density, the subscript *P* indicates the constant pressure and β is thermal expansion coefficient.

$$\rho = \frac{NM_{Na_2CO_3}}{nN_A V_E} \quad (5)$$

$$\beta = \frac{1}{V}\left(\frac{\partial V}{\partial T}\right)_P = -\frac{1}{\rho}\left(\frac{\partial \rho}{\partial T}\right)_P \quad (6)$$

2.2.3. The RDF and ADF of Na_2CO_3

The RDF and the ADF [28], which are defined by Equation (7) and Equation (10), can be used to describe eutectic salt's local microstructure.

The RDF expression is shown in Equation (7), where $N_{ij}(r)$ was the number of *j*-type ions inside a spherical cavity with a radius of *r* centered on *i*-type ion, and ρ_j was the ideal number density of *j*-type ion around *i*-type ion, and r_{min} was the position of the first peak valley of the RDF. It could be calculated from the RDF through the following formula as Equation (8),

$$g_{ij}(r) = \frac{1}{4\pi r^2 \rho_j}\left[\frac{dN_{ij}(r)}{dr}\right] \quad (7)$$

$$N_{ij}(r) = 4\pi\rho_j \int_0^r g_{ij}(r) r^2 dr \quad (8)$$

Because RDF and coordination number can only describe the probability of atomic pairing, they do not contain the orientation information between atoms. To analyze the local structure further, the ADF of Na_2CO_3 can be characterized. The ADF expression is shown in Equations (9) and (10), where r is used to represent the atomic spacing, $n(\theta)$ is the number of the angle *i-j-k* in the range between θ and $\theta + \Delta\theta$, and N is the total number of the angle *i-j-k* in the sphere, and based on the corresponding RDF, r_{min} represents the position of the first peak valley.

$$\theta_{ijk} = \arccos\left(\frac{r_{ij} + r_{ik} - r_{jk}}{2r_{ij}r_{ik}}\right) \tag{9}$$

$$p(\theta) = \frac{1}{N}\frac{dn(\theta)}{d\theta} \times 100 \tag{10}$$

2.3. *Experimental details*

The phase change temperature and C_p were measured by differential scanning calorimetry (DSC) and thermal gravity analyzer (STA449F5, NETZSCH, Selb, Germany). The weight of sample was 15 ± 2 mg. The phase change temperature was tested with an alumina crucible, and the Cp was tested with a platinum–rhodium crucible. The test environment is N2, and the temperature was heated from 500 K to 1100 K at the heating rate of 10 K/min.

3. Results and Discussion

3.1. *Thermophysical Properties of Na_2CO_3*

3.1.1. Phase Change Temperature and Specific Heat Capacity of Na_2CO_3

There is a corresponding relationship between the value of MSD and the atomic self-diffusion coefficient (D). The MSD has a linear relationship with time, and its slope has a relationship with D as Equation (2). The MSD curves of Na_2CO_3 at different temperatures are shown in Figure 2a, from which we can know that the slope of MSD increases gradually as the temperature rises, which shows that the movement of Na_2CO_3 atom increases gradually with the increase of temperature. Figure 2b shows that the self-diffusion coefficient (D) has a turning point at 1200 K obviously. Figure 3 shows the phase change temperature tested by Na_2CO_3, it shows that the phase change temperature is 1135 K, and the simulation error is 5.42% compared with the experimental test.

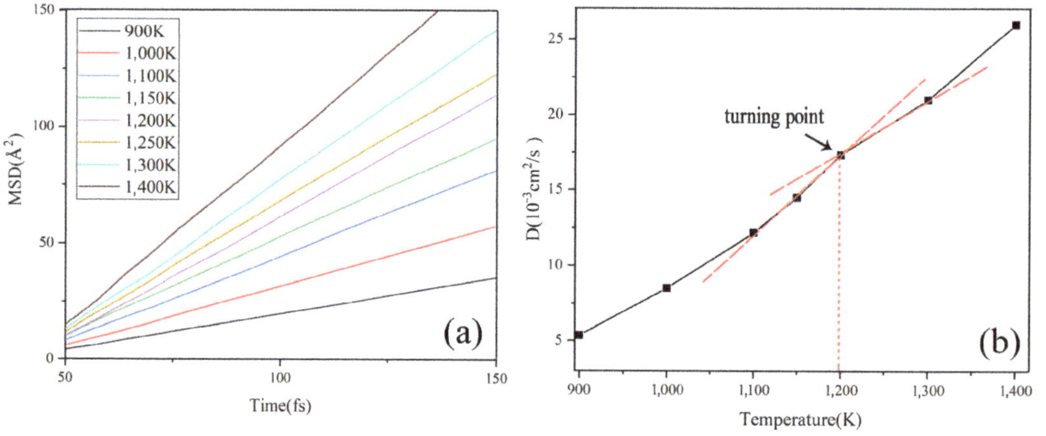

Figure 2. Simulation results of Na_2CO_3: (**a**) MSD change with time; (**b**) D change with temperature.

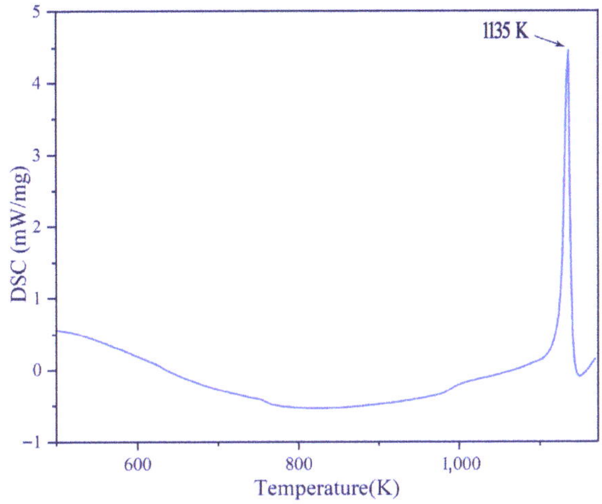

Figure 3. Phase change temperature tested by Na_2CO_3.

By simulating the enthalpy and temperature, the C_p of Na_2CO_3 can be determined. Figure 4a shows the average C_p between 500–1100 K is 1.45 J/(g·K). As shown in Figure 4b, the C_p of Na_2CO_3 (500~1100 K) obtained by simulation calculation is basically consistent with the results of experimental tests and literature reports [42], the C_p of Na_2CO_3 has a peak value between 700~800 K, and this peak may be caused by crystal transformation. The C_p of liquid Na_2CO_3 is shown in Figure 5. It can be seen that the C_p is 1.79 J/(g·K). As shown in Table 3, the minimum error between the simulated and experimental values of C_p is 2.8%, and the maximum error is 16.26%, which has good fitting results. The simulation shows that the enthalpy of Na_2CO_3 increases as temperature rises, and the specific heat capacity (C_p) is higher in liquid than in solid.

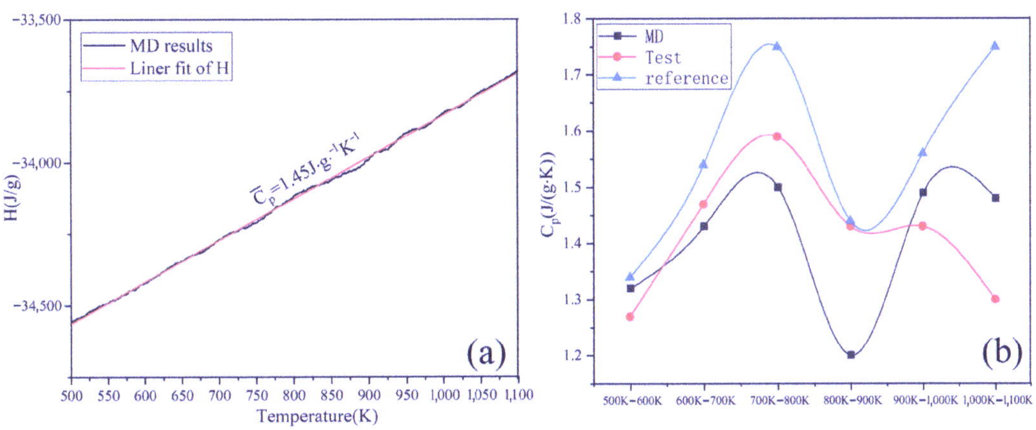

Figure 4. Heat capacity at constant pressure (C_p) of solid Na_2CO_3: (**a**) simulation; (**b**) experimental (reference and MD).

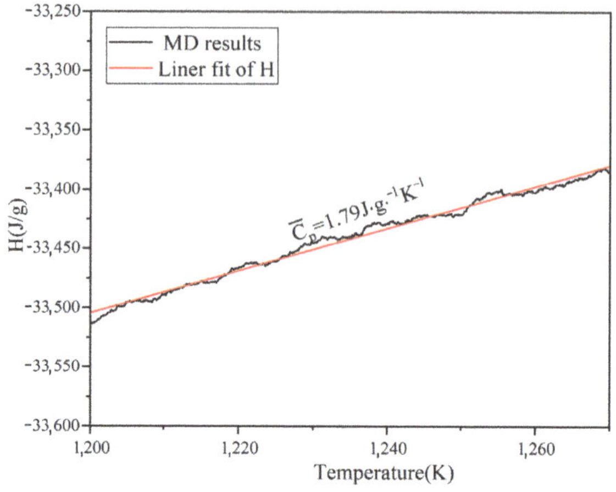

Figure 5. Simulation results of heat capacity at constant pressure (C_p) of liquid Na_2CO_3.

Table 3. The solid C_p of Na_2CO_3.

Temp. K	Simulation C_p J/(g·K)	Test C_p J/(g·K)	Error (with Test) %
500–600	1.34	1.27	5.22
600–700	1.43	1.47	2.80
700–800	1.50	1.59	6.00
800–900	1.23	1.43	16.26
900–1000	1.49	1.43	4.03
1000–1100	1.48	1.30	12.16
500–1100	1.45	1.42	2.07

3.1.2. Density and Thermal Expansion Coefficient of Na_2CO_3

The density–temperature curve in the Na_2CO_3 simulation system is shown in Figure 6. It shows that the density is about 1.8 g/cm^3 at 500 K. In the range of 500 K to 1100 K, the density decreases linearly with the temperature. Its density decreases rapidly and tends to be stable after 1400 K when the temperature is greater than 1200 K.

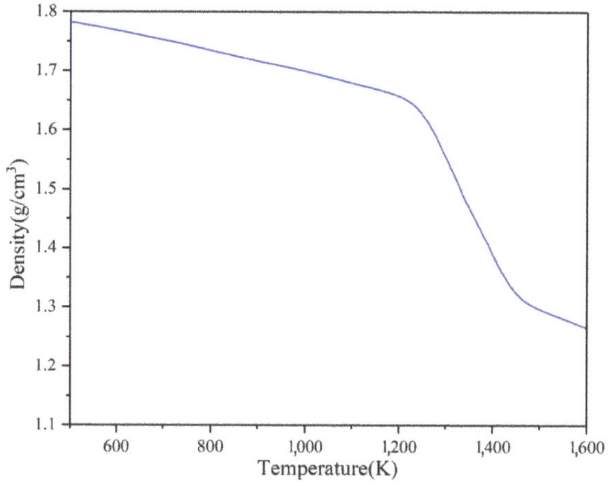

Figure 6. The density of Na_2CO_3 at different temperatures.

Table 4 shows the simulated calculation values of the thermal expansion coefficient of Na_2CO_3 in different temperature ranges. The thermal expansion coefficient of Na_2CO_3 is 2×10^{-4} K^{-1} between 500~900 K. The thermal expansion coefficient of Na_2CO_3 is 3.5×10^{-4} K^{-1} between 1000~1200 K. When the temperature is higher than 1200 K, the change of thermal expansion coefficient intensifies, and the maximum thermal expansion coefficient appears between 1200~1400 K and is 22×10^{-4} K^{-1}. The thermal expansion coefficient drops to 7×10^{-4} K^{-1} when the temperature is greater than 1400 K. It shows that the thermal expansion coefficient increases sharply, and the density decreases rapidly during phase transition.

Table 4. Coefficient of thermal expansion of Na_2CO_3 at different temperatures.

Temp. (K)	500~900	900~1000	1000~1200	1200~1400	1500~1600
β (10^{-4} K^{-1})	2	1	3.5	22	7

3.2. Local Microstructure

3.2.1. The RDF of Na_2CO_3

The relationship between the microstructure and temperature of Na_2CO_3 is the key to the study, and RDF can be used to reveal the change law of Na_2CO_3 in the process of solid–liquid phase change. The variation laws of RDF function and coordination number of Na_2CO_3 at different temperatures are illustrated in Figure 7, and Table 5. As the temperature rises, the long-range peak (the peak after the fourth peak) of the RDF curve weakens and disappears, and the short-range peak (the first four peaks) of the RDF curve widens. The long-range ordered structure of Na_2CO_3 crystal will gradually collapse to disappear as the temperature rises. Whenever the temperature falls below 1200 K, there is a long-range peak in the RDF curve, and the solid Na_2CO_3 crystal presents a long-range ordered structure; when the temperature is 1200~1500 K, the long-range peak of the RDF curve disappears and only the short-range peak (the first four peaks), and the liquid Na_2CO_3 presents a long-range disordered and short-range ordered structure. During the solid-liquid transition of sodium carbonate, the RDF curve changes in growth range peak and short-range peak. When the solid-state changes to the liquid state, the long-range peak disappears, and the short-range peak widens. Moreover, the degree of short-range peak broadening increases, and the order of short-range structure becomes worse as the liquid temperature rises.

Table 5. Characteristic parameters for local structures of Na_2CO_3.

	Temp.(K)	Na-Na	Na-C	Na-O	C-C	C-O	O-O
r_{max} (Å)	500	3.47	3.65	2.57	3.74	2.48	3.56
	700	3.56	3.65	2.57	3.74	2.48	3.56
	900	3.56	3.65	2.48	3.74	2.39	3.56
	1100	3.56	3.65	2.48	3.83	2.39	3.65
	1200	3.65	3.65	2.48	3.83	2.39	3.65
	1300	3.56	3.74	2.48	3.83	2.39	3.65
	1400	3.65	3.74	2.48	3.83	2.39	3.65
	1500	3.65	3.74	2.48	3.83	2.39	3.65
r_{min} (Å)	500	5.18	5.09	3.38	5.00	3.29	5.09
	700	5.09	5.18	3.47	5.18	3.38	5.09
	900	5.09	5.18	3.47	5.09	3.38	5.18
	1100	5.18	5.18	3.56	5.09	3.38	5.18
	1200	5.18	5.18	3.56	5.18	3.38	5.18
	1300	5.18	5.18	3.56	5.18	3.38	5.18
	1400	5.27	5.27	3.56	5.18	3.38	5.18
	1500	5.36	5.27	3.56	5.18	3.47	5.18

Table 5. *Cont.*

	Temp.(K)	Na-Na	Na-C	Na-O	C-C	C-O	O-O
$N\ (r_{min})$	500	9.83	4.88	4.86	4.53	5.12	14.42
	700	9.48	5.12	5.09	4.75	5.31	14.47
	900	9.03	4.94	4.86	4.31	5.07	14.33
	1100	8.98	4.70	4.86	4.22	4.91	13.71
	1200	8.97	4.51	4.77	4.48	4.87	13.47
	1300	8.68	4.44	4.69	4.35	4.82	13.17
	1400	8.80	4.55	4.59	4.15	4.73	12.85
	1500	8.26	4.40	4.50	4.09	4.78	12.52

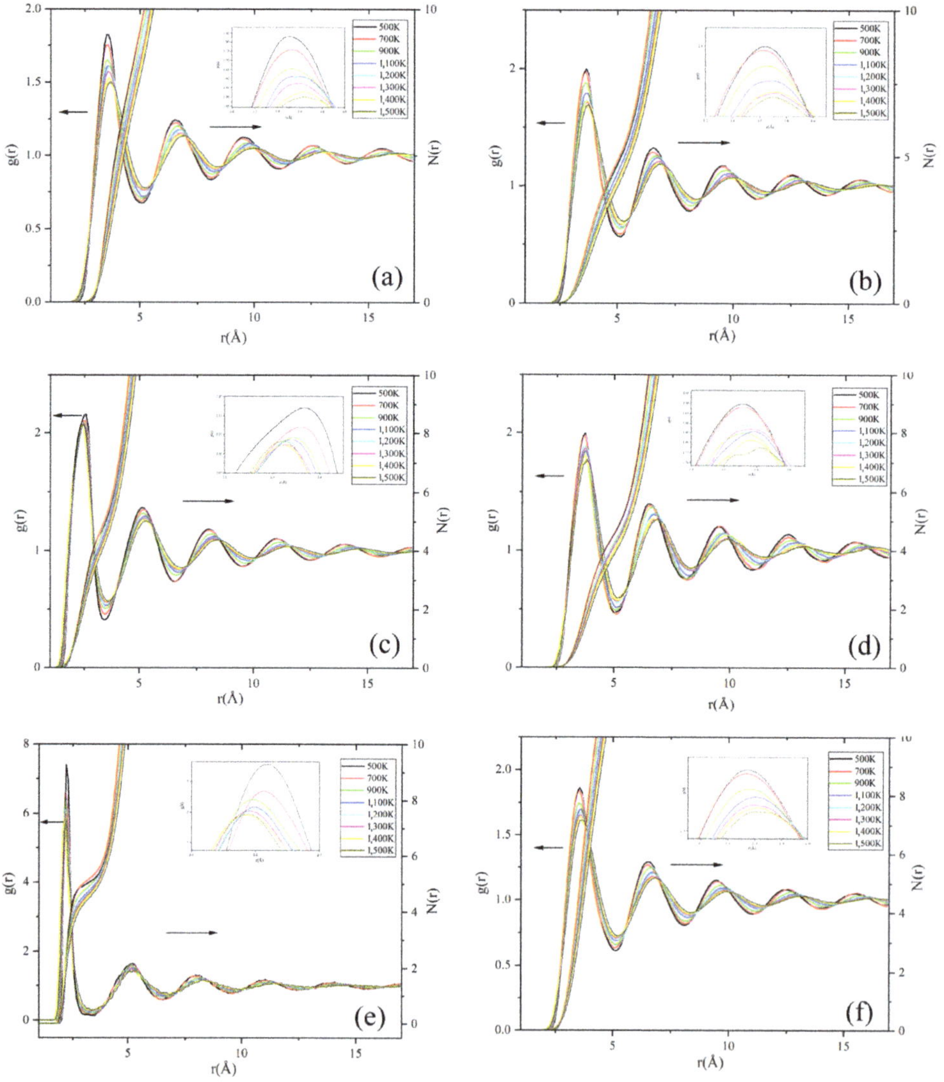

Figure 7. The RDF and coordination number of Na_2CO_3: (**a**) Na-Na; (**b**) Na-C; (**c**) Na-O; (**d**) C-C; (**e**) C-O; (**f**) O-O.

As shown in Figure 7 and Table 5, the r_{max} at the first peak of the RDF curve of Na_2CO_3 increases as the temperature rises, which shows the distances between Na-C, Na-Na, C-C, and O-O in the Na_2CO_3 system increase as the temperature increases, and in the process of solid–liquid transition, the coordination number curve shows an obvious right shift trend, and the coordination number of atoms decreases. This also explains that the molar volume of Na_2CO_3 increases as the temperature rises, resulting in the decrease of its density, the increase of thermal expansion coefficient and the change of other thermophysical properties. When the temperature is higher than 900 K, r_{max} hardly changes. As the temperature increases, the peak width becomes wider, which shows that the C-O stretching vibration and the out-of-plane bending vibration of C are increasing, and it means that the bond energy and the kinetic energy of CO_3^{2-} also increase as the temperature rises.

3.2.2. The ADF of Na_2CO_3

To learn more about the microstructure of sodium carbonate, the ADF of Na_2CO_3 at different temperatures was calculated and characterized on the basis of RDF simulation analysis, and the interatomic angular distribution and temperature variation of sodium carbonate were studied.

The Na-C-Na angular distribution function of Na_2CO_3 at different temperatures is shown in Figure 8a. At 500 K and 700 K, there are four obvious peaks in the angular function of Na-C-Na at 50°, 90°, 140° and 170°. When the temperature rises to 900 K, the angular function becomes two peaks (about 50° and 100°) and the peak shape becomes wider, and especially the intensity of the second peak decreases. It shows that the angle between Na-C-Na has changed significantly in this process. According to the literature [43], the solid-state phase transition of Na_2CO_3 occurs at 723 K (phase β to phase α), which may cause the peak value of the angular distribution function to be changed.

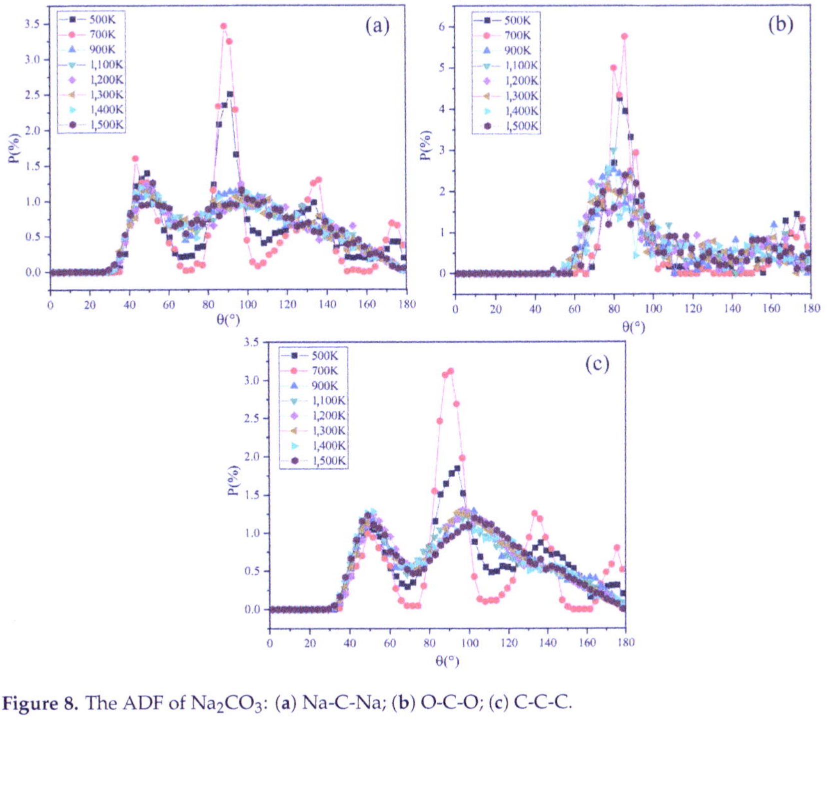

Figure 8. The ADF of Na_2CO_3: (**a**) Na-C-Na; (**b**) O-C-O; (**c**) C-C-C.

When Na_2CO_3 is solid, the angle distribution range of Na-C-Na is 40–180°. It transforms from solid-state to liquid state as the temperature rises, the angle distribution range of Na-C-Na expands to 30–180°, and the peak shape becomes wider and weakens. After the solid–liquid transformation, the angle distribution range of Na-C-Na of sodium carbonate expands significantly, which is also consistent with the report in the literature [24].

Figure 8b shows the angular distribution function of O-C-O in Na_2CO_3 at different temperatures. The angular function has two peaks in the range of 60–100° and 150–180°. When the temperature increases from 500 K to 900 K, the peak shape and peak intensity become significantly wider and weaken, and the change of peak shape and peak intensity becomes weaker when the temperature is over 900 K. In the solid-state, the included angle distribution range of O-C-O is 70–100°, while when the system temperature rises and becomes liquid, the included angle distribution range of O-C-O extends to 50–100°. As shown in Figure 8a,c, the change of angle between Na-C-Na is basically consistent with C-C-C.

This phenomenon shows that the atomic motion in Na_2CO_3 becomes more and more intense as the temperature rises, the activity range expands, the periodicity of the arrangement becomes worse and leading to the widening of the angular distribution, which is consistent with the change of the angular distribution function of Na_2CO_3 at different temperatures. The angular distribution of Na_2CO_3 in the liquid state is wider than that in the solid state. With the increase of liquid temperature, the order of its short-range structure becomes worse. When it becomes the liquid phase, the atomic activity range is larger, the periodicity of the arrangement becomes worse, and the distribution of the included angle between atoms continues to expand, but the change of the angular distribution in the liquid phase gradually slows down with the increase of temperature.

4. Conclusions

In this paper, molecular dynamics (MD) software LAMMPS was used for modeling and simulation. Thermodynamic properties and local structure of alkali carbonate Na_2CO_3 have been simulated by means of MD method with an effective pair potential model. The main conclusions are as follows:

(1) The self-diffusion coefficient (D) of Na_2CO_3 increases as the temperature rises, and the D changes significantly with temperature during solid-liquid transformation. The simulated solid-liquid phase change temperature is 1200 K, and the error is 5.4%.

(2) The enthalpy of Na_2CO_3 increases as temperature rises, and the specific heat capacity (C_p) is higher in liquid than in solid. The C_p has a peak in the temperature range of 700~800 K. The average specific heat capacity of solid is 1.45 J/g and that of liquid is 1.79 J/g, the minimum error is 2.8%.

(3) When the Na_2CO_3 transforms from solid to liquid, the thermal expansion coefficient changes suddenly and the density decreases rapidly. The calculated expansion coefficient shows that the expansion coefficient was the largest in the solid-liquid phase transformation (1200~1400 K), and the maximum expansion coefficient was $22 \cdot 10^{-4}\ K^{-1}$.

(4) The RDF and ADF results show that the atomic spacing of Na_2CO_3 increases, the coordination number decreases, and the angle distribution between atoms becomes wider as the temperature rises. When the solid state changes to the liquid state, the long-range peak disappears and the short-range peak widens in the RDF curve, the distribution of the included angle between atoms widens and the periodicity of the arrangement worsens. The degree of short-range peak broadening increases, and the order of short-range structure becomes worse, as the temperature of the liquid rises.

(5) The microscopic changes of ions during the phase transition of Na_2CO_3 from solid to liquid shows that the angle change of CO_3^{2-} in the liquid state is sharper.

Author Contributions: Conceptualization, H.L. and J.H.; methodology, H.L. and Y.L.; validation, H.L. and Y.L.; formal analysis, H.L. and H.Z.; investigation, H.L., J.Z.; resources, L.C. and G.Z.; data curation, H.L. and Y.L.; writing—original draft preparation, H.L.; writing—review and editing, H.L.; visualization, Y.L.; supervision, J.H; project administration, J.H. All authors have read and agreed to the published version of the manuscript.

Funding: This research received no external funding.

Data Availability Statement: Not applicable.

Conflicts of Interest: The authors declare no conflict of interest.

References

1. Zhang, H.; Baeynes, J.; Caceres, G.; Degrève, J.; Lv, Y. Thermal energy storage: Recent developments and practical aspects. *Prog. Energ. Combust.* **2016**, *53*, 1–40. [CrossRef]
2. Goods, S.H.; Bradshaw, R.W. Corrosion of stainless steels and carbon steel by molten mixtures of commercial nitrate salts. *J. Mater. Eng. Perform.* **2004**, *13*, 78–87. [CrossRef]
3. Mozafari, M.; Lee, A.; Cheng, S. Simulation Study of Solidification in the Shell-And-Tube Energy Storage System with a Novel Dual-PCM Configuration. *Energies* **2022**, *15*, 832. [CrossRef]
4. Mohammadpour, J.; Lee, A.; Timchenko, V.; Taylor, R. Nano-Enhanced Phase Change Materials for Thermal Energy Storage: A Bibliometric Analysis. *Energies* **2022**, *15*, 3426. [CrossRef]
5. Ge, Z.; Ye, F.; Ding, Y. Composite Materials for Thermal Energy Storage: Enhancing Performance through Microstructures. *Chemsuschem* **2014**, *7*, 1318–1325. [CrossRef]
6. Hasan, A. Phase change material energy storage system employing palmitic acid. *Sol. Energy* **1994**, *52*, 143–154. [CrossRef]
7. Wei, G.S.; Xing, L.J.; Du, X.Z.; Yang, Y. Research status and selection of phase change thermal Energy storage materials for CSP systems. *Proc. Csee.* **2014**, *34*, 325–335. [CrossRef]
8. Aneke, M.; Wang, M. Energy storage technologies and real life applications-a state of the art review. *Appl. Energy* **2016**, *179*, 350–377. [CrossRef]
9. Lu, Y.; Peng, G.W.; Wang, Z.P.; Wang, K.Z.; Ma, R. A review on research for molten salt as a phase change material. *Mater. Rev.* **2011**, *25*, 38–42. [CrossRef]
10. Peng, Q.; Ding, J.; Wei, X.L.; Yang, J.P.; Yang, X.X. Research progress in application of molten nitrate salt in energy utilization. *Mod. Chem. Ind.* **2009**, *29*, 17–24. Available online: http://refhub.elsevier.com/S0306-2619(18)30454-9/h0015 (accessed on 20 September 2022).
11. Oró, E.; Gracia, A.; Castell, A.; Farid, M.M.; Cabeza, L.F. Review on phase change materials (PCMs) for cold thermal energy storage applications. *Appl. Energy* **2012**, *99*, 513–533. [CrossRef]
12. Zhou, D.; Zhao, C.Y.; Tian, Y. Review on thermal energy storage with phase change materials (PCMs) in building applications. *Appl. Energy* **2012**, *92*, 593–605. [CrossRef]
13. Vignarooban, K.; Xu, X.; Arvay, A.; Hsu, K.; Kannan, A.M. Heat transfer fluids for concentrating solar power systems—A review. *Appl. Energy* **2015**, *146*, 383–396. [CrossRef]
14. Wang, T.; Mantha, D.; Reddy, R.G. Novel high thermal stability LiF–Na_2CO_3–K_2CO_3 eutectic ternary system for thermal energy storage applications. *Sol. Energ. Mater. Sol. Cells* **2015**, *140*, 366–375. [CrossRef]
15. Forsberg, C.W.; Peterson, P.F.; Zhao, H. High-Temperature Liquid-Fluoride-Salt Closed-Brayton-Cycle Solar Power Towers. *J. Sol. Energy Eng.* **2007**, *129*, 141–146. [CrossRef]
16. Cunha, J.P.; Eames, P. Thermal energy storage for low and medium temperature applications using phase change materials—A review. *Appl. Energy* **2016**, *177*, 227–238. [CrossRef]
17. Ismail, K.A.R.; Lino, F.A.M.; Machado, P.L.O.; Teggar, M.; Arıcı, M.; Alves, T.A.; Teles, M.P.R. New potential applications of phase change materials: A review. *J. Energy Storage.* **2022**, *53*, 105202. [CrossRef]
18. Wu, Y.T.; Ren, N.; Ma, C.F. Research and application of molten salts for sensible heat storage. *Energy Storage Sci. Technol.* **2013**, *2*, 586–592. Available online: http://refhub.elsevier.com/S0306-2619(18)30454-9/h0040 (accessed on 20 September 2022).
19. Shen, X.Y.; Ding, J.; Peng, Q.; Yang, J. Application of high temperature molten salt to solar thermal power. *Guangdong Chem. Ind.* **2007**, *34*, 49. Available online: http://refhub.elsevier.com/S0306-2619(18)30454-9/h0035 (accessed on 20 September 2022).
20. Gokon, N.S.; Nakamura, S.; Hatamachi, T.; Kodama, T. Steam reforming of methane using double-walled reformer tubes containing high-temperature thermal storage Na_2CO_3/MgO composites for solar fuel production. *Energy* **2014**, *68*, 773–782. [CrossRef]
21. Ye, F.; Ge, Z.; Ding, Y.; Yang, J. Multi-walled carbon nanotubes added to Na_2CO_3/MgO composites for thermal energy storage. *Particuology* **2014**, *15*, 56–60. [CrossRef]
22. Wang, J.; Wu, J.; Lu, G.; Yu, J. Molecular dynamics study of the transport properties and local structures of molten alkali metal chlorides. Part III. Four binary systems LiCl-RbCl, LiCl-CsCl, NaCl-RbCl and NaCl-CsCl. *J. Mol. Liq.* **2017**, *238*, 236–247. [CrossRef]

23. Suleiman, B.; Yu, Q.; Ding, Y.; Li, Y. Fabrication of form stable NaCl-Al_2O_3 composite for thermal energy storage by cold sintering process. *Front. Chem. Sci. Eng.* **2019**, *13*, 1–9. [CrossRef]
24. Papapetrou, M.; Kosmadakis, G.; Cipollina, A.; La Commare, U.; Micale, G. Industrial waste heat: Estimation of the technically available resource in the EU per industrial sector, temperature level and country. *Appl. Therm. Eng.* **2018**, *138*, 207–216. [CrossRef]
25. Kenisarin, M.M. High-temperature phase change materials for thermal energy storage. *Renew. Sustain. Energy Rev.* **2010**, *14*, 955–970. [CrossRef]
26. Kourkova, L.; Sadovska, G. Heat capacity enthalpy and entropy of Li_2CO_3 at 303.5–563.15 K. *Thermochim. Acta* **2007**, *452*, 80. [CrossRef]
27. Lantelme, F.; Turq, P.; Quentrec, B.; Lewis, J.W. Application of the molecular dynamics method to a liquid system with long range forces (Molten NaCl). *Mol. Phys.* **1974**, *28*, 1537–1549. [CrossRef]
28. Du, L.; Xie, W.; Ding, J.; Lu, J.; Wei, X.; Wang, W. Molecular dynamics simulations of the thermodynamic properties and local structures on molten alkali carbonate Na_2CO_3. *Int. J. Heat Mass Transfer.* **2019**, *131*, 41–51. [CrossRef]
29. Ding, J.; Du, L.; Pan, G.; Lu, J.; Wei, X.; Li, J.; Wang, W.; Yan, J. Molecular dynamics simulations of the local structures and thermodynamic properties on molten alkali carbonate K_2CO_3. *Appl. Energy* **2018**, *220*, 536–544. [CrossRef]
30. Fumi, F.G.; Tosi, M.P. Ionic sizes and born repulsive parameters in the NaCl-type alkali halides—I: The Huggins-Mayer and Pauling forms. *J. Phys. Chem. Solids* **1964**, *25*, 31–43. [CrossRef]
31. Tosi, M.P.; Fumi, F.G. Ionic sizes and born repulsive parameters in the NaCl-type alkali halides—II: The generalized Huggins-Mayer form. *J. Phys. Chem. Solids* **1964**, *25*, 45–52. [CrossRef]
32. Ding, J.; Pan, G.; Du, L.C.; Lu, J.; Wang, W.; Wei, X.; Li, J. Molecular dynamics simulations of the local structures and transport properties of Na_2CO_3 and K_2CO_3. *Appl. Energy* **2018**, *227*, 555–563. [CrossRef]
33. Du, L.; Ding, J.; Wang, W.; Pan, G.; Lu, J.; Wei, X. Molecular dynamics simulations on the binary eutectic system Na_2CO_3-K_2CO_3. *Energy Procedia* **2017**, *142*, 3553–3559. [CrossRef]
34. Janssen, G.J.M.; Tissen, J. Pair potentials from ab initio calculations for use in MD simulations of molten alkali carbonates. *Mol. Simulat.* **1990**, *5*, 83–98. [CrossRef]
35. Arakcheeva, A.; Bindi, L.; Pattison, P.; Meisser, N.; Chapuis, G.; Pekov, I. The incommensurately modulated structures of natural natrite at 120 and 293 K from synchrotron X-ray data. *Am. Miner.* **2010**, *95*, 574–581. [CrossRef]
36. Rao, R.P.; Seshasayee, M. Molecular dynamics simulation of ternary glasses Li_2O–P_2O_5–LiCl. *Solid State Commun.* **2004**, *131*, 537–542. [CrossRef]
37. Que, Y.; Chao, L.; Chen, Q.Y. Molecular Dynamics Simulation of Phase Change in Heptadecane-Pentadecane Binary System. *Mater. Rep.* **2010**, *24*, 77–80.
38. Naghizadeh, J.; Rice, S.A. Kinetic theory of dense fluids. X. Measurement and interpretation of self-diffusion in liquid Ar, Kr, Xe, and CH_4. *J. Chem. Phys.* **1962**, *36*, 2710–2720. [CrossRef]
39. Jolly, D.L.; Bearman, R.J. Molecular dynamics simulation of the mutual and self-diffusion coefficients in Lennard-Jones liquid mixtures. *Mol. Phys.* **1981**, *41*, 137–147. [CrossRef]
40. Lu, Y.K.; Zhang, G.Q.; Hao, J.J. Molecular dynamics simulation of thermodynamic properties and local structure of Na_2CO_3-K_2CO_3 eutectic salt during phase transition. *J. Chem. Phys.* **2021**, *43*, 103221. [CrossRef]
41. Feng, T.X.; Yang, B.; Lu, G.M. Deep learning-driven molecular dynamics simulations of molten carbonates: 1. Local structure and transport properties of molten Li_2CO_3-Na_2CO_3 system. *Ionics* **2022**, *28*, 1231–1248. [CrossRef]
42. Karad, S.K.; Jones, F.R.; Attwood, D. Moisture absorption by cyanate ester modified epoxy resin matrices. Part I. Effect of spiking parameters. *Polymer* **2002**, *43*, 5209–5218. [CrossRef]
43. Ye, D.L. *Handbook of Inorganic Thermodynamic Data*, 1st ed.; Metallurgical Industry Press: Beijing, China, 2002; pp. 659–660.

MDPI
St. Alban-Anlage 66
4052 Basel
Switzerland
www.mdpi.com

Energies Editorial Office
E-mail: energies@mdpi.com
www.mdpi.com/journal/energies

www.ingramcontent.com/pod-product-compliance
Lightning Source LLC
Chambersburg PA
CBHW061435210926
51850CB00008B/1515